Taxonomy of Angiosperms

(Fourth Revised Edition)

The Authors

Prof. T. Pullaiah obtained his M. Sc. (1973) and Ph. D. (1976) degrees in Botany from Andhra University. He was a Post-doctoral Fellow at Moscow State University, Russia, during 1976-1978. He travelled widely in Europe and visited Universities and Botanic Gardens in about 17 countries. He joined Sri Krishnadevaraya University as Lecturer in 1979 and became Professor in 1993. He held several positions in the University as Dean, Faculty of Life Sciences; Head, Department of Botany; Chairman, Board of Studies in Botany; Head, Department of Sericulture; Co-ordinator and Chairman, Board of Studies in Biotechnology; Principal, Sri Krishnadevaraya University College. He retired from active Service in 2011 and was selected by UGC as UGC-BSR Faculty Fellow in the Department of Botany, Sri Krishnadevaraya University. He has published 65 books, 330 research papers and 35 popular articles. His books include Encyclopaedia of World Medicinal Plants (5 volumes, revised edition 2018), Flora of Andhra Pradesh (5 volumes, revised edition 2017), Biodiversity in India (8 volumes), Flora of Eastern Ghats (4 volumes), Flora of Telangana (3 volumes) and Encyclopaedia of Herbal Antioxidants (3volumes) etc. Apple Academic Press, USA brought out 5 volume edited book on Ethnobotany of India, and they are being distributed by CRC Press of Taylor & Francis. Apple Academic Press is also bringing another four volume book Global Biodiversity and the book is in press. He was the Principal-Investigator of 20 major Research Projects totalling more than ten million Rupees funded by DBT, DST, CSIR, UGC, BSI, WWF, GCC etc. Under his guidance, 53 students obtained their Ph. D. degrees and 34 students their M. Phil. Degrees. He is the recipient of Best Teacher Award from the Government of Andhra Pradesh, Prof. P. Maheshwari Gold Medal and Dr. G.Panigrahi Memorial Award of Indian Botanical Society and Prof. Y. D. Tiagi Gold Medal of the Indian Association for Angiosperm Taxonomy. He was President of Indian Association for Angiosperm Taxonomy (2013) and President of the Indian Botanical Society (2014). He was a member of Species Survival Commission of International Union for Conservation of Nature and Natural Resources (IUCN).

Dr. S. Karuppusamy obtained his M.Sc. (1996) degree from Madurai Kamaraj University and Ph.D (2002) from Gandhigram University, Tamil Nadu. He was a Post Doctoral Fellow of Sri Krishndevaraya University (2000 – 2004), Anantapur. At present he is working as Assistant Professor, Department of Botany, The Madura College, Madurai since 2008. He has published 120 scientific research papers in nationally and internationally reputed journals. He is the co-author of Flora of Andhra Pradesh (Vol. 5), Flora of Eastern Ghats (Vol. 3 & 4) and *Caralluma* (sensu lato) – Antiobesity plants (2012). He is the recipient of Young Botanist Award 2006 from Indian Botanical Society and Fellow award from Indian Botanical Society (2013). He also recipient of Rolla S. Rao award from Indian Association for Angiosperm Taxonomy (IAAT) (2012) and Fellow of Indian Angiosperm Taxonomy (2012). He also has received a prestigious Prof. Y.S.Murthy gold medal award from Indian Botanical Society for contribution of Angiosperm Taxonomy. He is the Principal Investigator of major research projects funded from UGC, New Delhi, NIF, Ahmedabad, SERB, New Delhi and TBGP-Tamil Nadu Biodiversity Greening Project, Forest Department of Tamil Nadu. He is life member of many scientific societies in India. His research interests in the areas of Angiosperms Taxonomy, Phytochemistry, Ethnopharmacology, Plant tissue culture and conservation biology.

Taxonomy of Angiosperms

(Fourth Revised Edition)

T. Pullaiah

Department of Botany
Sri Krishnadevaraya University
Anantpur 515 003. A.P.

and

S. Karuppusamy

Department of Botany
The Madura College
Madurai 625011 T.N

2018

Regency Publications

A Division of

Astral International Pvt. Ltd.

New Delhi - 110 002

ISBN: 9789387057807 (Internatinal Edition)

Published by : **Regency Publications**
A Division of
Astral International Pvt. Ltd.
– ISO 9001:2015 Certified Company –
4736/23, Ansari Road, Darya Ganj
New Delhi-110 002
Ph. 011-43549197, 23278134
E-mail: info@astralint.com
Website: www.astralint.com

Preface

(Fourth Edition)

The book was out of print long back. The senior author was busy in publication of other books, hence was the delay in revising this book. This is also the reason for co-opting the second author for this revision.

Recent development in the field of Plant Taxonomy, coupled with new and exciting discoveries contributing to the progress in the disciplines that provide information used in new classification, have created the need for a fourth edition of Taxonomy of Angiosperms. The fourth edition is designed to present the updated APG-IV angiosperm classification, describing the developing taxonomic fields of DNA barcoding, cybertaxonomy or digital taxonomy and diversity informatics. Earlier we included orders but now we changed the pattern and selected some important angiosperm families for the taxonomic account. The aim of this book is to provide a broad over view of angiosperm taxonomy with updated data for undergraduate and post graduate students. Also, this book will prove stimulating to serious amateurs, taxonomy aspirants, teachers and professionals who specialise in other fields but utilize updated classification and other recent taxonomic information about flowering plants.

We are indebted to several colleagues, friends and students for fruitful discussion over the years which have crystallized many of the ideas incorporated in this book. We also express sincere gratitude to all those who have helped for conceiving the ideas for preparation of this book.

We take responsibility for errors that remain, we would be grateful if they are brought to our notice.

Anantapur

T. PULLAIAH
pullaiah.thammineni@gmail.com

Madurai

S. KARUPPUSAMY
ksamytaxonomy@gmail.com

Preface

(Third Edition)

I would like to share my feeling of satisfaction in the fact that the second edition of this book is out of print in such a short period. Instead of going for reprint I thought it good to revise the book. Most of the chapters have been thoroughly revised. The chapter on Biodiversity has been revised by Dr. B.Ravi Prasad Rao. Takhtajan's latest classification of 1997 has been included. Two more examples of lowest chromosome numbers have also been included.

I thank all those teachers who gave suggestions for improvement of the book. My Post-Doctoral student Dr. Karuppusamy deserves special mention for stimulating discussion and also a source of inspiration.

3-11-2006 **T. PULLAIAH**

Preface

(Second Edition)

It is heartening to note that the first edition of this book got good response. Several teachers and readers gave suggestions for improvement of this book. Most of these suggestions have been incorporated in the present edition.

Recent developments in sequencing of genetic material has its echo in Taxonomy. A chapter on Molecular taxonomy has been included. The Internet and computer revolution has necessitated the inclusion of a chapter on Computer applications in Taxonomy. A separate chapter on Biodiversity in India with all statistical details has also been given. Information on Taxonomic literature and taxonomists in India has been added. All other chapters have been thoroughly revised.

23.03.2003 **T. PULLAIAH**

Preface

(First Edition)

Taxonomy was a much neglected science in the Twentieth Century. This may be due to advancement of technology and development of newer branches like Biotechnology, Molecular Biology etc. This has resulted in such a situation where there are not enough people to identify plants. Taxonomy teaching in many Universities is much neglected, more so in India. This trend has to be reversed.

Biodiversity and its proper exploitation is gaining ground since last decade. Because of this there is a revival of interest of Plant Taxonomy. There are not enough books to cater to the needs of the students of Botany in India. The present book is an attempt in this direction.

Apart from traditional aspects of plant taxonomy, modern methods have also been discussed. In many universities Biodiversity, forest types of India, Botanical Survey of India, Ethnobotany and Economic Botany has been included in Plant Taxonomy paper, hence these aspects have also been included in this Book.

I thank my teachers and also my students who helped me in preparing this text. I am thankful to Mr. K. Sri Rama Murthy for drawing the diagrams.

I request all the teachers and users of this book to give suggestions for improvement of this Book.

Anantpur

10.12.1997 **T. PULLAIAH**

Contents

	Preface (Fourth Edition)	*v*
	Preface (Third Edition)	*vii*
	Preface (Second Edition)	*ix*
	Preface (First Edition)	*xi*
1.	Introduction	1
2.	History of Plant Classification	7
3.	Systems of Classification	37
4.	APG IV Classification	63
5.	Origin of Angiosperms	75
6.	Plant Nomenclature	83
7.	Plant Identification	101
8.	Field and Herbarium Methods	115
9.	Botanical Gardens	131
10.	Modern Methods in Plant Taxonomy	135
11.	Molecular Plant Taxonomy	183
12.	Computer Applications in Plant Taxonomy	195
13.	Biosystematics	205
14.	Botanical Survey of India	211
15.	Selected Families of Angiosperms	217
16.	Economic Botany	303
17.	Botanical Explorations in India and Taxonomists of India	311
	Index	319

1

Introduction

Taxonomy, the science of classifying plants into groups and identifying them from each other, may be considered as one of the oldest sciences in the world. When the primitive man distinguished the plants that he can eat safely, from those he cannot, he laid the foundation of taxonomy. When he found that the fruits and seeds that he identified as useful to him are not available everywhere he was wandering, he thought of settling in one place and cultivating them, thus heralding the advent of civilization.

The first classification of plants was thus according to whether they are useful or not. When man came to realize that most of the plants are of some use or other, he attempted classification according to their size and habit — into herbs, shrubs and trees. This was followed by various "Natural" systems of classification. The discovery of the presence of male and female parts in the flowers and that the seeds are the products from the union of these two, generated various theories about the origin and evolution of the flower as well as many "Evolutionary" and "Phylogenetic" systems of classification.

What is Taxonomy?

The literal meaning of the term taxonomy is "lawful arrangement" or arrangement by rules (from Greek, *taxis* = arrangement; *nomous* = law, rule). This term can be used in any branch of science e.g., Taxonomy of plants, Taxonomy of Animals, Taxonomy of rocks etc. This means in this chaotic world of objects, taxonomy brings an orderly classification of all the things.

The term classification is applied for both the process of classification as well as for the product of this practice, i.e., the system of classification. Biological classification is the process of grouping together the like organisms and the subsequent placing together of these groups into bigger ones. Therefore, the system of classification contains taxonomic categories like species, genus, family, order, class and division) into which the units or groups of units are placed. The system of classification thus requires a hierarchy of categories for these placements. Therefore, classification, in a biological sense, is the *ordering of the plants into groups that are arranged in a hierarchy.*

The taxonomist has learnt by experience, as principles of classification, two aspects: (i) in process of classification in which the taxonomist should assess the totality of similarities and differences of the organisms studied, and (ii) the characters are to be weighted with respect to their value in circumscribing the groups recognized and differentiating them. Different systems of classifications, such as practical, artificial, natural and evolutionary and phylogenetic systems have been proposed with the needs, levels of understanding the relationships of the groups, data means, and changing perspectives. This developmental process can be stated as the evolution of the current systems of classification. The refinement of the existing classification will go on; it is a continuous process. However, as means, the taxonomists have in mind, the ideal classification. It is a kind of *master classification,* in the structure of which all known facts are incorporated and in the design of which the doctrine of evolution plays a fundamental part.

The term taxonomy was first introduced to plant science by A.P. de Candolle in 1813. According to him plant Taxonomy means theory of plant classification.

The term taxonomy is based on the term taxon. The word 'taxon' was first used by a German Biologist Adolf Meyer in 1926 for animal groups. In Botany it was proposed by Herman J. Lam in 1948.

According to Radford *et al.* (1974) taxonomy includes identification, characterisation and classification (Fig. 1.1).

Phases of Development in Plant Taxonomy

According of Davis (1963) there are four phases in Plant Taxonomy. They are

1. Exploratory or Pioneer Phase

The discovery, description, naming, identification and classification of plants.

2. Consolidation or Systematic Phase

The synthesis, mostly based on gross morphology, of field and herbarium knowledge in the preparation of floras, manuals, monographs, and form-based classification systems.

3. Experimental or Biosystematic Phase

The analysis of breeding systems, variation patterns, evolutionary potential and pertinent work in the chemical, numerical, cytological, anatomical, embryological and palynological aspects of systematics.

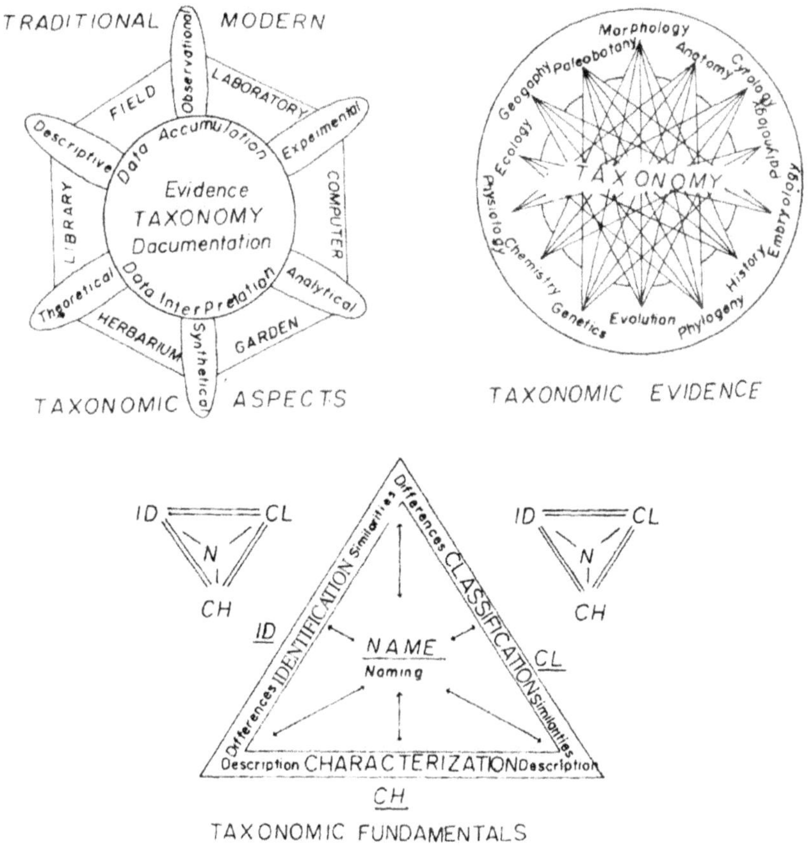

Fig. 1.1: Phases of Development in Plant Taxonomy

Valentine and Love (1958) recognised the first three phases while Davis and Heywood (1963) added the fourth one. The first two phases, which are mainly descriptive and based on gross morphological features, correspond to the "Alpha" taxonomy by Turrill (1938). The classifications of de Candolle and Bentham and Hooker are of this category. The last two phases correspond to the "Omega" classification by Turrill (1938). Today, the systems of plant classification of the world crossed the 'alpha-stage' and progressing nearer to the goal of 'omega' classification (Davis and Heywood, 1963).

Aims and Objectives

The three main aims of plant taxonomy are classification, identification and nomenclature.

Classification

Man has to distinguish objects around him. For this purpose he has to classify them. Early man classified objects around him as animate and inanimate, useful and unwanted, plants and animals etc. In plants he classified them as those that are useful to him and not useful to him. Then he classified them based on habit; herbs, shrubs and trees. When he found the sexual parts, the stamens and the ovary he tried to classify them based on these sexual parts. Later when the number of plants increased he tried to classify them based on Natural relationships. After the Theory of Evolution by Darwin Botanists are classifying the plants according to the phylogenetic relationships.

Identification

There are millions of species around man and these include several thousands of plant species. Man devised various methods to identify them. Modern botanists identify the plants with the help of keys.

Nomenclature

For communicating about the plants around him man has given names to plants. Since there are different languages botanists have devised a method of naming plants known as Binomial nomenclature. These names are in Latin.

Identification, nomenclature and classification are the three important aspects of plant taxonomy. These are the main aspects of modern taxonomy also. Evidences obtained from different branches and their utilization is the important part of modern plant taxonomy. The relationship of these different aspects and taxonomy are shown in Figure 1.1.

Suggested Further Reading

Davis, P.H. and V.H. Heywood. 1963. *Principles of Angiosperm Taxonomy.* Oliver and Boyd. London.

Dikshit, A., M.O.Siddiqui and A. Pathak. 2016. *Taxonomy of Angiosperms.* Studera Press, New Delhi.

Jones, S.B. Jr. and A.E. Luchsinger. 1986. *Plant Systematics.* Prentice Hall, New Jersey.

Judd, W.S., C.S. Campbell, E.A. Kellog, P.F. Stevens and M.J.Donughe. 2002. *Plant Systematics: a phylogenetic approach.* Sinauer Associates Inc. Sunderland, Massachusets.

Lawrence, G.H.M. 1951. *Taxonomy of Vascular Plants.* MacMillan, New York.

Naik, V.N. 1984. *Taxonomy of Angiosperms.* Tata McGraw Hill, New Delhi.

Radford, A.E., W.C. Dickison, J.R. Massey and C.R. Bell. 1976. *Vascular Plantsystematics.* Harper and Row, New York. .

Sivarajan, V.V. 1991. *Introduction to the Principles of Plant Taxonomy.* Oxford & IBH, New Delhi.

Shukla, P. and S.P. Misra. 1985. *An Introduction to Taxonomy of Angiosperms*. Vani Educational Books, New Delhi.

Singh, Gurcharan. 1999. *Plant Systematics Theory and Practice*. Oxford & IBH, New Delhi.

Singh, Gurcharan 2016. *Plant Systematics – An Integrated Approach*. 3rd edition. Science Publishers, Plymouth, UK.

Stace, C.A. 1980. *Plant Taxonomy and Biosystematics*. Edward Arnold, London.

Stuessy, Tod F. 1990. *Plant Taxonomy. The systematic evaluation of comparative data*. Columbia Univ. Press, New York.

Woodland, D.W. 1991. *Contemporary Plant Systematics*. Prentice Hall, New Jersey.

2

History of Plant Classification

Current plant classification systems are labors and insights of many workers throughout the centuries. According to Radford et al. (1976), although there are no sharp lines dividing the various periods, the history of plant classification can be broken down as follows:

- Period of Ancients: to ca. 1500
- Period of Herbalists: 1500-ca. 1580
- Period of Mechanical Systems: 1580-ca. 1760
- Period of Natural Systems: 1760 to ca. 1880
- Period of Phylogenetic Systems: 1880 to date

Period of Ancients

Theophrastus (370-285 B.C.)

Theophrastus, a student of Aristotle is considered as the Father of Botany. His two works about plants are *Enquiry into plants* and the *Causes of Plants*. Among the things noted by Theophrastus were: (1) Distinction between external structures (as organs) and internal structures (as tissues), (2) distinction of different kinds of tissues, (3) classification of plants into four major groups: trees, shrubs, under shrubs and herbs, (4) distinction of flowering plants (Phanerogams) and non flowering plants (Cryptogams), (5) recognition of various kinds of sexual and asexual reproduction, (6) a basic understanding of gross anatomy i.e. that the calyx and corolla are modified leaves etc., (7) recognition of the fruit in its modern technical sense (i.e. pericarp).

Theophrastus also differentiated between centripetal (indeterminate) and centrifugal (determinate) inflorescences, recognized differences in ovary position and in polypetalous and gamopetalous corollas. The plants were further recognized into annuals, biennials and perennials. Theophrastus wrote about nearly 500 different kinds of plants and he called them by the names then in common usage. The generic names *Cratagus, Daucus, Asparagus* and *Narcissus* are among those derived directly from the Greek of Theophrastus and they are used today in the same sense that Theophrastus used them.

Theophrastus was fortunate to have Alexander the Great as his patron. During his conquests Alexander made arrangements to send back materials to Athens enabling Theophrastus to write about exotic plants such as cotton, cinnamon and bananas.

Parasara (250-120 B.C.)

Parasara was an Indian scholar who compiled *Vrikshayurveda* (Science of Plant Life), one of the earliest works dealing with plant life from a scientific stand point. The book has separate chapters on morphology, properties of soil, forest types of India and details of internal structure. He also described the existence of cells (rasakosa) in the leaf, transportation of soil solution from the root to the leaves where it is digested by means chlorophyll (ranjakena pacyamanat) becoming a nutritive substance and yielding by-products. Plants were classified into numerous families (ganas) on the basis of morphological features not known to the European classification until 18th century. Samiganyan (Leguminosae) were distinguished by hypogynous flowers, five petals of different sizes, gamosepalous calyx and a fruit, actually a legume. Swastikaganyan (Cruciferae) similarly has a calyx resembling a swastika, ovary superior, 4 free sepals, 4 free petals, six stamens of 2 are shorter and 2 carpels forming bilocular fruit.

Caius Plinius Secundus (A.D. 23-79)

"Pliny the Elder" , a Roman admiral wrote *Historia Naturalis* (A.D . 77) a 37 volume work where he described about the universe, plants and animals. 9 volumes of this work were devoted to medicinal plants.

Pedanios Dioscorides (First Century A.D.)

Dioscorides was a contemporary of Pliny but we do not know exactly when he lived. He was a physician of Roman Emperor Nero. He traveled extensively to study plants. In his *Materia Medica* (A.D. 1), Dioscorides described about 600 plant species and their uses. In this book Dioscorides grouped plants together with at least a superficial degree of "natural" relationships, e.g., the mints (Lamiaceae) are grouped together and the umbels (Apiaceae) are grouped together. This work was illustrated and served to be one of the most authentic books for Europeans for next 1500 years. *De Historia stirpium* included both good descriptions and illustrations. Linnaeus took several plant names from Dioscorides, and these are in use now. These include: *Anemone, Dipsacum, Aloe, Aristolochia, Phaseolus* etc.

Charaka (First Century A.D.)

Charaka wrote *Charaka Samhita* in which he recognized trees without flowers, trees with flowers, herbs which wither after fructification and other herbs with spreading stems as separate groups. This huge treatise on Indian medicine, containing eight divisions, is largely based on a much earlier treatise published by Agnivesh.

Albertus Magnus (1193-1280)

He wrote on many subjects. The botanical work *De Vegetabilis* dealt with medicinal plants and provided descriptions of plants based on first hand information. Magnus was the first to recognize, on the basis of stem structure, the difference between Monocotyledonous and Dicotyledonous plants. In other major respects he accepted the classification of Theophrastus. The criteria used by Magnus for major categories are: leafy vs. non-leafy; leafy plants divided into Monocots and Dicots, Dicots divided into herbaceous and woody kinds.

Period of Herbalists

Herbal is an illustrated account of medicinal plants. In the late 1400's shortly after the advent of printing, a series of herbals was published under the name of *Gart der Gesundheit* or *Hortus Sanitatis*. These books were often more compilation of the local medical folklore, with crude illustrations, and they were usually published without attribution to any author.

Fifteenth and sixteenth century botany experienced a great general awakening in the field of medicine and classification. This was the period of plant description and illustration. Many of the world famous herbals appeared during these 200 years.

Otto Brunfels (1464-1534)

He was a practicing physician, *Herbarium Vivae Eicones* (1530-1536), a 3 volume work, written by Otto Brunfels is not much valuable from the classification point of view but is well known for illustration of living plants. He was the first person to recognize the Perfect (flowers visible when held at arm's length) and Imperfect (flowers not visible).

Jerome Bock (1498-1554)

Bock, a German, wrote under Latinisation his name as *Hieronymus Tragus*. He divided plants into trees, shrubs and herbs but endeavored to bring together related plants within these categories. He gave concise descriptions. In addition he provided notes on the natural distribution of many plants. He described all this in his book *Neue Kreuterbuch* (1539). He described 567 species.

Leonhart Fuchs (1501-1566)

Fuchs, a Professor of medicine, grew many of his medicinal plants and observed their characteristics. He was a German. He gave fine engravings. He described New world plants. He was a Bavarian physician. His herbal, *De Historia stirpium* (1542) included both good descriptions and illustrations of 487 species of medicinal plants.

Pierandrea Mattioli (1501-1577)

An Italian herbalist, Mattioli gave excellent illustrations.

Valerius Cordus (1515-1544)

He travelled in the forests of Germany and Italy. His work *Historia Plantarum* published in 1561, many years after his death, contained accurate descriptions of 502 species, 66 apparently new. He gave excellent descriptions.

Rembert Dodoens

Dutch botanist. He gave excellent descriptions and illustrations in 1583.

Charles de l'Ecluse (1526-1609)

A Flemish Physician, Ecluse treated plants of America, Spain, Portugal, Austria and Hungary.

Mathias de l'Obel (1538-1616)

Dutch herbalist. Classification based upon general form, growth habit, economic uses and width of leaf (separated monocots and dicots)

Grass like plants	*broader leaved*				
with narrow leaves→	*bulbous and→*	*dicot→*	*shrubs→*	*trees*	
(supposed simple	*rhizomatous*		*herbs*		*(most*
forms)	*monocots*				*perfect)*

Rycharde Bankes

Bankes in 1525 dealt with medicinal plants in his herbal.

William Turner

Plants were arranged, in 1561, 1562, 1568 in alphabetical sequence by Latin name.

The quality of description and illustration improved greatly during the period of Herbalists. There was a large increase in the number of plants known; the discovery of new world had an important impact during this period. There was some attempt to group closely related plants together. Some large natural families were recognized, such as composites, umbels, grasses, rushes etc.

Many genera commemorate herbalists names, as for example, *Brunfelsia* for Brunfels, *Fuchsia* for Leonard Fuchs, *Lobelia* for Mathias de l'Obel, *Gerardia* for John Gerard, *Clusia* for Charles L'Ecluse.

Period of Mechanical Systems

It is from the sixteenth and seventeenth centuries onwards that the science of botany developed as an independent discipline. As a result, there were attempts to study more and more plants and a large number of characters in order to arrive at a satisfactory classification.

Andraea Caesalpino (1519-1603)

Caesalpino, an Italian botanist who studied botany under Luca Ghini and became the Director of the Botanical Garden and later Professor of Botany and

Medicine at Bologna. He went to Rome in 1592 as the personal physician to Pope Clement VIII. He prepared a herbarium of 768 well-mounted plants in 1563 which is still preserved in the Museum of Natural History at Florence. He was the first botanist to try to base a taxonomic scheme upon reason and logic rather than purely on utilitarian concepts. He was philosophically an Aristotelian in his approach and he based his classification on the assumption that certain features of plant structures were intrinsically more useful than others. Such an approach is termed a priori reasoning and while it may lead to erroneous conclusions it does provide a way of testing a hypothesis.

Caesalpino's (1583) famous book, *De plantis Libri* contains descriptions of about 1520 plant species. The most fundamental drawback of his system is total denial of sex in spite of observation of presence of ovary, seed and fruit.

His classification is based upon woodiness vs. herbaceousness, then on characters of fruits, seeds and embryos. Fruits were classified on basis of position, locule number and seed number. Classification stresses number, position and form of parts. He recognized inferior/ superior ovary, bulbs present/absent, sap milky/ watery, activities of plants. He recognized Fagales, legumes, umbels, crucifers, composites and borages as natural groups.

Jean Bauhin (1541-1631)

Jean Bauhin, a French and Swiss physician, is important for his excellent illustrated *Historia Plantarum universalis* published posthumously (1650) in three volumes. This was a comprehensive work and dealt with synonymy of about 5000 plants. In addition, it included description, which for the first time in botanical history, were good diagnoses of the species he treated. He recognized crucifers and labiates.

Gaspard Bauhin (1560-1624)

Gaspard Bauhin, a Swiss botanist and brother of Jean Bauhin, put out a monumental opus in his *Pinax theatri botanici* (1623). This is a register of 6000 different kinds of plants known to exist up to that time and it contains an account of what name the various different botanists and herbalists had used for each plant. In other words, Gaspard Bauhin accounted for the rapidly developing synonymy.

Gaspard Bauhin classified plants on the basis of texture and form. He is to be remembered also as one of the first to distinguish nomenclaturally between species and genera. To many of the plants classified and described by him, he gave a generic name and specific (trivial) epithet. This binary nomenclature, with which Linnaeus is usually credited, was founded, therefore, by Bauhin more than a century before its use by Linnaeus.

His other books include *Phytopinax* (1596), *Prodromus theatri botanici* (1620) and *Pinax theatri botanici*.

Robert Morrison (1620-1683)

Morrison took his first appointment at Blois on the Loire as the Keeper of the gardens for Duke of Orleans. A few years later he was appointed as Physician for the King Charles II and Regius Professor of Botany. Lastly he took over as in-charge of the Physic Garden at Oxford. In his *Plantarum Historia Universalis* (1623) and Introductio Generalis in Rem Herbarium Morrison made usage of habit (woody/ herbaceous) fruit and seed characters for distinguishing the class.

John Ray (1628-1705)

Another man who made a noteworthy contribution to the growth of systematic botany throughout the 17th century was the Englishman, John Ray. He published numerous works, but most significant for us are his *Methodusplantarum nova* (1628) and the *Historia plantarum* (3 volumes, 1686-1704). His classification is as follows: *Methodus plantarum* (1703).

- Herbae
  - Imperfectae (basically Cryptogams)
  - Perfectae
    - Dicotyledones
    - Monocotyledones
- Arborae
    - Monocotyledones
    - Dicotyledones

Ray proposed a classification accounting for nearly 18000 species, and under the main divisions of woody and herbaceous plants, he recognized the taxa of monocots and dicots, the classes based on fruit type (cone bearing, nut bearing, bacciferous, pomiferous, pruniferous and siliquous), and subdivided these on the basis of leaf and flower characters. Some of the groups in Ray's classification correspond to the present day families such as Umbelliferae, Verticillatae (Labiatae), Tetrapetalae (mostly Cruciferae), Leguminosae, Stamineae (grasses) etc. Ray produced a system of classification in which plants that looked alike were grouped together, and he recognized the fact that individual characters in themselves may or may not be worthwhile taxonomic features.

Pierre Magnol (1638-1715)

French botanist Magnol is the botanist to use the term family in the classification of plants. In 1689 he listed 76 families. He used conspicuous characters of roots, stems, flowers and seeds.

Joseph Pitton de Tournefort (1656-1708)

Born in Aix, France, Tournefort developed an interest in the collection of medicinal plants right from his school days. He worked as a Professor of Botany at the Jardin des Plantis in Paris. He collected plants from Greece, Asia minor, France and Spain. His publications include *Elements de Botanique* (1694) and *Institutiones Rei Herbariae* (1700). He is the author of modern genus concept and is considered to have had a good concept of genera. He gave generic descriptions and recognized 698 genera and 10,146 species. Tournefort is also the first botanist to use the term Herbarium. Many of his generic names were validated by later botanists and remain well known to the present. These include *Salix, Populus, Fagus, Betula, Lathyrus, Acer, Verbena, Castanea, Quercus, Ulmus* and *Abutilon*.

Tournefort divided flowering plants into 2 large categories, trees and herbs. Each of these was subdivided into groups based on characters such as apetalous or petalous and then on polypetalous or gamopetalous and flowers regular or irregular. His system was widely used until about 1760, when it was superseded by that of Linnaeus while in France it was used until 1780 when it was superseded by de Jussieu.

Carolus Linnaeus (1707-1778)

Carolus Linnaeus is regarded as father of modern taxonomic botany and zoology. Linnaeus was born in Rashault in Sweden. He was educated at the University of Uppsala and eventually obtained M.D. degree at the University of Harderwijk in Netherlands.

Rudolf Camerarius in 1694 discovered sexuality in plants. Linnaeus was impressed by the ideas of Camerarius and considered the male and female sex organs (stamens and gynoecium) as important for classification of plants. He classified all plants based on sexual characters and hence his classification is known as sexual system.

While working under the guidance of Dr. Rudbeck, Linnaeus published his first paper in 1729 on sexuality of plants. In 1730 he published, under the latinized title of *Hortus Unpandicus*, an enumeration of plants in Uppsala Botanic Garden. In this book he listed the plants according to Tournefort system. But in the revised edition of *Hortus Uplandicus* published in 1732 he arranged the plants according to his own system, his so-called sexual system. This served as the basis for his later publication *Systema Naturae* (1735), where he classified plants, animals and minerals. In 1736 he published *Fundamenta Botanica*. In 1737 four of his valuable publications — *Flora Lapponica, Hortus Cliffortianus, Critica Botanica* and *Genera Plantarum* have been published.

He proposed his sexual system of classification in *Genera plantarum* (1737) where he gave descriptions to 935 genera. This was published in five editions and two supplements and in all a total of 1336 genera were diagnosed.

Linnaeus' systematic contributions did not arrive on the scene in one piece. They were proposed in several publications over the years, but the place where he got it all together is his '*Species Plantarum*', a two volume catalogue of the plant kingdom, published in May, 1753. This book marks the first consistent use of binomial nomenclature, and it has subsequently been adopted by the botanical community as the starting point for modern botanical nomenclature. In *Species Plantarum* (1753) Linnaeus diagnosed 6000 species and 1000 genera.

Linnaeus recognized 24 classes, which are based on number, union and length of stamens. These classes were subdivided into orders on the basis of number of styles and unisexuality. All the non-flowering plants were lumped together in a single class, the Cryptogamia. Although this system is highly artificial it was accepted at that time because it provided a very easy means of identification.

In *Philosophia Botanica* (1751), Linnaeus enumerated 67 "natural orders" . Some of these represent natural groups like Palmae, Orchidaceae, Graminae, Coniferae,

Compositae, Boraginaceae etc, but some orders are quite mixed, the monocots and dicots appearing together.

Linnaeus had a large number of students who explored several parts of the world. He died in 1778 and his collections were sold by his widow to an English botanist, J.E. Smith. Later, these became the property of the Linnaen society of London, founded in 1788.

Table 2.1: Linnaeus' Sexual System (Systema Naturae, 1735; Genera Plantarum, 1737; Species Plantarum, 1753)

Class	(Ordo)	Examples of Species
1. Monandria	Monogynia	*Canna indica*
(Stamen one)	Digynia	*Cinna arundinacea*
II. Diandria	Monogynia	*Ligustrum vulgare, Chionanthus virginica*
(Stamens two)	Digynia	*Anthoxanthum odoratum*
	Trigynia	*Piper nigrum*
III. Triandria	Monogynia	*Valeriana officinalis, Melothria pendula*
(Stamens three)	Digynia	*Phalaris canariensis, Triticum aestivum*
	Trigynia	*Mollugo verticillata*
IV. Tetrandria	Monogynia	*Dipsacus fullonum, Plantago major*
(Stamens four)	Digynia	*Cuscuta americana*
	Tetragynia	*Ilex cassine*
V. Pentandria	Monogynia	*Heliotropium indicum, Lysimachia quadrifolia*
(Stamens five)	Digynia	*Asclepias syriaca, Carum carvi*
	Trigynia	*Rhus vernix, Alsine media*
	Tetragynia	*Parnassia plaustris*
	Pentagynia	*Aralia spinosa, Suriana maritima*
VI. Hexandria	Monogynia	*Tillandsia utriculata, Pontederia cordata*
(Stamens six)	Digynia	*Oryza sativa*
	Trigynia	*Rumex verticillatus, Medeola virginiana*
	Tetragynia	*Petiveria alliacea*
	Polygynia	*Alisma cordifolia*
VII. Heptandria(7)	Monogynia	*Aesculus pavia*
VIII. Octandra	Monogynia	*Tropaeolum major, Gaura biennis*
(Stamens 8)	Digynia	*Galenia africana*
	Trigynia	*Polygonum persicaria*
	Tetragynia	*Paris quadrifolia*
IX. Enneandria	Monogynia	*Laurus benzoin*
(Stamens 9)	Trigynia	*Rheum rhaponticum*
	Hexagynia	*Butomus umbellatus*

Class	(Ordo)	Examples of Species
X. Decandria	Monogynia	*Cercis canadensis, Melia azedarach*
(Stamens 10)	Digynia	*Tiarella cordifolia*
	Trigynia	*Silene virginica*
	Pentagynia	*Oxalis stricta, Agrostemma githago*
	Decagynia	*Phytolacca americana*
XI. Dodecandria	Monogynia	*Asarum canadense, Rhizophora mangle*
(Stamens	Digynia	*Agrimonia eupatoria*
12-19)	Pentagynia	*Glinus lotoides*
	Dodecagynia	*Sempervivum arboreum*
XII. Icosandria	Monogynia	*Cactus opuntia, Prunus virginiana*
(>19, attached	Digynia	*Crataegus crusgalli*
to calyx)	Trigynia	*Sorbus domestica*
	Pentagynia	*Pyrus communis*
	Polygynia	*Rosa canina*
XIII. Polyandria	Monogynia	*Capparis spinosa, Podophyllum peltatum*
(Stamens >19,	Digynia	*Paeonia officinalis*
attached to receptacle)	Trigynia	*Delphinium ajacis*
	Tetragynia	*Tetracera volubulis*
	Pentagynia	*Aquilegia canadensis*
	Hexagynia	*Magnolia virginiana*
	Polygynia	*Annona glabra, Anemone virginiana*
XIV. Didynamia	Gymnospermia	*Ajuga reptans*
(Stamens	Angiospemia	*Scrophularia marilandica*
didynamous)		
XV. Tetradynamia	Siliculosa	*Draba verna*
(Stamens tetra-	Siliquosa	*Erisymum officinale*
dynamous)		
XVI. Monadelphia	Pentandria	*Waltheria indica*
(Stamens in	Decandria	*Geranium maculatum*
one bundle)	Polyandra	*Sida rhombifolia*
XVII. Diadelphia	Hexandria	*Fumaria officinale*
(Stamens in two bundles)	Octandria	*Polygala incarnata*
	Decandria	*Erythrina herbacea*
XVIII. Polyadelphia	Pentandria	*Theobroma cacao*
(Stamens in	Icosandria	*Citrus aurantium*
several bundles)	Polyandria	*Hypericum perforatum*

Class	(Ordo)	Examples of Species
XIX. Syngenesia	Polygamia Aequalis	*Tragopogon virginicum*
(Stamens with united anthers)	Polygamia Superflua	*Artemisia annua*
	Polygamia Frustranea	*Helianthus annuus*
	Polygamia Necessaria	*Chrysogonum virginianum*
	Monogamia	*Lobelia cardinalis, Viola canadensis*
XX. Gynandria	Diandria	*Orchis spectabilis*
(Stamens adnate to pistil)	Triandria	*Sisyrinchium bermudiana*
	Tetrandria	*Nepenthes distillatoria*
	Pentandria	*Passiflora incarnata*
	Hexandria	*Aristolochia serpentaria*
	Decandria	*Helicteres angustifolia*
	Polyandria	*Arum triphyllum*
XXI. Monoecia	Monandria	*Zannichellia palustris*
(Plants	Diandria	*Lemna minor*
monoecious)	Triandria	*Typha latifolia, Zea mays*
	Tetrandra	*Urtica diocia*
	Pentandria	*Ambrosia trifida*
	Hexandria	*Zizania aquatica*
	Heptandria	*Guettarda speciosa*
	Polyandria	*Myriophyllum spicatum, Quercus phellos*
	Monadelphia	*Pinus taeda, Ricinus communis*
	Syngenesia	*Momordica cylindrica*
	Gynandria	*Andrachne fruticosa*
XXII. Dioecia	Monandria	*Najas marina*
(Plant's	Diandria	*Salix babylonica*
dioecious)	Triandria	*Empetrum nigrum*
	Tetrandria	*Viscum album, Myrica cerifera*
	Hexandria	*Smilax lanceolata*
	Octandria	*Populus alba*
	Ennaeandria	*Mercurialis procumbens*
	Decandria	*Carica papaya*
	Polyandria	*Cliffortia trifoliata*
	Monadelphia	*Juniperus virginiana*
	Syngenesia	*Ruscus aculeatus*
	Gynandria	*Clutia retusa*
XXIII. Polygamia	Monoecia	*Celtis occidentalis, Veratrum nigrum*
(Plants	Dioecia	*Nyssa aquatica*
polygamous)	Polyoecia	*Ficus carica*

Class	(Ordo)	Examples of Species
XXIV. Cryptogamia (Flowers concealed)	Filices Musci	*Equisetum arvense, Adiantum pedatum* *Lycopodium inundatum, Hypnum cuspidatum*
	Algae Fungi	*Marchantia polymorpha, Ulva canfervoides* *Hydnum imbricatum*

Period of Natural Systems

During this period plants were classified based on their relationships and plants that looked alike were kept together.

Michel Adanson (1727-1806)

Adanson, a French botanist, while exploring the plants wealth in Africa, realized the inadequacy of the Linnaen system and developed his own system in *Families des Plantes* (1763). He rejected artificial classification. He emphasized equal weighing of characters and believed that all organs should be taken into account. He rejected a priori reasoning. Adanson is considered as "Grand Father of Numerical Taxonomy" .

De Jussieu

The de Jussieu family of France had four members who made noteworthy contributions to botany. Antonie (1686-1758), Bernard (1699-1996), Joseph (1704-1779) and Antonie Laurent (1748-1836). The first three of them are brothers while Antonie Laurent is nephew of Bernard. All four of them were connected with the Jardin des Plantes in Paris. Antonie and Bernard studied under Pierre Magnol.

Bernard never published his system during his lifetime due to personal dissatisfaction over perfection of his system. He rearranged the plants in a garden at La Trianon, Versailles according to his own system, which was not based on habit and was not purely artificial.

Antonie Laurent de Jussieu published his uncle's system, with some improvements in '*Genera Plantarum*' (1789). This is the first major work to be natural in its approach, which in practice meant that plants that looked alike were grouped together.

Antonie Laurent de Jussieu divided the flowering plants into 15 classes and these classes were subdivided into 100 natural orders corresponding to most major families. Each of these was clearly differentiated, named and provided with a description. Some related families such as Palmae, Liliaceae, Amaryllidaceae and Iridaceae were grouped together. Monopetalae contained families now known as Gamopetalae, while his class Diclines Irregulares (an unnatural assemblage) contained the Coniferae, Amentiferae, Urticae, Cucurbitaceae and Euphorbiae.

System of A.de Jussieu (*Genera Plantarum*, 1789)

Class	Order (Family) Examples
I. Acotyledones	I. Fungi, Algae, Hepaticae
II. Monocotyledones	
Stamina hypogyna	II. Aroideae, Typhae
Stamina perigyna	III. Palmae, Junci
Stamina epigyna	IV. Musae, Orchides
III. Dicotyledones	
Apetalae	
Stamina Epigyna	V. Aristolochiae
Stamina perigyna	VI. Elaeagni, Lauri
Stamina hypogyna	VII. Amaranti
Monopetalae	
Corolla hypogyna	VII. Acanthi, Vitices
Corolla perigyna	IX. Guaiacanae, Ericae
Corolla epigyna	
Antheris Connatus	X. Cichoraceae
Antheris distinctis	XI. Dipsacaceae
Polyepetalae	
Stamina Epigyna	XII. Araliae, Umbelliferae
Stamina Hypogyna	XIII. Cruciferae, Acera
Stamina perigyna	XIV. Cacti, Myrti
Diclines irregulares	XV. Euphorbiae, Urticae

de Candolle

Three generations of de Condolles have contributed much to the science of systematic Botany. The first of this Swiss-French family is Augustin Pyramus de Candolle (1778-1841).

A.P. de Candolle published in 1813 a book entitled *Theorie Elementaire de la Botanique* in which the basic principles of plant classification were given. He introduced the term Taxonomy to designate the theory of plant classification.

A.P. de Candolle's monumental work *Prodromus Systematis Naturalis Regni Vegetabilis* was an attempt to write a flora for the whole world. This work began in 1816 and before A.P. de Candolle died in 1841 seven volumes have been published. The next 10 volumes, written by specialists were published under the editorship of his son, Alphonse de Candolle (1806-1893). In this book A.P. de Candolle, proposed to classify and describe every species of seed plants then known to science. The Prodromus included descriptions of 58,975 species of Dicotyledons and Gymnosperms known at that time.

De Candolle's system of classification was an expansion and elaboration of de Jussieu's system. In this system the arrangements of Dicots has improved (united Apetalae and Diclines). The families were arranged in a linear sequence starting from the polypetalous Ranunculaceae, through gamopetalous groups and apetalous groups. The choice of Ranalian group as a starting point in linear sequence of the families had done much to popularize the system. The de Candolle system dominated until 1860, despite the fact that nearly 24 systems were proposed during the period 1825-1845.

System of Augustin Pyramus and Alphonse de Candolle. 1816 and 1873.

Prodromus Systematis Naturalis Regni Vegetabilis

I. Vasculares (Vascular plants with cotyledons):
  Class I Dicotydedoneae (includes Conifers):
    Subclass I. Thalamiflorae (Polypetalous and hypogynous):
      Orders (Families): 1-54. Ranunculaceae, Dilleniaceae, Magnoliaceae, Rutaceae, Simarubeae, Ochnaceae etc.
  Subclass II. Calycifloreae (Perigynous or epigynous and polypetalous or sympetalous):
      Orders (Families): 55-118. Celastrineae, Rhamneae, Bruniaceae, Samudeae, Homalineae, Pyrolaceae, Monotropeae etc.
  Subclass III. Corollifloreae (Gamopetalous and hypogynous):
      Orders (Families):119-155. Lentibulariaceae, Primulaceae, Aegiceraceae, Solanaceae, Plantaginaceae etc.
  Subclass IV. Monochlamydeae (Calyx only present):
      Orders (Families): 156-207 Phytolaccaceae, Salsolaceae, Basellaceae, Amaranthaceae, Euphorbiaceae etc.
  Class II Moncotyledoneae (includes cycads):
    Subclass I. Phanerogamae (Liliaceae, Iridaceae, Orchidaceae etc.).
    Subclass II. Cryptogamae (Ferns etc.)
II Cellulares (Plants without cotyledons):
    A. Mosses, Liverworts
    B. Lichens, Fungi, Algae
(Ferns placed in Cellulares in last arrangement of de Candolle system)

After Prodromus a monograph series was published by Alphonse and later his son Anne Casimir (1836-1918). About 100 such monographs have been published which, even now, provide a lot of information.

Robert Brown (1773-1858)

Robert Brown of Scotland did not propose any system of classification but contributed to better understanding of floral morphology and to classification problems. He pointed out the nature of Gymnosperms (naked seeds) and their distinctness from Angiosperms.

John Lindley (1799-1865)

Lindley, an English man, in his *Introduction to the Natural System of Botany* (1830) proposed the first comprehensive natural system published in English. He recognized the difference between Gymnosperms and Angiosperms. The Dicots were divided into Polypetalae, Apetalae, Achlamydeae and Monopetalae. In his *Vegetable Kingdom* (1846), Lindley overemphasized the importance of the parasitic habit of the dicots and the importance of net vs. parallel venation among monocots.

Stephan Endlicher (1805-1849)

Endlicher, a Viennese Botanist divided the plants into the Thallophytes (algae, fungi and lichens) and Cormophytes (mosses, ferns and seed plants). He presented this system in his book *Genera Platarum* (1840). Lower dicots were apetalous groups, a practice later followed in later (Englerian) systems as opposed to the beginning with Ranalian groups in de Candolle, Bessey and Hallier systems. Dicots were divided into Apetalae, Gamopetalae and Dialypetalae. Some parasitic Angiosperms were placed with Cryptogams.

Adolphe Theodore Brongniart (1770-1847)

Brongniart, a French botanist, divided the plants into Cryptogams and Phanerogams (seed plants). His 'dicots' included both Gymnosperms and Angiosperms. He divided the Angiospermous dicots into Gamopetalae and Dialypetalae. He did not consider the taxon often referred to as apetalae to be one of closely related elements but distributed the species among those petaliferous families to which they appeared to be more closely related.

Alexander Braun

A German taxonomist, Alexander Braun in his '*Flora der Provinz Bradenburg* (1850) placed Monocots and Dicots together and separated from Gymnosperms. The Monocots were arranged to show a progression from the simple to the complex—began with Lemnaceae and ended with Orchidaceae. Similarly he began the Dicotyledons with Apetalae followed by Sympetalae, Eleutheropetalae and ending with Leguminosae. This classification influenced his successors Eichler, Engler, Wettstain and others.

George Bentham (1800-1884) and Joseph Dalton Hooker (1817-1911)

George Bentham, an Englishman, was an amateur botanist until almost middle age, after which time he gave the subject of systematic botany all his attention. In addition to being most critical, discrimination and analytical systematist he also was an accomplished linguist and Latinist. Prior to his joint publication with Hooker of the Genera Plantarum, Bentham published world monographs of the families Labiatae, Ericaceae, Polemoniaceae, Scrophulariaceae and Polygonaceae. Bentham was the author of the 7-volume *Flora of Australia*.

Sir Joseph Dalton Hooker, son of the botanist Sir William Jackson Hooker (1785-1865), was more the plant explorer and plant geographer than was Bentham.

He collected plants from Himalayas, Lebanon, Antarctic and Atlas mountains. He succeeded his father as Director of Royal Botanical Gardens, Kew. Joseph Dalton Hooker also wrote *Flora of British India* (1872-1897), *Student's Flora of British Isles* (1870) and also revised later editions of Hand- book of British Flora. He also supervised the publication of *Index Kewensis.*

Bentham and Hooker published their 3 volume work *Genera Plantarum* in Latin, at intervals between 1862 and 1883. This work comprised the names and descriptions of all genera of seed plants then known, classified according to their own system.

This system was patterned directly on that of de Candolle. But every genus was studied anew from the material of British and continental herbaria. Full and complete descriptions were prepared from studies and dissections of the plants themselves and did not represent a compilation made from literature.

Bentham and Hooker divided all seed plants into 3 classes, 3 subclasses, 21 series, 25 cohorts and 202 orders. Originally it was designed to include 200 orders and each order was given a definite number. Orders Vochysiaceae and the Cyrilleae were incorporated later. These were not given separate numbers but were included as 20 a and 46 a respectively.

The seed plants were considered by them to number about 97, 205 species distributed among the major taxa and categories as indicated in Table 2.1.

Table 2.2: Distribution of taxa in Bentham and Hooker's *Genera Plantarum*

	Orders (Families)	Genera	Species (Estimated)
Dicotyledons	84	2,610	31,874
Gamopetalae	45	2,619	34,556
Monochlamydeae	36	801	11,784
Gymnospermae	3	44	415
Monocotyledons	34	1,495	18,516
Total	202	7,569	97,205

In this work the family of contemporary systems was termed as order while that of an order was termed as cohort. The name cohort was first used by Endlicher for a taxon of higher level.

This system was accepted throughout the British Empire and in the United States and was adapted to a lesser extent by some continental botanists.

Bentham and Hooker's system is Post-Darwinian in chronology but Pre- Darwinian in concept. It is interesting to know that the publication of Darwin's theories of evolution and Origin of Species coincided with the time of production of the first volume of Bentham and Hooker's *Genera Pantarum*. Hooker then favoured a complete reorganization of their classification but was deterred from effecting it by Bentham, who did not accept the essentials of Darwin's work, although he did so about a decade later.

The system of Bentham and Hooker is given in detail in the next chapter.

Period of Phylogenetic Systems

The theory of Evolution proposed by Darwin in 1858 influenced taxonomy in many ways. Several systems conforming to this theory have been proposed since then and still continue to be proposed.

Post Darwinian systems can be broadly categorized into two groups: the Englerian school and the Ranalian school. The Englerian school considered simplicity as primitive and complexity as advanced and is based on the concept of progressive evolution. The Ranalean School considered the Ranales group as the primitive and evolution proceeded in both progressive and retrogressive manner.

Englerian School

August Wilhelm Eichler (1839-1887)

Eichler, a German, proposed his classification in his book *Bluthendiagramme* (1875-1878). This was not a Phylogenetic system in the modern sense, but Eichler did accept the concept of evolution. In 1883 he elaborated his earlier treatise into a unified system accounting for all major groups of the entire plant kingdom

Eichler's system is given briefly below:

A. Cryptogamae

I. Divison: Thallophyta

1. Class: Algae
2. Class: Fungi

II. Division: Bryophyta

1. Class: Hepaticae
2. Class: Musci

III. Division: Pteridophyta

1. Class: Equisetinae
2. Class: Lycopodinae
3. Class: Filicinae

B. Phanerogamae

I. Division: Gymnospermae

II. Division: Angiospermae

1. Class: Monocotyleae
2. Class: Dicotyleae

(i) Subclass: Choripetalae

(ii) Subclass: Sympetalae

Eichler began Dicots with Amentiferae and united apetalous and polypetalous dicots. He considered simplicity as primitive.

Adolf Engler (1844-1930)

Engler was a Professor of Botany at the University of Berlin and Director of Berlin Botanical Garden. In 1892 he published Guide to Breslau Botanic Garden. Engler in collaboration with Karl Prantl (1849-1893) presented a 20 volume work *Die Naturlichen Planzefamilien* (1897-1915). This was an illustrated work, with modern keys. It provided a means for identification of all the known genera of plants from Algae to the most advanced seed plants. A revised edition of this work was published by Engler in collaboration with Gilg in 1924 and with Diels in 1936 under the Syllabus der pflanzenfamilien.

Engler considered Monocotyledons primitive than Dicotyledons. According to Engler and his associates the most primitive flowering plants in both Monocotyledons and Dicotyledons have apetalous simple flowers in strobilus like inflorescences.

Engler considered unisexual flowers as primitive and bisexual flowers as advanced. The Englerian school further maintains that these unisexual flowers were wind pollinated like the cone of Gymnosperms and that the different groups of Angiosperms have been derived from various groups of Gymnosperms as well as from Pteridophytes, thus proposing a polyphyletic origin.

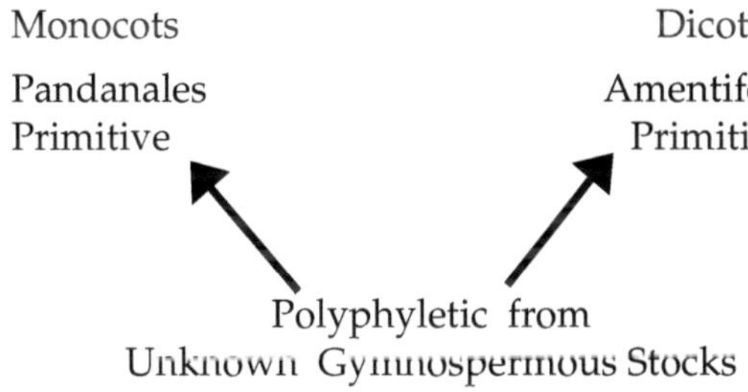

The seed plants (termed Embryophyta Siphonogama by Engler) were divided into Gymnospermae and Angiospermae. The Angiospermae was divided into two classes Monocotyledonae and Dicotyledonae, the latter into two sub- classes Archichlamydeae (composed of Apetalae without petals and Choripetalae with separate petals) and Metachlamydeae (with united petals).

Englerian evolutionary principles were challenged by Bessey (1897), Hallier (1905), Arber and Parkin (1907) and others. Among the objectionable features was the acceptance by Engler of the Dichlamydeous flowers as derived from Monochlamydeous flowers, the derivation of all parietal placentation in syncarpous ovaries from axile placentation, of free central from parietal, and the interpretation of the majority of simple unisexual flowers as primitive.

The materials in the American and continental herbaria were arranged according to Engler's system.

Richard von Wettstein (1862-1931)

Wettstein, an Austrian Botanist, presented his system of classification in 1901 in *Handbuch der Systematischen Botanik.* This underwent revisions, the fourth and the last one being published posthumously in two volumes (1930-35).

Wettstein believed that Angiosperms have evolved from Gymnosperms, which were very much similar to the present day Gnetales. His system is similar to that of Engler but Wettstein rearranged the relative positions of many Dicots families. Wettstein also differed with Engler in considering Dicotyledons as more primitive than Monocotyledons and the later to have been derived from the Ranalean stocks. He considered the Casuarinales as the primitive most Angiospermous group and believed that *Casuarina* was derived from *Ephedra.*

Alfred Barton Rendle (1865-1938)

An Englishman, Rendle presented his system in his 2-volume *Classification of flowering plants* (1904, 1925). A revision of this has been published in 1930 and 1938. This system is basically Englerian type and considered Amentiferae as primitive group. The Monocotyledons were treated primitive than Dicotyleddons. Wind pollination and woodiness were considered by Rendle as primitive features. Rendle does not claim that his classification is phylogenetic in the complete sense. He divided the Dicotyledons into 3 groups. Monochlamydeae, Dialypetalae and Sympetalae.

August A. Pulle (1879-1950)

Pulle of The Netherlands in 1938 published a system in his book *Compedium van de Terminologie, Nomenclatur en Systematiek.* This book was revised in 1950. Pulle's system was a modification of Engler's classification. He accepted Spermatophyta as a division but differed from Engler in treating Spermatophyta into four subdivisions-Pteridospermae, Gymnospermae, Chlamydospermae and Angiospermae. The Angiospermae were divided into Monocotyledonae and Dicotyledoneae. The Dicotyledoneae were divided into 8 series. In the revised edition published in 1950 he merged the Cactales with Centrospermae, a view accepted by most of the recent taxonomists. Pulle did not consider the Monochlamydeae to be a natural group. In the opinion of the Pulle the Monochlamydeae and Sympetalae are not natural groups.

Bunzo Hayata (1921)

Hayata produced a dynamic system of Angiosperm classification, which was basically a rearrangement of Engler's system. Hayata's arrangement points out the reticulate nature of character variation in the Angiosperms.

Carl Skottsberg (1940)

Skottsberg of Sweden modified Englerian classification. He considered that the Monocots have been derived from an unknown primitive Dicot. He thought that apocarpy and polycarpy as primitive, Syncarpy and monocarpy as advanced in both Monocots and Dicots. The apetalous forms were considered polyphyletic. The Amentiferae were placed after Rosales, but not redistributed among petaliferous families to which they appear to be more closely related. He began the Dicotyledons with Casuarinaceae but did not consider them to be ancestral to other Dicotyledons.

The Ranalian School

Arber and Parkin (1907) proposed that the extinct Gymnospermous group Bennettitales has given rise to *Magnolia* like Angiosperms. Bennettitales have bisexual strobilus and *Magnolia* also has bisexual flower. This view has been supported by several authors like Bessey (1915), Hutchinson (1959, 1969), Cronquist (1968, 1981), Takhtajan (1967, 1980) and others.

Charles Edwin Bessey (1845-1915)

C.E.Bessey was a Professor of Botany at the University of Nebraska, and student of Asa Gray. Bessey in 1893 read a paper presenting the nucleus of what was to become his well-known classification; this was amplified in 1897 and his final treatment published in 1915. This, in many respects, was a modification of Bentham and Hooker's system realigned according to evolutionary principles.

Bessey considered the seed plants to have had polyphyletic origin and to be composed to three separate phyla of which he dealt only with the Anthophyta (Angiosperms). He derived the Anthophyta from his Cycadophyta and the latter from implied Bennettitalian ancestry. He divided the Angiosperms into the Oppositifoliae (Dicotyledons) and Alternifoliae (Monocotyledons). His classification of these two taxa was based on evidence obtained from Paleobotanical, ontogenetical and morphological studies of homologies.

Bessey considered the Ranales or their ancestors as primitive Angiosperms. The evolutionary lines of development of various orders were summarized in a way of branched prickly pear plant, which is popularly known as Bessey's Cactus (Fig.2.1). There are three major series derived from a single stock. The branch on the left represent monocots, the other two branches indicate the two major groupings, Cotyloideae (the Rosalean line) and Strobiloideae (the Ranalian line) within the Dicotyledons.

Bessey derived a set of "dicta" or propositions based upon his own experience and sources in the literature as to what structural feature might be considered primitive, and what might be advanced, primitive, in this case, refers to being present in the most ancient plants; advanced refers to a feature more recently evolved. Besseyan dicta have influenced the evolutionary thoughts of Angiosperm taxonomists to a very great extent. Subsequent studies have corroborated most of his conclusions. Bessey's contribution probably lies more in his set of "Phylogenetic dicta" than in his system per se. The phylogenetic dicta, which Bessey propounded, have, with modifications, provided the theoretical base and guiding principles for most of contemporary systems of Angiosperms classification.

Hans Hallier (1868-1932)

Hallier of Hamburg and Leyden presented a system of classification designed to be phylogenetic. He considered the seed plants to comprise 2 phyla, each monophyletic and derived the Angiosperms from an unknown and extinct tribe of cycads situated near or descended from the Bennettitales admitting with the Marattiales. The Monocots were treated as more advanced than the Dicots and to have been derived from extinct stocks ancestral to the Lardizabalaceae. In the classification of Monocots and Dicots Hallier followed independently the dicta established by Bessey that recognized the strobiloid type of flower as primitive and held polycarpy and spiral arrangement of parts to be primitive over syncarpy and cyclic arrangements. Hallier placed large significance on ovule morphology and position. He defended use of many kinds of characters for classification. His realignment of families displayed many insights, such as the alliance of Salicaceae with Flacourtiaceae, Cactaceae with Aizoaceae, Cucurbitaceae with Passifloraceae etc.

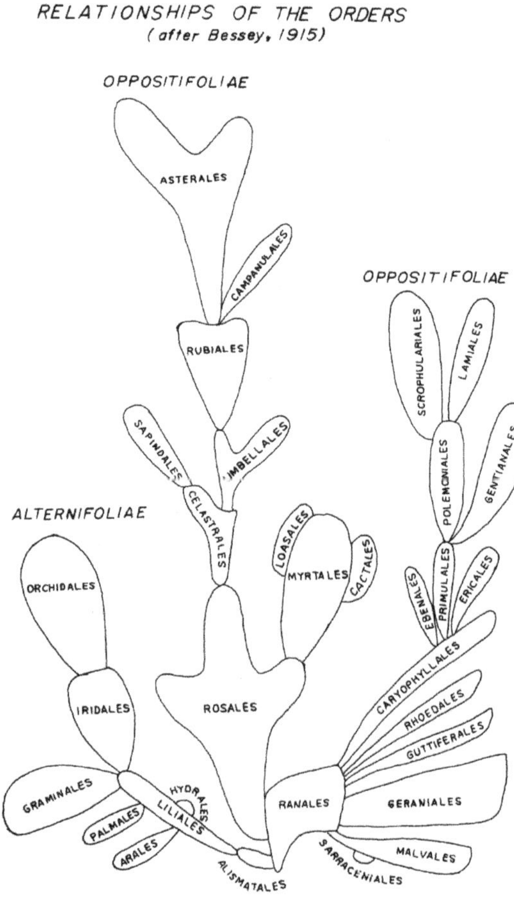

Fig. 2.1: Bessey's (1915) Cactus showing relationships of the orders.

John Hutchinson (1884-1972)

Hutchinson of England proposed a phylogenetic system in his 2-volume work. The Families of Flowering Plants (1926, 1934). He published his revised concepts in his book British Flowering Plants (1948). The revised edition of The Families of Flowering Plants was published in 1959 and third edition in 1973. In 1969 he published Evolution and Phylogeny of Flowering Plants. He based his system on 22 phylogenetic principles or dicta and these appear below in condensed form.

1. Evolution is both upwards and downwards, the former tending toward preservation and the latter to their reduction and suppression (of character)..
2. Evolution does not necessarily involve all organs at the same time
3. Broadly speaking, trees and shrubs are more primitive than herbs in any one family or genus.
4. Trees and shrubs are older than climbers in any one family or genus.
5. Perennials are older than biennials and annuals;
6. Aquatic flowering plants are derived from terrestrial ancestors, and epiphytes, saprophytes, and parasites are more recent than plants of normal habit;
7. Dicotyledons are more primitive than monocots.
8. Spiral arrangement is more primitive than cyclic.
9. Simple leaves are usually more primitive than compound leaves.
10. Unisexual flowers are more advanced than bisexual; dioecious plants are more recent than monoecious.
11. The solitary flower is more primitive than inflorescence.
12. Aestivation types are evolved from contorted to imbricate to valvate.
13. Apetalous flowers are derived from petaliferous flowers.
14. Polypetaly is more primitive than gamopetaly.
15. Actinomorphy is more primitive than zygomorphy.
16. Hypogyny is usually more primitive than perigyny and epigyny is the most advanced.
17. Apocarpy is more primitive than syncarpy.
18. A gynoecium of many pistils preceded one of few pistils.
19. Seeds with endosperm and small embryo are older than seeds without endosperm and a large embryo;..
20. Numerous stamens, in general, indicate greater primitiveness than does an androecium of a few stamens (exception, Malvaceae).
21. Separate anthers, in general, indicate greater primitiveness than does an androecium of either fused anthers or filaments.
22. Aggregate fruits are more highly evolved than single fruits; as a rule the capsule precedes the berry or drupe.

A comparision of these principles with the dicta of Bessey shows that, of the above, numbers 1-3, 7-10, 13-21 are enumerated also by Bessey with the same context and, for the most part, in the same sequence.

Hutchinson's system is basically in the Bentham- Bessey tradition. He considered the Angiosperm to be monophyletic possibly from an ancestral group of Gymnosperms related to the Cycadeoids. He divided the Angiosperms into two subphyla Dicotyledons and Monocotyledons The Dicotyledons were divided into those considered "fundamentally and predominantly" woody (Lignosae) and those considered "fundamentally and predominantly" herbaceous (Herbaceae). Parallel lines of development were considered to have occurred within two main groups. The Monocotyledons were derived from the stocks ancestral to Ranales.

The Division Lignosae started with Magnoliales and ended with Verbenales while Herbaceae started with Ranales and ended with Lamiales. The apocarpous Monocotyledons — Alismataceae and Butomaceae — were considered to be the most primitive members of the group because of their close resemblance to Ranales. There are three lines of monocot evolution-Calyciflorae, Corolliflorae and Glumiflorae.

In the last revision in 1969 Hutchinson divided the Angiosperms into 411 families. This includes 82 orders and 342 families in Dicotyledons and 29 orders and 69 families in Monocotyledons. The Lignosae was divided into 54 orders and Herbaceae into 28 orders.

Verbenales (5)
Bignoniales (4)
Rubiales (2)
Apocynales (4)
Loganiales (7)
Sapindales (11)
Meliales (1)
Rutales (4)
Ebenales (3)
Myrsinales (1)
Rhamnales (4)
Santalales (5)
Olacales (6)
Celastrales (10)
Myrtales (11)
Guttiferales (4)
Ericales (9)
Ochnales (7)
Theales (10)
Euphorbiales (12)
Malpighiales (12)
Malvales (1)
Tiliales (5)
Cactales (1)
Cucurbitales (4)

Passiflorales (3)
Loasales (2)
Lamiales (4)
Graminales (1)
Polygalales (4)
Boraginales (1)
Cyperales (1)
Violales
Polemoniales (3)
Juncales (4)
Tamaricales (3)
Geraniales (5)
Orchidales (1)
Capparidales (3)
Personales (7)
Burmanniales (3)
Pittosporales (5)
Solanales (4)
Haemodorales (6)
Proteales (1)
Asterales (1)
Cyclanthales (1)
Thymelaeales (6)
Goodeniales (3)
Pandanales (1)

Bixales (9)
Urticales (6)
Casuarinales (1)
Juglandales (3)
Fagales (3)
Balanopsidales (1)
Myricales (1)
Leitneriales (1)
Salicales (1)
Hamamelidalies (8)
Araliales (6)
Chenopodiales (11)
Polygonales (2)
Caryophyllales (5)
Resedales (1)
Brassicales (3)
Rhoedales (2)
Piperales (3)
Aristolochiales (4)
Berberidales (6)
Ranales (7)

Campanales (2)
Valerianales (3)
Umbellales (1)
Podostemales (2)
Sarraceniales (2)
Saxifragales (9)
Plantaginales (1)
Primulales (2)
Gentianales (2)
Onagrales (4)
Styracales (3)
Eriocaulales (1)
Xyridales (2)
Commelinales (4)
Najadales (2)
Potamogetonales (2)
Aponogetonales (2)
Juncaginales (3)
Triuridales (1)
Alismatales (3)
Butomales (2)

Palmales (1)
Agavales (2)
Dioscoreales (4)
Iridales (1)
Amaryllidales (1)
Typhales (2)
Arales (2)
Alstroemeriales (3)
Liliales (6)
Zingiberales (6)
Bromeliales (1)
Styracales (3)
Cunoniales (10)
Leguminales (3)
Rosales (3)
Coriariales (1)
Dilleniales (4)
Laurales (7)
Annonales (2)
Magnoliales (9)

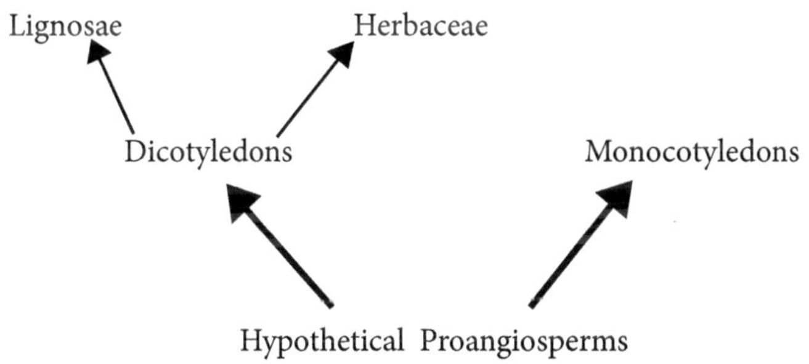

Fig. 2.2: Linear sequence of orders in Hutchinson's system (1973) (Figures in parenthesis indicate the number of families)

The division of Dicotyledons into Lignosae and Herbaceae is considered unnatural by most workers. By this division he kept the related families widely separated e.g.; Magnoliaceae and Ranunculaceae, Umbelliferae and Araliaceae, Labiatae and Verbenaceae, Bignoniaceae and Scrophulariaceae. Hutchinson divided Monocots into 3 subphyla and considered criteria based on inflorescence to be phylogenetically more fundamental than those based on ovary position. In the phylogenetic tree, Hutchinson did not derive one order directly from the other but from its ancestral stock. This is more safe and acceptable than to believe any present day order to be directly ancestral to any other.

Although Hutchinson's system has not been followed widely, it has provided a sound basis for the later phylogenetic systems by Oswald Tippo, Cronquist, Takhtajan and others. The arrangement of families within the Monocotyledons has been appreciated all over the world.

Hutchinson is generally considered as a "Splitter" as he accepted a large number of orders and families.

Oswald Tippo (1942)

Oswald Tippo of the University of Illinois published an outline (1942) of a system based on the earlier schemes proposed by G.M. Smith (1938) for Non Vascular plants and A.J. Eames (1936 for Vascular plants. Tippo considered Magnoliales to be primitive order while Amentiferae were considered to be advanced. The broad classification of Tippo (1942) is given below.

Kingdom: Plantae
- *Subkingdom: Thallophyta*
- *Subkingdom: Embryophyta*
  - *Phylum: Bryophyta*
  - *Phylum: Tracheophyta*
    - *Subphylum: Psilopsida*
    - *Subphylum: Lycopsida*
    - *Subphylum: Sphenopsida*
    - *Subphylum: Pteropsida*
      - *Class: Filicinae*
      - *Class: Gymnospermae*
      - *Class: Angiospermae*
        - *Subclass: Dicotyledonae*
        - *Subclass: Monocotyledone*

Carl Christian Mez (1866-1944)

Mez, Professor of Botany at the University of Koenigsberg, Germany, presented a paper in 1926, which was modified in 1936. This system is based on serodiagnostic data. According to this theory the relationships between larger groups of genetically related plants could be determined by study and analysis of their protein reactions. This physiological approach sometimes known as the serum diagnosis, consisted of mixing an extracted plant protein with serum, either, in animal or *in vitro* and following the formation of antibodies in the inoculated serum, of adding a protein extract prepared from the plant whose relationship is being studied. If a precipitation occurred when this second protein extract was added to the serum, a genetical relationship was believed to be indicated to exist between the plants involved. The proximity of relationship was considered to be indicated by the abundance and character of the precipitate and the degree of dilution of the serum and extract at which the precipitation would occur.

Lyman Benson

Lyman Benson of Pomona college, California, designed a system of classification, which is basically a Besseyan-type. In his book *Plant Classification* (1957) he presented his system in the form of a two dimensional chart (Fig.2.3) thus developing an essentially horizontal arrangement. Benson did not think any flowering plant order to have been derived from any other living order and believed that the related orders have evolved from preexisting groups with some members similar to the species now composing the living taxa. Benson divided the Angiosperms into Dicotyledons and Monocotyledons. The Dicotyledons were divided into 5 informal polyphyletic grades namely Thalamiflorae, Corolliflorae, Calyciflorae, Ovariflorae and Amentiferae.

Armen L. Takhtajan (1910-2009)

Armen LeonovichTakhtajan was Head of the Department of Higher plants at the Komarov Botanical Institute in Leningrad (now called as St. Petersburg), Russia. He was an international authority on Phytogeography, origin and phylogeny of flowering plants. He proposed an evolutionary system of classification of flowering plants in 1959 in his book *Die Evolution der Angiospermen* (in German). This was explained in 1966 in System and Phylogeny of Flowering Plants whose revised edition appeared in 1987 (both in Russian). An English version of his system appeared in 1969 in Flowering Plants-Origin and Dispersal. He revised his system in 1980 in Botanical review, in 1987 *Systema Magnoliophytorum*, in 1997 in *Diversity and Classification of Flowering Plants* and lastly in 2009 in *Flowering Plants.*

Takhtajan system is basically Bessey-Hallier tradition. Takhtajan took all available evidence-Morphological, Anatomical, Embryological, Cytological, Chemical, Ultrastructural, Palynological and Paleobotanical into consideration in classifying Angiosperms. His latest classification is according to International Code of Botanical Nomenclature. Accordingly Takhtajan divided the Magnoliophyta (Angiosperms) into two classes-Magnoliopsida (Dicotyledones) and Liliopsida (Monocotyledones). Magnoliopsida is considered to be primitive. Takhtajan system is Ranalean in tradition and considered that Angiosperms have evolved from Bennettitalean ancestors and the Magnoliales were considered primitive among Angiosperms (Fig. 2.4).

Takhtajan (2009) in his last classification divided the Magnoliopsida into 8 subclasses-Magnoliidae (17 orders 37 families), Ranunculidae (12 orders 20 families), Hamamelididae (12 orders 18 families), Caryophyllidae (6 orders 32 families), Dilleniidae (30 orders 108 families), Rosidae (28 orders 110 families), Asteridae (14 orders 62 families) and Lamiidae (6 orders, 55 families) while class Liliopsida was divided into four subclasses — Alismatidae (5 orders 22 families), Liliidae (16 orders 63 families), Arecidae (1 order and 1 family) and Commelinidae (9 orders 35 families). His subclasses were divided into super orders and orders. The Magnoliopsida includes 8 subclasses 32 super orders, 125 orders and 443 families while Liliopsida is categorized into 4 subclasses, 12 super orders, 31 orders and 121 families. So a total of 564 families are seen in the classification of Angiosperms by Takhtajan. Takhtajan considered the Liliopsida (Monocotyledones) to have been derived from Magnoliales.

Takhtajan's system is based on his 67 phyletic principles. He considered Angiosperms to be monophyletic and Monocotyledons to have been derived from primitive Dicotyledons. Among the subclass Takhtajan considered Magnoliidae to be the primitive most group from which all other subclasses have been derived (Fig. 2.4). He considered the Asteridae the most Advanced among Magnoliopsida. He considered the vesselless Dicotyledonous family Winteraceae as the primitive

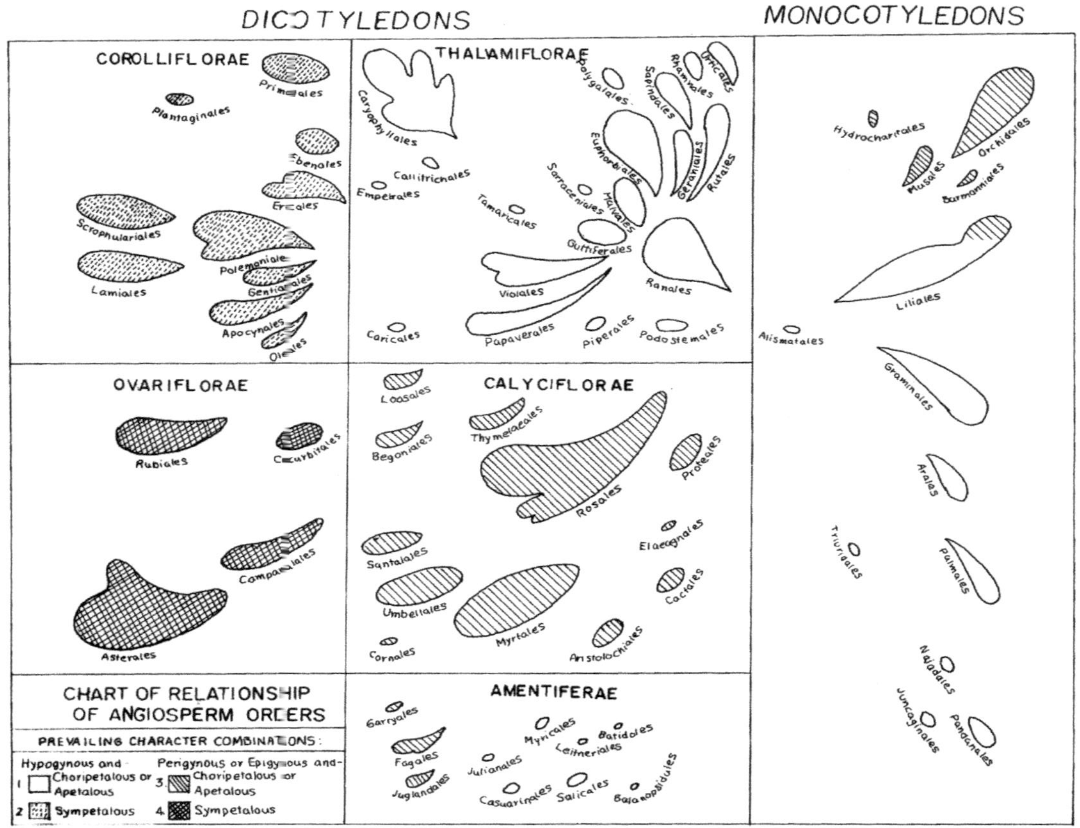

Fig. 2.3: Benson's (1957) chart showing relationships of the orders.

most family among Angiosperms and Poaceae the most advanced family among Angiosperms.

For many years Takhtajan considered Winteraceae along with Degeneriaceae to represent the most primitive angiosperms. Finally, however, he chose Degeneriaceae as the most primitive family, placed under the order Magnoliales. He shifted Winteraceae to a separate order Winterales higher up in the phyletic line after order Eupomatiales after Magnoliales. This is interesting in light of the fact that many contemporary authors, including Cronquist (1988) and Thorne (1983, 1992) regard Winteraceae as the primitive most family of the living angiosperms.

Arthur Cronquist (1919-1992)

Arthur Cronquist of New York Botanical Garden presented a system in his book *Evolution and Classification of Flowering Plants* (1968). His revised system appeared in 1981 in his book *An Integrated System of Classification of Flowering Plants*. He further revised this in 1988 in his book *TheEvolution and Classification of Flowering Plants* (ed. 2).

Cronquist's system is similar to that of Takhtajan except for minor details. Cronquist also took all the evidences-Morphology, Anatomy, Embryology, Palynology, Chemistry, Serology, Cytology, Ultrastructure-into consideration while classifying the flowering plants. Cronquist's system is also according to International Code of Botanical Nomenclature with regard to names of taxa.

Cronquist divided the Angiosperms into Mognolipsida and Liliopsida. The Magnoliosida is divided into six subclasses Magnoliidae, Hamamelidae, Caryophyllidae, Dillenidae, Rosidae and Asteridae. The Liliopsida is divided into 5 subclasses-Alismatidae, Arecidae, Commelinidae, Zingiberidae and Liliidae. There are no super orders. The Magnoliideae is divided into 8 orders and 39 families, Hamamelidae into 11 orders and 25 families, Caryophyllidae into 3 orders and 14 families, Dillenidae into 13 orders and 78 families, Rosidae into 18 orders and 116 families and Asteridae into 11 orders and 49 families. In the class Liliopsida Alismatidae is divided into 4 orders and 16 families, Arecidae into 4 orders and 6 families, Commelinidae into 7 orders and 16 families, Zingiberidae into 2 orders and 9 families and Liliidae into 2 orders and 19 families.

A total of 386 families are there in Angiosperms in Cronquist's (1988) system of which 320 are under Magnoliopsida (Dicotyledones) and 66 under Liliopsida (Monocotyledones).

Conquist's system is also Ranalian and considered the Angiosperms to have been derived from Bennettitalean ancestors. Cronquist also considered the Magnoliidae the primitive group among Angiosperms and the Monocotyledons have been derived from primitive Ranales (Fig. 2.5). Cronquist (1981) provided keys to orders and families.

C.R. de Soo

From Budapest, Hungary, de Soo (1975) has presented a modified system of classification of flowering plants essentially similar to Takhtajan. His system is largely based on phytochemical information. He used the name Malvidae for Dillenidae

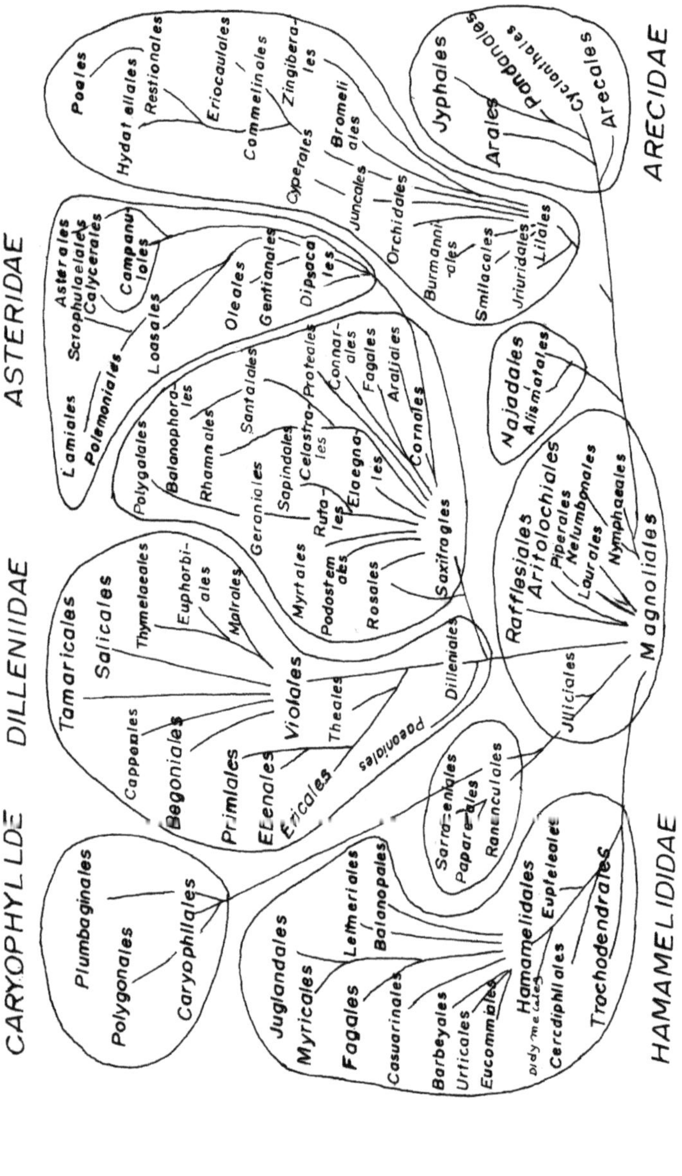

Fig. 2.4: Takhtajan's (1980) chart showing putative relationships among classes, sub-classes and orders.

and Eucomiidae for Hamamelididae. He prefered the name Angiospermophyta for Angiosperms, Dicotyledonopsida for dicots, Monocotyledonopsida for monocots.

Rolf Dahlgren (1932-1987)

Rolf Dahlgren of Copenhagen, Denmark presented the Angiosperm evolution in the form of a shrub in transection (Dahlgren, 1975, 1977, 1980). Dahlgren divided the Angiosperms into two classes Magnoliopsida and Liliopsida. The Magnoliopsida have been divided into 24 super orders, 80 orders and 346 families while Liliopsida are divided into 7 super orders, 26 orders and 92 families.

Robert Thorne (1920-)

Robert Thorne (1976, 1977) presented a system of classification, which is Ranalean in tradition. He divided the Angiosperms into 21 super orders, 50 orders, 74 suborders and 321 families. The Angiospermae was named as Annonopsida and divided into two subclasses Dicotyledonae and Monocotyledonae. Thorne's system is also in the Bessey-Hallier tradition.

APG Classification

The details of latest classification APG IV is given in chapter 4.

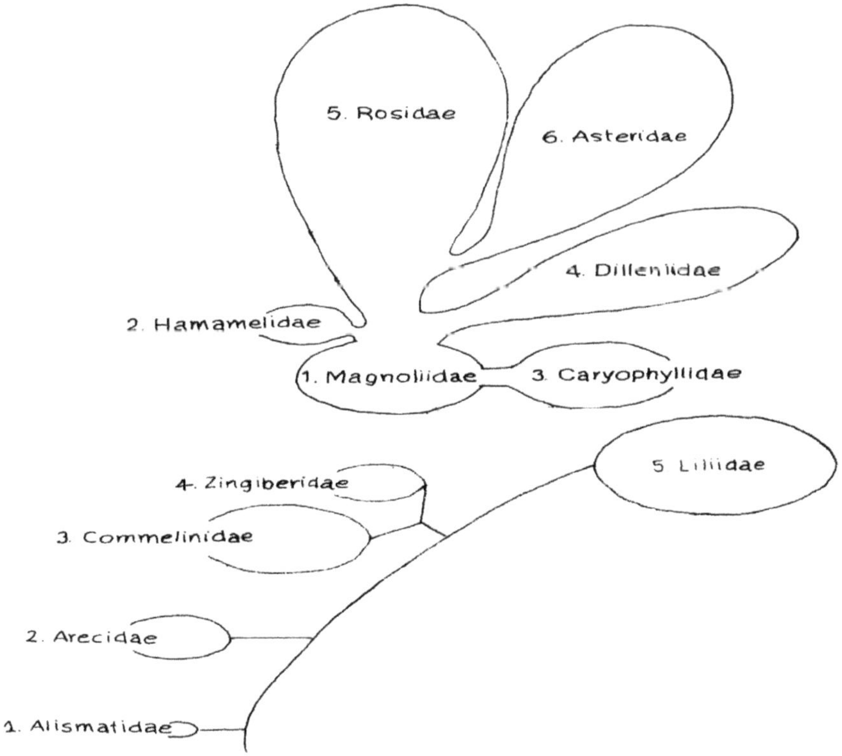

Fig. 2.5: Putative evolutionary relationships among subclasses Mangnoliopsida and Liliopsida of Cronquist System.

References

Bentham, G. and J.D. Hooker. 1862-1883. *Genera Plantarum*. London. 3 vols.

Bessey, C.E. 1915. Phylogenetic Taxonomy of flowering Plants. Ann. Mo. Bot. Gard. 2:109-164.

Cronquist, A. 1968. *The Evolution and classification of flowering plants*. New York.

Cronquist, A. 1981. *An Integrated system of classification of Angiosperms*. New York.

Cronquist, A. 1988. *The Evolution and Classification of Flowering Plants*. 2nd ed. New York.

Engler, A. and L. Diels. 1936. *Syllabus der Pflanzenfamilien*. 11th ed. Berlin

Engler, A. and K. Prantl. 1897-1915. *Die Narturlichen Pflanzenfamilien*. Leipzig. 23 vols.

Hutchinson, J. 1969. *Evolution and phylogeny of flowering Plants*. London.

Hutchinson, J. 1973. *The Families of Flowering Plants*. (3rd ed.). Oxford.

Takhtajan, A. 1959. *Die Evolution der Angiospermen*. Jena.

Takhtajan, A. 1969. *Flowering Plants: Origin and Dispersal*. Washington.

Takhtajan, A. 1980. Outline of the classification of flowering plants. Bot. Rev. 46:255-369.

Takhtajan, A. 1997. *Diversity and Classification of Flowering Plants*. New York.

Takhtajan, A. 2009. *Flowering Plants*, Springer Science.

3

Systems of Classification

Bentham and Hooker's System: *Genera Plantarum* (1862-1883)

Many of the Indian herbaria have been arranged according to Bentham and Hooker's system with modifications to incorporate recent changes. Many Indian Universities still follow this system although it is obsolete.

The system is given below in detail. Seed plants were classified into 3 classes. The characters are also given for the benefit of the readers.

Class I. Dicotyledons (Cotyledons two, tap root system, reticulate venation, perianth usually pentamerous or tetramerous).

Subclass (1) Polypetalae (Perianth dichlamydeous, heterochlamydeous, petals free).

Series A: Thalamiflorae (Calyx consists of usually distinct sepals which are free from ovary, ovary superior, petals and stamens hypogynous, disc absent):

Order 1 : Ranales (Floral parts, especially stamens and carpels numerous, carpels free, floral parts arranged in spiral, cyclic or spirocyclic manner):

Families: 1. Ranunculaceae 2. Dilleniaceae

3. Calycanthaceae 4. Magnoliaceae

5. Annonaceae 6. Menispermaceae

7. Berberidaceae 8. Nymphaeaceae

Order 2 : Parietales (Carpels united to form a unilocular ovary with parietal placentation):

Families: 9. Sarraceniaceae

10. Papaveraceae 11. Cruciferae

12. Capparidae 13. Resedaceae

14. Cistineae 15. Violarieae

16. Canellaceae 17. Bixieae

***Order 3* :** Polygalineae (Ovary syncarpous, 2-locular):

Families: 18. Pittosproeae 19. Tremandreae

20. Polygaleae 21. Vochysiaceae

***Order 4* :** Caryophyllinae (Placentation free central):

Families: 22. Frankniaceae 23. Caryophyllaceae

24. Portulacaceae 25. Tamariscinae

***Order 5* :** Guttiferales (Stamens many, sepals imbricate):

Families: 26. Elatineae 27. Hypericineae

28. Guttiferae 29. Ternstroemiaceae

30. Dipterocarpae 31. Chlaenaceae

***Order 6* :** Malvales (Ovary 3-8-carpellary with axile placentation, stamens usually numerous, sepals valvate):

Families: 32. Malvaceae 33. Sterculiaceae

34. Tiliaceae

Series B: Disciflorae (Calyx consists of either distinct or united sepals which may be free or adnate to ovary. A prominent ring or cushion-shaped disc usually present below the ovary. Sometimes it is broken up into glands. Stamens usually definite in number, inserted upon or at the outer or inner base of the disc. Ovary superior).

***Order 7* :** Geraniales (Flowers pentamerous, disc usually annular or broken into glands, androecium obdiplostemonous, ovary multicarpellary, syncarpous with axile placentation, ovules pendulous and raphe usually ventral).

Families: 35. Lineae 36. Humiriaceae

37. Malpighiaceae 38. Zygophyllaceae

39. Geraniaceae 40. Rutaceae

41. Simarubeae 42. Ochnaceae

43. Burseraceae 44. Meliaceae

45. Chailletiaceae

***Order 8* :** Olacales (Androecium diplostemonous, ovules pendulous, raphe dorsal):

Families: 46. Olacineae 47. Ilicineae

48. Cyrillaceae

Order 9 : Celastrales (Androecium haplostemonous, ovules erect, raphe ventral

Families: 49. Celastrineae 50. Stackhousiaceae

51. Rhamneae 52. Ampelidae

Order 10 : Sapindales (Androecium diplostemonous, ovule ascending, raphe ventral or inverted, leaves usually compound, exstipulate, flowers usually in panicles, ovules often 1-2 in each locule):

Families: 53. Sapindaceae 54. Sabiaceae

55. Anacardiaceae

Ordines anomali:

Families: 56. Coriareeae 57. Moringaceae

Series (C): Calyciflorae (Calyx consists of united sepals, rarely free and adnate to ovary, petal one or two seriate, peri- or epigynous):

Order 11 : Rosales (Stamens indefinite, often twice or more the number of petals, styles distinct):

Families: 58. Connaraceae 59. Leguminosae

60. Rosaceae 61. Saxifragaceae

62. Crassulaceae 63. Droseraceae

64. Hamamelidae 65. Bruniaceae

66. Haloragaceae

Order 12: Myrtales (Stamens definite, rarely indefinite, flowers perigynous or epigynous, placentation axile)

Families: 67. Rhizophoracea 68. Combretaceae

69. Myrtaceae 70. Melastomaceae

71. Lythraceae 72. Onagraceae

Order 13 : Passiflorales (Placentation parietal, tendril climbers, leaves often much lobed and divided, stamens 5):

Families: 73. Samydaceae 74. Loasaceae

75. Turneraceae 76. Passifloraceae

77. Cucurbitaceae 78. Begoniaceae

79. Datiscaceae

Order 14 : Ficoidales (Perianth undifferentiated, ovules on axile or basal placentation)

Families: 80. Cactaceae 81. Ficoideae

Order 15 : Umbellales (Inflorescence umbel):

Families: 82. Umbelliferae 83. Araliaceae

84. Cornaceae

Subclass (2) Gamopetalae (Perianth dichlamydeous, heterochlamydeous, petals united).

Series (A): Infereae (Ovary inferior, stamens as many as corolla lobes and alternating with them):

Order 1 : Rubiales: Stamens epipetalous, anthers distinct, ovary 2-many-locular, locule 1-many-ovuled):

Families: 85. Caprifoliacee 86. Rubiaceae

Order 2 : Asterales (Stamens epipetalous, anthers distinct or united, ovary 1-locular, 1-ovuled).

Families: 87. Valerianeae 88. Dipsacaceae

89. Calyceraceae 90. Compositae

Order 3 : Campanales (Stamens not adnate to corolla, ovary 2-6-locular, locule many-ovuled):

Families: 91. Stylidaceae 92. Goodenovieae

93. Campanulaceae

Series (B): Heteromereae (Ovary superior, carpels more than 2, stamens epipetalous or free from corolla, either equal to the number of petals or double this number or indefinite):

Order 4 : Ericales (Stamens twice as many as corolla lobes or isomerous and alternating with them):

Families: 94. Ericaceae 95. Vaccineae

96. Monotropeae 97. Epacrideae

98. Diapensiaceae 99. Lennoaceae

Order 5 : Primulales (Stamens as many as corolla lobes and opposite them, ovary 1-locular):

Families: 100. Plumbagineae 101. Primulaceae

102. Myrsinaceae

Order 6: Ebenales (Stamens as many as corolla lobes and opposite them or many, ovary 2-many-locular).

Families: 103. Sapotaceae 104. Ebenaceae

105. Styraceae

Series (C): Bicarpellatae (Ovary superior, carpels 2, stamens as many as corolla lobes and alternating with them or fewer):

Order 7 : Gentianales (Corolla actinomorphic, leaves opposite):

Families: 106. Oleaceae 107. Salvadoraceae

108. Apocynaceae   109. Asclepiadaceae
110. Loganiaceae   111. Gentianaceae

Order 8 : Polemoniales (Corolla actinomorphic, leaves alternate):

Families: 112. Polemoniaceae   113. Hydrophyllaceae
114. Boragineae   115. Convolvulaceae
116. Solanaceae

Order 9 : Personales (Corolla zygomorphic, posterior stamen often reduced to staminode, ovules often more than 4):

Families: 117. Scrophularaceae   118. Orobanchaceae
119. Lentibulariaceae   120. Columelliaceae
121. Gesneriaceae   122. Bignoniaceae
123. Pedaliaceae   124. Acanthaceae

Order 10 : Lamiales (Corolla zygomorphic, posterior stamen reduced, ovules 4):

Families: 125. Myoporineae   126. Selaginae
127. Verbenaceae   128. Labiatae

Ordines anomali:

Family 129. Plantaginaceae

Subclass (3): Monochlamydeae (Flowers usually with only one or incomplete whorl of perianth which is usually sepaloid, sometimes perianth altogether absent). Series (A): Curvembryeae (Embryo curved, lateral or peripheral, ovary usually

1-ovuled, seeds mostly with mealy endosperm):

Families: 130: Nyctaginaceae   131. Illecebraceae
132. Amaranthaceae   133. Chenopodiaceae
134. Phytolaccaceae   135. Batideae
136. Polygonaceae

Series (B): Multiovulatae Aquaticae (Submerged aquatic herbs, ovary syncarpous with many ovules):

Familiy: 137. Podostemaceae

Series (C): Multiovulatae Terrestris (Terrestrial herbs, ovary syncarpous with numerous ovules):

Families: 138. Nepenthaceae   139. Cytinaceae
140. Aristolochiaceae

Series (D): Micrembryeae (Embryo minute, ovary either apocarpous or syncarpous, with a single ovule):

Families: 141. Piperaceae   142. Chloranthaceae
143. Myristichaceae   144. Monimiaceae

Series (E): Daphnales (Ovary monocarpellary, rarely syncarpous and 2-4-ovuled, plants woody or herbaceous, perianth mostly sepaloid, stamens perigynous).

Families: 145. Laurineae, 146. Proteaceae, 147. Thymeleaceae, 148. Penaeaceae, 149. Elaegnaceae

Series (F): Achlamydosporeae (Ovary 1-locular, ovules undifferentiated, endospermic):

Families: 150. Loranthaceae, 151. Santalaceae, 152. Balanophoraceae

Series (G): Unisexuales (Flowers unisexual, perianth usually absent):

Families: 153. Euphorbiaceae, 154. Balanopseae, 155. Urticaceae, 156. Platanaceae, 157. Leitneriaceae, 158. Juglandaceae, 159. Myricaceae, 160. Casuarinaceae, 161. Cupuliferae

Ordines anomali

Families: 162. Salicineaae, 163. Lacistemaceae, 164. Empetraceae, 165. Ceratophyllaceae

Class II: Gymnospermae

Families: 166. Gnetaceae, 167. Coniferae, 168. Cycadaaceae

Class III: Monocotyledons

Series (A): Microspermae (Inner perianth lobes petaloid, ovary inferior, ovules numerous, minute, exalbuminous):

Families: 169. Hydrocharitaceae, 170. Burmanniaceae, 171. Orchidaceae

Series (B): Epigynae (Perianth biseriate, inner petaloid, ovary inferior, ovules large, few to many):

Families: 172. Scitaminae, 173. Bromeliaceae, 174. Haemodoraceae, 175. Iridaceae, 176. Amaryllidaceae, 177. Taccaceae, 178. Dioscreaceae

Series (C): Coronarieae (Perianth biseriate, inner whorl petaloid, ovary free, seeds albuminous, ovary superior):

Families: 179. Roxburghiaceae, 180. Liliaceae

181. Pontederiaceae 182. Philydraceae

183. Xyridaceae 184. Mayaceae

185. Commelinaceae 186. Rapateaceae

Series (D): Calycineae (Perianth biseriate, sepaloid, rigid or herbaceous and mem- branous, seeds albuminous, ovary superior):

Families: 187. Flagellarieae 188. Juncaceae

189. Palmae

Series (E): Nudiflorae (Perianth absent or reduced, seeds albuminous):

Families: 190. Pandaceae 191. Cyclanthaceae

192. Typhaceae 193. Aroideae

194. Lemnaceae

Series (F): Apocarpae (Ovary superior, carpels one or more, apocarpous, seeds exalbuminous, perianth either absent or 1 or 2-seriate):

Families: 195. Triuridaceae 196. Alismataceae

197. Naidaceae

Series (G): Glumaceae (Flowers minute, subtended by bracts or glumes, perianth reduced or absent, ovary or locules 1-ovuled):

Families: 198. Eriocaulaceae 199. Centrolepidiaceae

200. Restiaceae 201. Cyperaceae

202. Gramineae

Merits

(1) Every species and genus was studied anew from the material of British and continental herbaria.

(2) Full and complete descriptions were prepared from studied and dissections of plants themselves and did not represent a compilation made from literature.

(3) The system has great practical value for identification of plants. It is very easy to follow for routine identification.

(4) The description of Families and genera are precise. Keys to the identification are very useful. Larger genera have been divided into subgenera to facilitate identification.

Demerits

(1) Keeping Gymnosperms between Dicotyledons and Monocotyledons.

(2) The Monochlamydeae is an unnatural assemblage. Keeping the families with single whorl of perianth and families without perianth, although they are showing similarity with polypetalous families, is unjustified.

(3) Liliaceae and Amaryllidaceae have been separated on the basis of superior or inferior ovary and are kept in different series, although they show similarities.

(4) Elimination of orders in Monochlamydeae.

(5) Although the system was proposed after Darwin's theory of Evolution the system was not in phylogenetic sequence.

(6) Unisexuales is a loose assemblage of diverse families, which share only one major character i.e. unisexual flowers.

System of Classification by Engler and Prantl (1897-1915) "*Die Naturlichen Pflanzenfamilien*"

I-Monocotyledonae

Order 1	: Pandanales	
Families	: 1. Typhaceae	2. Pandanaceae
	3. Sparganiceae	
Order 2	: Helobieae	
Families	: 4. Potamogetonaceae	5. Najadaceae
	6. Aponogetonaceae	7. Scheuchzeriaceae
	8. Alismataceae	9. Butomaceae
	10. Hydrocharitaceae	
Order 3	: Triuridales	
Family	: 11. Triuridaceae	
Order 4	: Glumiflorae	
Families	: 12. Gramineae	13. Cyperaceae
Order 5	: Principes	
Family	: 14. Palmae	
Order 6	: Synanthae	
Family	: 15 Cyclanthaceae	
Order 7	: Spathiflorae	
Families	: 16. Araceae	17. Lemnaceae
Order 8	: Farinosae	
Families	: 18. Flagellariaceae	19. Restionaceae
	20. Centrolepidiaceae	21. Mayaceae
	22. Xyridaceae	23. Eriocaulacee
	24. Thurniaceae	25. Rapateacae
	26. Bromeliaceae	27. Commelinaceae

	28. Pontederiaceae	29. Cynastraceae
	30. Philydraceae	
Order 9	: Liliflorae	
Families	: 31. Juncaceae	32. Stemonacee
	33. Liliaceae	34. Haemodoraceae
	35, Amarylliedaceae	36. Velloziaceae
	37 Taccaceae	38. Dioscoreaceae
	39. Iridaceae	
Order 10	: Scitamineae	
Families	: 40. Musaceae	41. Zingiberaceae
	42. Cannaceae	43. Marantaceae
Order 11	: Microspermae	
Families	: 44. Burmanniaceae	45. Orchidaceae

II. Dicotyledonae

A. Archichlamydae

Order 1	: Verticillatae	
Family	: 46. Casuarinaceae	
Order 2	: Piperales	
Families	: 47. Saururaceae	48. Piperaceae
	49. Chloranthaceae	50. Lacistemaceae
Order 3	: Salicales	
Family	: 51. Salicaceae	
Order 4	: Garryales	
Family	: 52. Garryaceae	
Order 5	: Myricales	
Family	: 53. Myricaceae	
Order 6	: Balanospidales	
Family	: 54. Balanopsidaceae	
Order 7	: Leitneriales	
Family	: 55. Leitneriaceae	
Order 8	: Juglandales	
Family	: 56. Juglandaceae	
Order 9	: Batidales	
Family	: 57. Batidaceae	

Order 10	: Julianales	
Family	: 58. Julianaceae	
Order 11	. Fagales	
Families	: 59. Betulaceae	60. Fagaceae
Order 12	: Urticales	
Families	: 61. Ulmacee	62. Moraceae
	63. Urticaceae	
Order 13	: Proteales	
Family	: 64. Proteaceae	
Order 14	: Santalales	
Family	: 65. Myzodendraceae	66. Santalaceae
	67. Opiliaceae	68. Grubbiaceae
	69. Olacaceae	70. Octoknemataceae
	71. Loranthaceae	72. Balanophoraceae
Order 15	: Aristolochiales	
Families	: 73. Aristolochiaceae	74. Rafflesiaceae
	75. Hydnoraceae	
Order 16	: Polygonales	
Family	: 76. Polygonaceae	
Order 17	: Centrospermae	
Families	: 77. Chenopodicaeae	78. Amaranthaceae
	79. Nyctaginaceae	80. Cynocrambaceae
	81. Phytolaccaceae	82. Aizoaceae
	83. Portulacaceae	84. Basellaceae
	85. Caryophyllaceae	
Order 18	: Ranales	
Families	: 86. Nymphaeaceae	87. Ceratophyllaceae
	88. Trochodendraceae	89. Cercidiphyllaceae
	90. Ranunculaceae	91. Lardizabalaceae
	92. Berberidaceae	93. Menispermaceae
	94. Magnoliaceae	95. Calycanthaceae
	96. Lactoridaceae	97. Annonaceae
	98. Eupomatiaceae	99. Myristichaceae
	100. Gomortegaceae	101. Monimiaceae
	102. Lauraceae	

Order 19	: Rhoedales	
Families	: 104. Papaveraceae	105. Capparidaceae
	106. Cruciferae	107. Tovariaceae
	108. Resedaceae	109. Moringaceae
Order 20	: Sarraceniales	
Families	: 110. Sarraceniaceae	111. Nepenthaceae
	112. Droseraceae	
Order 21	: Rosales	
Families	: 113. Podostemaceae	114. Tristichaceae
	115. Hydrostachyaceae	116. Crassulaceae
	117. Cephalotaceae	118. Saxifragaceae
	119. Pittosporaceae	120. Brunelliaceae
	121. Cunoniaceae	122. Myrothamnaceae
	123. Bruniaceae	124. Hamamelidaceae
	125. Eucommiaceae	126. Platanaceae
	127. Crassomataceae	128. Rosaceae
	129. Connaraceae	130. Leguminosae
Order 22	: Pandales	
Family	: 131. Pandaceae	
Order 23	: Geraniales	
Families	: 132. Geraniaceae	133. Oxalidaceae
	134. Tropaeolaceae	135. Linaceae
	136. Humiriaceae	137. Erythroxylaceae
	138. Zygophyllaceae	139. Cneoraceae
	140. Rutaceae	141. Simarubaceae
	142. Burseraceae	143. Meliaceae
	144. Malpighiaceae	145. Trigoniaceae
	146. Vochysiaceae	147. Tremandraceae
	148. Polygalaceae	149. Dichapetalaceae
	150. Euphorbiaceae	151. Callitrichaceae
Order 24	: Sapindales	
Families	: 152. Empetraceae	153. Coriariaceae
	154. Limnanthaceae	155. Anacardiaceae
	156. Cyrillaceae	157. Buxaceae

158. Pentaphylacaceae 159. Corynocarpaceae
160. Aquifoliaceae 161. Celastraceae
162. Hippocrateaceae 163. Salvadoraceae
164. Stackhousiaceae 165. Staphylaceae
166. Icacinaceae 167. Aceraceae
168. Hippocastanaceae 169. Sapindaceae
170. Sabiaceae 171. Melianthaceae
172. Balsaminaceae

Order 25 : Rhamnales

Families : 173. Rhmnaceae 174. Vitaceae

Order 26 : Malvales

Families : 175. Elaeocarpaceae 176. Chlaenaceae
177. Gonystilaceae 178. Tiliaceae
179. Malvaceae 180. Bombacaceae
181. Sterculiaceae 182. Scytopetalaceae

Order 27 : Parietales

Families : 183. Dilleniaceae 184. Eucryphiaceae
185. Ochnaceae 186. Caryocaraceae
187. Marcgraviaceae 188. Quinaceae
189. Theaceae 190. Guttiferae
191. Dipterocarpaceae 192. Elatinaceae
193. Frankeniaceae 194. Tamaricaceae
195. Fouquiriaceae 196.Cistaceae
197. Bixaceae 198. Cochlospermaceae
199. Winteraceae 200. Violaceae
201. Flacourtiaceae 202. Stachyuraceae
203. Turneraceae 204. Malesherbiaceae
205. Passifloraceae 206. Achariaceae
207. Datiscaceae 208. Begoniaceae
209. Ancistrocladaceae

Order 28 : Opuntiales

Families : 210. Cactaceae

Order 29 : Myrtiflorae

Families : 211. Geissolomataceae 212. Penaeaceae

	213. Oliniaceae	214. Thymeleaceae
	215. Elaegnaceae	216. Lythraceae
	217. Sonneretiaceae	218. Punicaceae
	219. Lecythidaceae	220. Rhizophoraceae
	221. Nyssaceae	222. Alangiaceae
	223. Combretaceae	224. Myrtaceae
	225. Melastomaceae	226. Onagraceae
	227. Haloragidaceae	228. Hippuridaceae
	229. Cynomoriaceae	
Order 30	: Umbelliflorae	
Families	: 230. Araliaceae	231. Umbellferae
	232. Cornaceae	

B. Sympetalae

Order 1	: Ericales	
Families	: 233. Clethraceae	234. Pyrolaceae
	235. Lennoaceae	236. Ericaceae
	237. Epacridaceae	238. Diapensiaceae
Order 2	: Primulales	
Families	: 239. Theophrastaceae	240. Myrsinaceae
	241. Primulaceae	
Order 3	: Plumbaginales	
Families	: 242. Plumbaginaceae	
Order 4	: 243. Sapotaceae	244. Ebenaceae
	245. Symplocaceae	246. Styraceae
Order 5	: Contortae	
Families	: 247. Oleaceae	248. Loganiaceae
	249. Gentinaceae	250. Apocynaceae
	251. Asclepiadaceae	
Order 6	: Tubiflorae	
Families	: 252. Convolvulaceae	253. Polemoniaceae
	254. Hydrophyllaceae	255. Boraginaceae
	256. Verbenaceae	257. Labiatae
	258. Nolanaceae	259. Solanaceae
	260. Scrophulariaceae	261. Bignoniaceae

	262. Pedaliaceae	263. Martyniaceae
	264. Orobanchaceae	265. Gesneriaceae
	266. Columelliaceae	267. Lentibulariaceae
	268. Globulariaceae	269. Acanthaceae
	270. Myoporaceae	271. Phrymaceae
Order 7	: Plantaginales	
Family	: 272. Plantaginaceae	
Order 8	: Rubiales	
Families	: 273. Rubiaceae	274. Caprifoliaceae
	275. Adoxaceae	276. Valerianaceae
	277. Dipsacaceae	
Order 9	: Cucurbitales	
Family	: 278. Cucurbitaceae	
Order 10	: Campanulatae	
Families	: 279. Campanulaceae	280. Goodeliaceae
	281. Brunoniaceae	282. Stylidaceae
	283. Calyceraceae	284. Compositae

Merits

1. It is a phylogenetic system.
2. The classification covers entire plant kingdom and provides description and identification keys down to the level of the family (in *Syllabus der Pflanzenfamilien*), genus (in *Die Naturlichen Pflanzenfamilien*) and even species for large number of families (in *Das Pflanzenreich*). Valuable illustrations and information on anatomy and geography are also provided.
3. Compositae and Orchidaceae are considered as advanced families
4. Merger of polypetalae and Monochlamydeae into a single subclass Archichlamydeae
5. Engler and Prantl have classified the whole plant kingdom

Demerits

1. Monocots are placed before Dicots.
2. Simplicity is considered primitive.
3. Amentiferae regarded as primitive group.
4. Angiosperms were considered a polyphyletic group. Most of the recent evidence points towards the monophyletic origin of Angiosperms.
5. Derivation of free central placentation from parietal placentation, and of

the latter from axile placentation is contrary to the evidence from floral anatomy. Free central placentation is now believed to have evolved from axile placentation through the disappearance of septa.

Hutchinson's System of Classification of Flowering Plants

Families of Flowering Plants (1959, 1973)

Evolution and Phylogeny of Flowering Plants (1969)

Subphylum : Dicotyledones

Division I : Lignosae

Hutchinson's sequence of orders and families

Order 1 : Magnoliales

Families : 1. Magnoliaceae 3. Winteraceae
2. Illiciaceae 4. Canellaceae
5. Schisandraceae 6. Himantandraceae
7. Lactoridaceae 8. Trochodendraceae
9. Cercidiphyllaceae

Order 2 : Annonales

Families : 10. Annonaceae 11. Eupomatiaceae

Order 3 : Laurales

Families : 12. Monimiaceae 13. Austrobaileyaceae
14. Trimeniaceae 15. Lauraceae
16. Gomortegaceae 17. Hernandinaceae
18. Myristichaceae

Order 4 : Dilleniales

Families : 19. Dilleniaceae 20. Connaraceae
21. Crossosomataceae 22. Brunelliaceae

Order 5 : Coriariales

Family : 23. Coriariaceae

Order 6 : Rosales

Families : 24. Rosaceae 25. Dichapetalaceae
26. Calycanthaceae

Order 7 : Leguminales

Families : 27. Caesalpiniaceae 28. Mimosaceae
29. Papilionaceae

Order 8 : Cunoniales

Families	: 30. Prostemonaceae	31. Cunoniaceae
	32. Pteladelphaceae	33. Hydrangeaceae
	34. Grossulariaceae	35. Oliniaceae
	36. Greyiaceae	37. Escallionaceae
	38. Baueraceae	39. Crypteroniaceae
Order 9	: Styrales	
Families	: 40. Lissocarpaceae	41. Styraceae
	42. Symplocaceae	
Order 10	: Arales	
Families	: 43. Cornaceae	44. Alangiceae
	45. Garryaceae	46. Nyssaceae
	47. Araliaceae	48. Caprifoliaceae
Order 11	: Hamamelidales	
Families	: 49. Tetracentraceeae	50. Hamamelidaceae
	51. Myrothamnaceae	52. Platanaceae
	53. Stachyuraceae	54. Buxaceae
	55. Daphniphyllaceae	56. Bruniaceae
Order 12	: Salicales	
Family	: 57. Salicaceae	
Order 13	: Leitneriales	
Family	: 58.Leineriaceae	
Order 14	: Myricales	
Family	: 59. Myricceae	
Order 15	: Balanopsidales	
Family	: 60. Balanopsidaceae	
Order 16	: Fagales	
Families	: 61. Betulaceae	62. Fagaceae
	63. Corylaceae	
Order 17	: Juglandales	
Families	: 64. Rhoipteleaceae	65. Juglandaceae
	66. Picrodendraceae	
Order 18	: Casuarinales	
Family	: 67. Casuarinaceae	
Order 19	: Urticales	

Families	: 68. Ulmaceae	69. Cannabinaceae
	70. Moraceae	71. Urticaceae
	72. Barbeyaceae	73. Eucommiaceae
Order 20	: Bixales	
Families	: 74. Bixaceae	75. Cistaceae
	76. Flacourtiaceae	77. Cochlospermaceae
	78. Hoplestigmataceae	79. Achatocarpaceae
	80. Lacistemaceae	
Order 21	: Thymeleales	
Families	: 81. Gonystylaceae	82. Aquilariaceae
	83. Geissolomataceae	84. Penaeaceae
	85. Thymeleaceae	86. Nyctaginaceae
Order 22	: Proteales	
Family	: 87. Proteaceae	
Order 23	: Pittosprales	
Families	: 88. Pittosporaceae	89. Byblidaceae
	90. Stegnospermataceae	91. Vivianiaceae
	92. Tremandraceae	
Order 24	: Capparidales	
Families	: 93. Capparidaceae	94. Moringaceae
	95. Tovariaceae	
Order 25	: Tamaricales	
Families	: 96. Frankeniaceae	97. Tamaricaceae
	98. Fouquieriaceae	
Order 26	: Violales	
Family	: 99. Violaceae	
Order 27	: Polygalales	
Families	: 100. Polygalaceae	101. Krameriaceae
	102. Trigoniaceae	103. Vochysiaceae
Order 28	: Loasales	
Families	: 104. Turneraceae	105. Loasaceae
Order 29	: Passiflorales	
Families	: 106. Malesherbiaceae	107. Passifloraceae
	108. Achariaceae	

Order 30	: Cucurbitales	
Families	: 109. Cucurbitaceae	110. Begoniaceae
	111. Datiscaceae	112. Caricaceae
Order 31	: Cactales	
Family	: 113. Cactaceae	
Order 32	: Tiliales	
Family	: 114. Dirachmaceae	115. Scytopetalaceae
	116. Tiliaceae	117. Sterculiaceae
	118. Peridiscaceae	119. Bombacaceae
Order 33	: Malvales	
Family	: 120. Malvaceeae	
Order 34	: Malpighiales	
Families	: 121. Ixonanthaceae	122. Malpighiaceae
	123. Humiriaceae	124. Linaceae
	125. Irvingiaceeae	126. Huaceae
	127. Ledocarpaceae	128. Erythroxylaceae
	129. Ctenolophonaceae	130. Lepidobotryaceae
	131. Balanitaceae	132. Zygophyllaceae
Order 35	: Euphorbiales	
Family	: 133. Euphorbiaceae	
Order 36	: Theales	
Families	: 134. Bonnetiaceae	135. Theaceae
	136. Sauraniaceae	137. Actinidiaceae
	138. Pellicieraceeae	139. Pentaphylacaceae
	140. Tetrameristaceae	141. Marcgraviaceae
	142. Caryocaraceae	143. Medusagynaceae
Order 37	: Ochnales	
Families	: 144. Strasburgheriaceae	145. Ochnaceae
	146. Rhodolaenaceae	147. Sphaerosepalaceae
	148. Dipterocarpaceae	149. Ancistrocladaceae
Order 38	: Ericales	
Families	: 150. Clethraceae	151. Pyrolaceae
	152. Ericaceae	153. Epacridaceae
	154. Diapensiaceae	155. Monotropaceae

	156. Lennoaceae	157. Vacciniaceae
Order 39	: Guttiferales	
Families	: 158. Hypericaceae	159. Clusiaceae
	160. Eucryphiaceae	161. Quiinaceae
Order 40	: Myrtales	
Families	: 162. Myrtaceae	163. Lecythidaceae
	164. Rhizophoraceae	165. Sonneretiaceae
	166. Punicaceae	167. Combretaceae
	168. Melastomataceae	
Order 41	: Celastrales	
Families	: 169. Pandaceae	170. Aquiifoliaceae
	171. Salvadoraceae	172. Koeberliniaceae
	173. Cneoraceae	174. Cardiopteridaceae
	175. Cyrillaceae	176. Icacinaceae
	177. Empetraceae	178. Aextoxicaceae
	179. Pentaeiplanoraceae	180. Celastraceae
	181. Corynocarpaceae	182. Stackhousiaceae
	183. Goupiaceae	184. Hippocrateaceae
	185. Erythropalaceae	186. Capusiaceae
	187. Scyphostegiaceae	
Order 42	: Olacales	
Families	: 188. Olacaceae	189. Opiliaceae
	190. Octoknemaceae	191. Aptandraceae
	192. Dipentodontaceae	193. Medusandraceae
Order 43	: Santalales	
Families	: 194. Loranthaceae	195. Grubbiaceae
	196. Santalaceae	197. Myzodendraceae
	198. Balanophoraceae	
Order 44	: Rhamnales	
Families	: 199. Heteropyxidaceae	200. Elaegnaceae
	201. Rhamnaceae	202. Vitaceae
Order 45	: Myrsinales	
Families	: 203. Myrsinaceae	204. Theophrastaceae
	205. Aegiceraceeae	

Order 46 : Ebenales

Families : 206. Ebenaceae 207. Sapotaceae
208. Sarcospermaceae

Order 47 : Rutales

Families : 209. Rutaceae 210. Simaroubaceae
211. Burseraceae 212. Averrhoaceae

Order 48 : Meliales

Family : 213. Meliaceae

Order 49 : Sapindales

Families : 214. Melianthaceae 215. Sapindaceae
216. Podoaceae 217. Sabiaceae
218. Anacardiaceae 219. Aceraceae
220. Hippocastanaceae 221. Staphyleaceae
222. Akaniaceae 223. Julianaceae
224. Didiereaceae

Order 50 : Loganiales

Families : 225. Potaliaceae 226. Loganiaceae
227. Buddleiaceae 228. Antoniaceae
229. Spigeliaceae 230. Strychnaceae
231. Oleaceae

Order 51 : Apocynales

Families : 232. Plocospermaceae 233. Apocynaceae
234. Periplocaceae 235. Asclepiadaceae

Order 52 : Rubiales

Families : 236. Dialypetalanthaceae 237. Rubiaceae

Order 53 : Bignoniales

Families : 238. Cobaeaceae 239. Bignoniaceae
240. Pedaliaceae 241. Martyniaceae

Order 54 : Verbenales

Families : 242. Ehretiaceae 243. Verbenaceae
244. Stilbeaceae 245. Chloanthaceae
246. Phrymaceae

Division II: Herbaceae

Order 55 : Ranales

Families : 247. Paeoniaceae 248. Hellboraceae
249. Ranunculaceae 250. Nymphaeaceae
251. Podophyllaceae 252. Ceratophyllaceae
253. Cabombaceae

Order 56 : Berberidales

Families : 254. Sargentodoxaceae 255. Lardizabalaceae
256. Menispermaceae 257. Nandiaceae
258. Circaestraceae 259. Berberidaceae

Order 57 : Aristolochiales

Families : 260. Aristolochiaceaee 261. Hydnoraceae
262. Cytinaceae 263. Nepenthaceae

Order 58 : Piperales
264. Piperaceae 265. Saururaceae
266. Chloranthaceae

Order 59 : Rhoedales
267. Papaveraceae 268. Fumariaceae

Order 60 : Cruciales

Family : 269. Cruciferae

Order 61 : Resedales

Family : 270. Resedaceae

Order 62 : Caryophyllales

Families : 271. Elatinaceae 272. Molluginaceae
273. Caryophyllaceae 274. Ficoidaceae
275. Portulacaceae

Order 63 : Polygonales

Families : 276. Polygonaceae 277. Illecebraceae

Order 64 : Chenopodiales

Families : 278. Barbeuiaceae 279. Phytolaccaceae
280. Gyrostemonaceae 281. Agdestiaceae
282. Petiveriaceae 283. Chenopodiaceae
284. Amaranthaceae 285. Cynocrambaceae
286. Batidaceae 287. Basellaceae

Order 65 : Lythrales

Families : 288. Lythraceae 289. Onagraceae

	290. Trapaceae	291. Haloragidaceae
	292. Callitrichaceae	
Order 66	: Gentianales	
Families	: 293. Gentianaceae	294. Menyanthaceae
Order 67	: Primulales	
Families	: 295. Primulaceae	296. Plumbaginaceae
Order 68	: Plantaginales	
Family	: 297. Plantaginaceae	
Order 69	: Saxifragales	
Families	: 298. Crassulaceae	299. Cephalotaceae
	300. Saxifragaceae	301. Eremosynaceae
	302. Vahliaceae	303. Francoaceae
	304. Donatiaceae	305. Parnassiaceae
	306. Adoxaceae	
Order 70	: Sarraceniales	
Families	: 307. Droseraceae	308. Sarraceniaceae
Order 71	: Podostemales	
	309. Podostemaceae	310. Hydrostachyaceae
Order 72	: Umbellales	
Family	: 311. Umbelliferae	
Order 73	: Valerianales	
Families	: 312. Valerianaceae	313. Dipsacaceae
	314. Calyceraceae	
Order 74	: Campanales	
Families	: 315. Campanulaceae	316. Lobeliaceae
Order 75	: Goodeniales	
Families	: 317.Goodeniaceae	318. Brunoniiaceae
	319. Stylidaceae	
Order 76	: Asterales	
Family	: 320. Compositae	
Order 77	: Solanales	
Families	: 321. Solanaceae	322. Convolvulaceae
	323. Nolanaceae	
Order 78	: Personales	

Families : 324. Scrophulariaceae 325. Acanthaceae
326. Gesneriaceae 327. Orobanchaceae
328. Lentibulariaceae 329. Columelliaceae

Order 79 : Geraniales

Families : 330. Geraniaceae 331. Limnanthaceae
332. Oxalidaceae 333. Tropaeolaceae
334. Balsaminaceae

Order 80 : Polemoniales

Families : 335. Polemoniaceae 336. Hydrophyllaceae
337. Cuscutaceae

Order 81 : Boraginales

Family : 338. Boraginaceae

Order 82 : Lamiales

Families : 339. Myoporaceae 340. Selaginaceae
341. Globulariaceae 342. Labiatae

Subphylum: Monocotyledons

Division I : Calyciferae

Order 83 : Butomales

Families : 343. Butomaceae 344. Hydrocharitaceae

Order 84 : Alismatales

Families : 345. Alismataceae 346. Scheuchzeriaceae
347. Petrosaviaceae

Order 85 : Triuridales

Family : 348. Triuridaceae

Order 86 : Juncaginales

Families : 349. Juncaginaceae 350. Lilaeaceae
351. Posidoniaceae (Heterostylaceae)

Order 87 : Aponogetonales

Families : 352. Aponogetonaceae 353. Zosteeraceae

Order 88 : Potamogetonales

Families : 354. Potamogetonaceae 355. Ruppiaceae

Order 89 : Najadales

Families : 356. Zannichelliaceae 357. Najadaceae

Order 90 : Commelinales

Families	: 358. Commelinaceae	359. Cartonemataceae
	360. Flagellariaceae	361. Mayaceae
Order 91	: Xyridales	
Families	: 362. Xyridaceae	363. Rapateaceae
Order 92	: Eriocaulales	
Family	: 364. Eriocaulaceae	
Order 93	: Bromeliales	
Family	: 365. Bromeliaceae	
Order 94	: Zingiberales	
Families	: 366. Musaceae	367. Strelitzeaceae
	368. Lowiaceae	369. Zingiberaceae
	370. Cannaceae	371. Marantaceae
Division II: Corolliferae		
Order 95	: Liliales	
Families	: 372. Liliaceae	373. Tecophilaeaceae
	374. Trilliaceae	375. Pontederiaceae
	376. Smilacaceae	377. Ruscaceae
Order 96	: Alstromeriales	
Families	: 378. Alstromeriaceae	379. Petermanniaceae
	380. Philesiaceae	
Order 97	: Arales	
Families	: 381. Araceae	382. Lemnaceae
Order 98	: Typhales	
Families	: 383. Sparganiaceae	384. Typhaceae
Order 99	: Amaryllidales	
Family	: 385. Amaryllidaceae	
Order 100	: Iridales	
Family	: 386. Iridaceae	
Order 101	: Dioscoreales	
Families	: 387. Stenomeridaceae	388. Trichopodaceae
	389. Roxburghiaceae	390. Dioscoreaceae
Order 102	: Agavales	
Families	: 391. Xanthorrhoaceae	392. Agavaceae
Order 103	: Palmales	

Family	: 393. Palmae	
Order 104	: Pandanales	
Family	: 394. Pandanaceae	
Order 105	: Cyclanthales	
Family	: 395. Cyclanthaceae	
Order 106	: Haemodorales	
Families	: 396. Haemodoraceae	397. Hypoxidaceae
	398. Velloziaceae	399. Apostasiaceae
	400. Taccaceae	401. Philydraceae
Order 107	: Burmanniales	
Families	: 402. Burmanniaceae	403. Thismiaceae
	404. Corsiaceae	
Order 108	: Orchidales	
Family	: 405. Orchidaceae	

Division III: Glumiflorae

Order 109	: Juncales	
Families	: 406. Juncaceae	407. Thurniaceae
	408. Centrolepidaceae	409. Restionaceae
Order 110	: Cyperales	
Family	: 410. Cyperaceae	
Order 111	: Graminales	
Family	: 411. Graminae	

Demerits

1. System based on Habit is considered as draw back. Due to this related families are kept far apart, e.g., Umbelliferae and Araliaceae, Labiatae and Verbenaceae, Cruciferae and Capparidae, Bignoniaceae and Acanthaceae.
2. Hutchinson derives angiosperms from proangiosperms, but does not provide information about the nature of this hypothetical ancestral group.
3. He regards Magnoliaceae as the most primitive family of living Dicotyledones, but most contemporary authors consider vesselless Winteraceae to be the most primitive

Merits

1. The system is more phylogenetic than that of Engler and Prantl, as it is based on phylogenetic principles generally recognized by most authors.
2. The treatment of Magnoliales as the starting point in the evolutionary series.

4

APG IV Classification

To establish a consensus view of the classification of flowering plants, an international group of flowering plant systematists, christened as the Angiosperm Phylogeny Group (APG), came together to propose a classification based on sound phylogenetic data sets in 1998. Latter three versions of classifications have been proposed in 2003, 2009 and 2016, each superseding the previous systems. The original APG System is unusual in being based not on total evidence, but on cladistic analysis of the DNA sequences of three genes - two chloroplast genes (*cpDNA; atpB* and *rbcL*) and one gene coding for ribosomes (nuclear ribosomal 18s DNA). Although based on molecular evidence only, its constitutent groups prove to be supported by other evidence as well, for example pollen morphology supports the split between eudicots and the rest of the former Dicotyledons.

The system is rather controversial in its decisions at the family level, splitting a number of long-established families and submerging some other families. It also is unusual in not using botanical names above the level of order that is an order is the highest rank, that will have a formal botanical name in this system. Higher groups are defined only as clades with names such as Monocots, Eudicots, Rosids and Asterids.

The term "Clade" was coined by Julian Huxley (1957) to refer the relationships between individual groups and the formation of organism groups in divergent evolutionary tree is called as Cladogenesis. A clade is a single "branch" or monophylum on the tree of life. Cladogram (family tree) is of a biological group.

APG (1998) recognized 462 families, grouped into 40 monophyletic orders and classified under few informal monophyletic higher groups like Monocots, Commelinids, Eudicots etc. The APG II (2003) recognized 457 families and 45 orders. APG III (2009) has 413 families, except 10 families, all other are assigned to 59 orders. A revised and updated version of APG IV (2016) has 416 families in 64 orders.

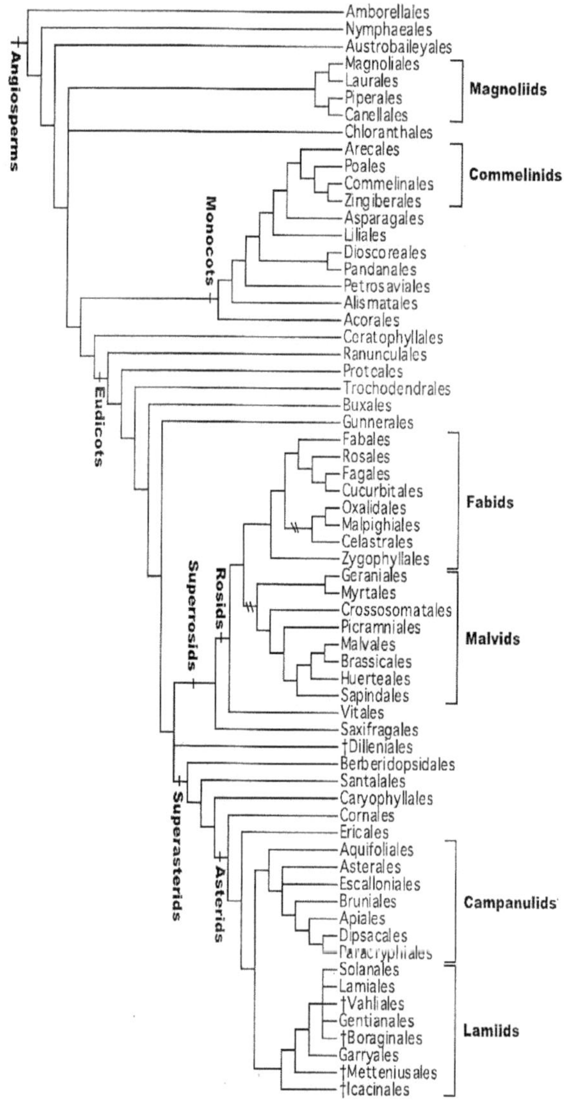

Figure 1. Interrelationships of the APG IV orders and some families supported by jackknife/bootstrap percentages >50 or Bayesian posterior probabilities >0.95 in large-scale analyses of angiosperms. See text for literature supporting these relationships. The alternative placements representing incongruence between nuclear/mitochondrial and plastid results for the Celastrales/Oxalidales/Malpighiales (COM) clade are indicated by slash marks (\\). †Orders newly recognized in APG.

Salient Features of APG IV

- Several new orders are recognized such as Boraginales, Dilleniales, Icacinales, Metteniusiales and Vahliales.
- Asclepiadaceae merged with Apocynaceae
- Bombacaceae, Sterculiaceae and Tiliaceae merged with Malvaceae
- Brassicaceae s.l. split into 3 families viz., Brassicaceae s.s., Capparaceae and Cleomaceae

- Chenopodiaceae merged with Amaranthaceae
- Euphorbiaceae s.l. is split into Euphoriaceae s.s., Phyllathaceae, Picrodendraceae, Putranjivaceae
- Fabaceae with 3 subfamilies: Faboideae, Caesalpinioideae and Mimosoideae
- Genera *Sopubia* and *Striga* (Scrophulariaceae) included in Orobanchaceae
- Liliaceae s.l. is split into 14 families and many taxa are included in Asparagaceae
- Molluginaceae and Gisekiaceae are recognized separately from Aizoaceae
- Some of the genera of Verbenaceae (e.g., *Tectona*) are included in Lamiaceae

Largest Families (Apg IV - 2016) (Species/Genera)

Orchidaceae (27,800/880)

Asteraceae (23,600/1620)

Fabaceae (19,560/745)

Rubiaceae (13,150/611)

Poaceae (11,337/707)

APG IV Classification considers Amborellaceae as the primitive most family of Angiosperms. *Amborella* is a monotypic genus of shrubs or small trees endemic to New Caledonia. The genus is the only member of the family Amborellaceae and the order Amborellales and contains a single species *Amborella trichopoda*. *Amborella* is of great interest to plant systematist because molecular phylogenetic analysis consistently place the genus at or near the base of the flowering plant lineage. That is, the Amborellaceae represent a line of flowering plants that diverged very early on (about 130 million years ago) from all the other extant species of flowering plants, and, among extant flowering plants, is the sister group to the other flowering plants. Comparing characteristics of this basal angiosperm, other flowering plants and fossils may provide clues about how flowers first appeared—what Darwin called the "abominable mystery". This position is consistent with a number of conservative characteristics of its physiology and morphology; for example, the wood of *Amborella* lacks the vessels characteristic of most flowering plants.

Because of its evolutionary position at the base of the flowering plant clade, there was support for sequencing the complete genome of *Amborella trichopoda* to serve as a reference for evolutionary studies. In 2010, the US National Science Foundation began a genome sequencing effort in *Amborella*, and the draft genome sequence was posted on the project website in December 2013.

In APG IV Classification families that made up potentially monofamilial orders, Dasypogonaceae and Sabiaceae, are instead referred to Arecales and Proteales respectively. Family names substitute for some like Asphodelaceae for Xanthorrhoeaceae (Asparagales) and Francoaceae for Melianthaceae (Geraniales). New families recognized in the APG IV system are Petenaeaceae (Huerteales), Kewaceae, Limeaceae, Macarthuriaceae and Microteaceae (all from Caryophyllales).

Several changes have been made in generic level composition of families in particularly Lamiales. These changes in APG IV on the basis of published studies include placement of *Petenaea* in its own family (Petenaeaceae), recognition of Kewaceae for the genus *Kewa* (previously included in Hypertelis in Molluginaceae) and reorganisation of some families that were known to be problematic. For example, several genera had been moved from Icacinaceae into the previously monogeneric Metteniusaceae.

Mostly molecular based classification for flowering plants is being developed by the APG and updated modern version was published more recently in 2016. In general, this classification system and taxonomic placements that are not compatible with published studies, but where further research is needed for the changes can be made in this system. Many botanic gardens and herbaria (including Kew) have adopted the APG system (Wearn *et al.*, 2013). APG IV will not be the final version, but the decreasing number of changes means that reorganising collections to reflect changes is becoming less arduous. The resulting system allows greater predictability than previous classifications because the groups that are recognised reflect evolutionary relationships of the flowering plants.

The detailed APG IV Classification is given below.

References

Angiosperm Phylogeny Group (1998). An ordinal Classification for the families of flowering plants. Annals of the Missouri Botanical Garden 85(4): 531-553.

Angiosperm Phylogeny Group II: Bremer, B., K. Bremer, M.W.Chase, J.L.Reveal, D.E.Soltis, P.S.Soltis and P.F.Stevens. 2003. An update of the Angiosperm Phylogeny Group Classification for the orders and families of flowering plants: APGII. Botanical Journal of the Linnean Society 141: 399-436.

Angiosperm Phylogeny Group III: Bremer, B., K. Bremer, M.W.Chase, M.F.Fay, J.L.Reveal, D.E.Soltis, P.S.Soltis and P.F.Stevens. 2009. An update of the Angiosperm Phylogeny Group classification for the orders and families of flowering plants APG III. Botanical Journal of the Linnean Society 161: 105-121.

Angiosperm Phylogeny Group IV (2016) An update of the Angiosperm Phylogeny Group classification for the orders and families of flowering plants APG IV. Botanical Journal of the Linnean Society 181: 1-20.

Wearn, J. A., Chase, M. W., Mabberley, D. J. & Couch, C. 2013. Utilizing a phylogenetic plant classification for systematic arrangements in botanic gardens and herbaria. Botanical Journal of the Linnean Society 172: 127–141.

Angiosperm Phylogeny Group Classification (APG IV) 2016

Order : Amborellales

Family 1. Amborellaceae

Order : Nymphaeales

Family 2. Hydatellaceae

Family 3. Cabombaceae

Family 4. Nymphaeaceae

Order : Austrobaileyales

Family 5. Austrobaileyaceae

Family 6. Trimeniaceae

Family 7. Schisandraceae

Mesoangiosperms Magnoliids

Order : Canellales

Family 8. Canellaceae

Family 9.Winteraceae

Order : Piperales

Family 10. Sauruaceae

Family 11. Piperaceae

Family 12. Aristolochiaceae

Order : Magnoliales

Family 13. Myristicaceae

Family 14. Magnoliaceae

Family 15. Degenereaceae

Family 16. Himantandraceae

Family 17. Eupomatiaceae

Family 18. Annonaceae

Order : Laurales

Family 19. Calycanthaceae

Family 20. Siparunaceae

Family 21. Gomortegaceae

Family 22. Atherospermataceae

Family 23. Hernandiaceae

Family 24. Monimiaceae

Family 25. Lauraceae

Independent Lineage : Unplaced To More Inclusive Clade

Order : Chloranthales

Family 26. Chloranthaceae

Monocots

Order : Acorales

Family 27. Acoraceae

Order : Alismatales

Family 28. Araceae

Family 29. Tofieldiaceae

Family 30. Alismataceae

Family 31. Butomaceae

Family 32. Hydrocharitaceae

Family 33. Scheuzeriaceae

Family 34. Aponogetonaceae

Family 35. Juncaginaceae

Family 36. Maundiaceae

Family 37. Zosteraceae

Family 38. Potamogetonaceae

Family 39. Posidoniaceae

Family 40. Ruppiaceae

Family 41. Cymodoceaceae

Order: Petrosaviales

Family 42. Petrosaviaceae

Order: Dioscoreales

Family 43. Nartheciaceae

Family 44. Burmanniaceae

Family 45. Dioscoreaceae

Order : Pandanales

Family 46. Triuridaceae

Family 47. Velloziaceae

Family 48. Stemonaceae

Family 49. Cyclanthaceae

Family 50. Pandanaceae

Order : Liliales

Family 51. Campynemataceae

Family 52. Corsiaceae

Family 53. Melanthiaceae

Family 54. Petermanniaceae

Family 55. Alstromeriaceae

Family 56. Colchicaceae

Family 57. Philesiaceae

Family 58. Ripogonaceae

Family 59. Smilacaceae

Family 60. Liliaceae

Order : Asparagales

Family 61. Orchidaceae

Family 62. Boryaceae

Family 63. Blandfordiaceae

Family 64. Asteliaceae

Family 65. Lanariaceae

Family 66. Hypoxidaceae

Family 67. Doryanthaceae

Family 68. Ixioliriaceae

Family 69. Tecophilaeaceae

Family 70. Iridaceae

Family 71. Xeronemataceae

Family 72. Asphodelaceae

Family 73. Amaryllidaceae

Family 74. Asparagaceae

Order : Arecales

Family 75. Dasypogonaceae

Family 76. Arecaceae

Order : Commelinales

Family 77. Hanguanaceae

Family 78. Commelinaceae

Family 79. Philydraceae

Family 80. Pontederiaceae

Family 81. Haemodoraceae

Order : Zingiberales

Family 82. Strelitziaceae

Family 83. Lowiaceae

Family 84. Heliconiaceae

Family 85. Musaceae

Family 86. Cannaceae

Family 87. Marantaceae

Family 88. Costaceae

Family 89. Zingiberaceae

Order : Poales

Family 90. Typhaceae

Family 91. Bromeliaceae

Family 92. Rapateaceae

Family 93. Xyridaceae

Family 94. Eriocaulaceae

Family 95. Mayaceae

Family 96. Thurniaceae

Family 97. Juncaceae

Family 98. Cyperaceae

Family 99. Restionaceae

Family 100. Flagellariaceae

Family 101. Joinvilleaceae

Family 102. Ecdeiocoeleaceae

Family 103. Poaceae

Probable Sister of Eudicots

Order : Ceratophyllales

Family 104. Ceratophyllaceae

EUDICOTS

Order : Ranunculales

Family 105. Eupteleaceae

Family 106. Papaveraceae

Family 107. Circaeasteraceae

Family 108. Lardizabalaceae

Family 109. Menispermaceae

Family 110. Berberidaceae

Family 111. Ranunculaceae

Order : Proteales

Family 112. Sabiaceae

Family 113. Nelumbonaceae

Family 114. Platanaceae

Family 115. Proteaceae

Order : Trochodendrales

Family 116. Trochodendraceae

Order : Buxales

Family 117. Buxaceae

Core Eudicots

Order : Gunnerales

Family 118. Myrothamnaceae

Family 119. Gunneraceae

Order: Dilleniales

Family 120. Dilleniaceae

Super Rosids

Order: Saxifragales

Family 121. Peridiscaceae

Family 122. Paeoniaceae

Family 123. Altingiaceae

Family 124. Hamamelidacceae

Family 125. Cercidiphyllaceae

Family 126. Daphniphyllaceae

Family 127. Itaceae

Family 128. Grossulariaceae

Family 129. Saxifragaceae

Family 130. Crassulaceae

Family 131. Aphanopetalaceae

Family 132. Tetracarpaeaceae

Family 133. Penthoraceae

Family 134. Haloragaceae

Family 135.Cynomoriaceae

Rosids

Order : Vitales

Family 136. Vitaceae

Order : Zygophyllales

Family 137. Krameriaceae

Family 138. Zygophyllaceae

Order : Fabalaes

Family 139. Quillajaceae

Family 140. Fabaceae

Family 141. Surianaceae

Family 142. Polygalaceae

Order : Rosales

Family 143. Rosaceae

Family 144. Barbeyaceae

Family 145. Dirachmaceae

Family 146. Elaeagnaceae

Family 147. Rhamnaceae

Family 148. Ulmaceae

Family 149. Cannabaceae

Family 150. Moraceae

Family 151. Urticaceae

Order : Fagales

Family 152. Nothofagaceae

Family 153. Fagaceae

Family 154. Myricaceae

Family 155. Juglandaceae

Family 156. Casuarinaceae

Family 157. Ticodendraceae

Family 158. Betulaceae

Order : Cucurbitales

Family 159. Apodanthaceae

Family 160. Anisophyllaceae

Family 161. Corynocarpaceae

Family 162. Coriariaceae

Family 163. Cucurbitaceae

Family 164. Tetramelaceae

Family 165. Datiscaceae

Family 166. Begoniaceae

Order : Celastrales

Family 167. Lepidobotryaceae

Family 168. Celastraceae

Order : Oxalidales

Family 169. Huaceae

Family 170. Connaraceae

Family 171. Oxalidaceae

Family 172. Cunoniaceae

Family 173. Elaeocarpaceae

Family 174. Cephalotaceae

Family 175. Brunelliaceae

Order : Malpighiales

Family 176. Pandaceae

Family 177. Irvingiaceae

Family 178. Ctenolophonaceae

Family 179. Rhizophoraceae

Family 180. Erythroxylaceae

Family 181. Ochnaceae

Family 182. Bonnetiaceae

Family 183. Clusiaceae

Family 184. Calophyllaceae

Family 185. Podostemaceae

Family 186. Hypericaceae

Family 187. Caryocaraceae

Family 188. Lophopyxidaceae

Family 189. Putranjivaceae

Family 190. Centroplacaceae

Family 191. Elatinaceae

Family 192. Malpighiaceae

Family 193. Balanopaceae

Family 194. Trigoniaceae

Family 195. Dichapetalaceae

Family 196. Euphroniaceae

Family 197. Chrysobalanaceae

Family 198. Humiriaceae

Family 199. Achariaceae

Family 200. Violaceae

Family 201. Goupiaceae

Family 202. Passifloraceae

Family 203. Lacistemataceae

Family 204. Salicaceae

Family 205. Peraceae

Family 206. Rafflesiaceae

Family 207. Euphorbiaceae

Family 208. Linaceae

Family 209. Ixonanthaceae

Family 210. Picrodendraceae

Family 211. Phyllanthaceae

Order : Geraniales

Family 212. Geraniaceae

Family 213. Francoaceae

Order : Myrtales

Family 214. Combretaceae

Family 215. Lythraceae

Family 216. Onagraceae

Family 217. Vochysiaceae

Family 218. Myrtaceae

Family 219. Melastomataceae

Family 220. Crypteroniaceae
Family 221. Alzateaceae
Family 222: Penaeaceae

Order : Crossosomatales

Family 223. Aphloiaceae
Family 224. Geissolomataceae
Family 225. Strasburgeriaceae
Family 226. Staphyleaceae
Family 227. Guamatelaceae
Family 228. Stachyuraceae
Family 229. Crossosomataceae

Order : Picramniales

Family 230. Picramniaceae

Order : Huerteales

Family 231. Gerrardinaceae
Family 232. Petenaeaceae
Family 233. Tapisciaceae
Family 234. Dipentodontaceae

Order : Sapindales

Family 235. Biebersteiniaceae
Family 236. Nitrariaceae
Family 237. Kirkiaceae
Family 238. Burseraceae
Family 239. Anacardiaceae
Family 240. Sapindaceae
Family 241. Rutaceae
Family 242. Simaroubaceae
Family 243. Meliaceae

Order : Malvales

Family 244. Cytinaceae
Family 245. Muntingiaceae
Family 246. Neuradaceae
Family 247. Malvaceae
Family 248. Sphaerosepalaceae
Family 249. Thymeleaceae
Family 250. Bixaceae
Family 251. Cistaceae
Family 252. Sarcolaenaceae
Family 253. Dipterocarpaceae

Order : Brassicales

Family 254. Akaniaceae
Family 255. Tropaeolaceae
Family 256. Moringaceae
Family 257. Caricaceae
Family 258. Limnanthaceae
Family 259. Setchellanthaceae
Family 260. Koeberliniaceae
Family 261. Bataceae
Family 262. Salvadoraceae
Family 263. Emblingiaceae
Family 264. Bataceae
Family 265. Pentadiplandraceae
Family 266. Gyrostemonaceae
Family 267. Resedaceae
Family 268. Capparaceae
Family 269. Cleomaceae
Family 270. Brassicaceae

Order : Berberidopsidales

Family 271. Aextoxicaceae
Family 272. Berberidopsidaceae

Order : Santalales

Family 273. Olacaceae
Family 274. Opiliaceae
Family 275. Balanophoraceae
Family 276. Santalaceae
Family 277. Misodendraceae
Family 278. Schoepfiaceae
Family 279. Loranthaceae

Order : Caryophyllales

Family 280. Frankeniaceae
Family 281. Tamaricaceae
Family 282. Plumbaginaceae
Family 283. Polygonaceae
Family 284. Droseraceae
Family 285. Nepenthaceae
Family 286. Drosphyllaceae
Family 287. Dioncophyllaceae
Family 288. Ancistrocladaceae
Family 289. Rhabdodendraceae
Family 290. Simmondsiaceae
Family 291. Physenaceae
Family 292. Asteropeiaceae
Family 293. Macarthuriaceae
Family 294. Microteaceae
Family 295. Caryophyllaceae
Family 296. Achatocarpaceae
Family 297. Amaranthaceae
Family 298. Stegnospermataceae
Family 299. Limeaceae
Family 300. Lophiocarpaceae
Family 301.Kewaceae
Family 302. Barbeuiaceae
Family 303. Gisekiaceae
Family 304. Aizoaceae
Family 305. Phytolaccaceae
Family 306. Petiveriaceae
Family 307. Sarcobataceae
Family 308. Nyctaginaceae
Family 309. Molluginaceae
Family 310. Montiaceae
Family 311. Didiereaceae
Family 312. Basellaceae
Family 313. Halophytaceae
Family 314. Talinaceae
Family 315. Portulacaceae
Family 316. Anacampserotaceae
Family 317. Cactaceae

Order : Cornales

Family 318. Nyssaceae
Family 319. Hydrostachyaceae
Family 320. Hydrangeaceae
Family 321. Loasaceae
Family 322. Curtisiaceae
Family 323. Grubbiaceae
Family 324. Cornaceae

Order : Ericales

Family 325. Balsaminaceae
Family 326. Marcgraviaceae
Family 327. Tetrameristaceae
Family 328. Fouquieriaceae
Family 329. Polemoniaceae
Family 330. Lecythidaceae
Family 331. Sladeniaceae
Family 332. Pentaphylacaceae
Family 333. Sapotaceae
Family 334. Ebenaceae
Family 335. Primulaceae
Family 336. Theaceae
Family 337. Symplocaceae
Family 338. Diapensiaceae
Family 339. Styracaceae
Family 340. Sarraceniaceae
Family 341. Roridulaceae
Family 342. Actinidiaceae
Family 343. Clethraceae
Family 344. Cyrillaceae

Family 345. Ericaceae
Family 346. Mitrastemonaceae

Order : Icacinales

Family 347. Oncothecaceae
Family 348. Icacinaceae

Order : Metteniusales

Family 349. Metteniusaceae

Order : Garryales

Family 350 Eucommiaceae
Family 351. Garryaceae

Order : Gentinalaes

Family 352. Rubiaceae
Family 353. Gentianaceae
Family 354. Loganiaceae
Family 355. Gelsemiaceae
Family 356. Apocynaceae

Order : Boraginales

Family 357. Boraginaceae

Order : Vahliales

Family 358. Vahliaceae

Order : Solanales

Family 359. Convolvulaceae
Family 360. Solanaceae
Family 361. Montiniaceae
Family 362. Sphenocleaceae
Family 363. Hydroleaceae

Order : Lamiales

Family 364. Plocospermataceae
Family 365. Carlemanniaceae
Family 366. Oleaceae
Family 367. Tetrachondraceae
Family 368. Calceolariaceae
Family 369. Gesneriaceae
Family 370. Plantaginaceae
Family 371. Scrophulariaceae
Family 372. Stilbaceae
Family 373. Linderniaceae
Family 374. Byblidaceae
Family 375. Martyniaceae
Family 376. Pedaliaceae
Family 377. Acanthaceae
Family 378. Bignoniaceae
Family 379. Lentibulariaceae
Family 380. Schlegeliaceae
Family 381. Thomandersiaceae
Family 382. Verbenaceae
Family 383. Lamiaceae
Family 384. Mazaceae
Family 385. Phrymaceae
Family 386. Paulowniaceae
Family 387. Orobanchaceae

Order : Aquifoliales

Family 388. Stemonuraceae
Family 389. Cardiopteridaceae
Family 390. Phyllonomaceae
Family 391. Helwingiaceae
Family 392. Aquifoliaceae

Order : Asterales

Family 393. Rousseaceae
Family 394. Campanulaceae
Family 395. Pentaphragmataceae
Family 396. Stylidaceae
Family 397. Alseuosmiaceae
Family 398. Phellinaceae
Family 399. Argophyllaceae
Family 400. Menyanthaceae
Family 401. Goodeniaceae

Family 402. Calyceraceae

Family 403. Asteraceae

Order : Escalloniales

Family 404. Escalloniaceae

Order : Brunniales

Family 405. Columelliaceae

Family 406. Bruniaceae

Order : Paracryphiales

Family 407. Paracryphiaceae

Order : Dipsacales

Family 408. Adoxaceae

Family 409. Caprifoliaceae

Order : Apiales

Family 410. Pennantiaceae

Family 411. Torricelliaceae

Family 412. Griseliniaceae

Family 413. Pittosporaceae

Family 414. Araliaceae

Family 415. Myodocarpaceae

Family 416. Apiaceae

5

Origin of Angiosperms

Among the plants inhabiting the earth, the Angiosperms or flowering plants have the greatest number of species and occupy more types of habitats than any other group. They include trees, shrubs, herbs, perennials, annuals and lianas. They range in size from tiny duck weeds to giant trees (Gums of Australia).

Charles Darwin once called the origin of Angiosperms an abominable mystery because little evidence was available to him to help trace the Angiosperms back in them. The origin and evolution of angiosperms continues to present an intriguing challenge to those paleobotanists who specialize in the study of fossil angiosperms. The collection of fossils, the accumulated literature and the experience available today provide much better understanding of angiosperm origin.

Polyphyletic or Monophyletic

The morphological and taxonomic evidence leads to the conclusion that the flowering plants are one natural monophyletic branch of development and are not made up of various unrelated parallel branches of different origins. The common origin of all the orders and families of angiosperms is demonstrated above all by their multitude of common morphological characters. It is demonstrated by the uniform staminal structure with the characteristic endothecial layer of anther wall, by the presence of specialized megasporophylls (carpels) with stigmas, by the constancy of the relative position of androecium and Gynoecium on the floral axis, by the characteristic male and female gametophytes, basically similar throughout the group and accompanied by double fertilization and the formation of triploid endosperm and by the presence of sieve tubes (Takhtajan, 1969). As Parkin (1923) pointed out, the independent origin of these characters in different taxonomic groups is statistically very unlikely.

Takhtajan (1969) is of the opinion that the orders Urticales, Casuarinales and Fagales so distinct at first sight from the Magnoliales are linked with them by an intermediate group, the Hamamelidales. Similarly the Lemnaceae and Araceae though do not resemble each other are connected through the somewhat intermediate genus *Pistia*. One may instance numerous such examples, and they show that those groups which at first sight seem completely isolated and wider comparison with other groups, in fact sooner or later fall into their natural place in the system of Angiosperm classification. All available evidences lead to the conclusion that the angiosperms form one natural monophyletic branch of development (=*monophyletic*) and not *polyphyletic*.

Simpson (1961) defined monophyly as the derivation of a taxon through one or more lineages from one immediately ancestral taxon of the same or lower rank. Such a definition would be true if say genus B evolved from genus A through one species of the latter, since in that case the genus would be monophyletic at the same rank (genus) as well as the lower (species) rank. On the other hand, if genus B evolved from two species of genus A, it would be monophyletic at the genus level but polyphyletic at the lower rank. Most authors, however, adhere to a stricter interpretation of monophyly, namely the group should have evolved from a single immediately ancestral species belonging to the group in question. There are thus two different levels of monophyly: a minimum monophyly (Melville, 1983) wherein one supraspecific taxon is derived from another of equal rank (Simpso's definition) and a strict monophyly wherein one higher taxon is derived from single evolutionary species (Hennig, 1966).

It is widely accepted principle that taxonomic groups (at least above the species level) should be monophyletic. If it is confirmed that the group is polyphyletic, it is advisable to split into monophyletic groups. Based on this principle, the genus *Paeonia* has been separated from Ranunculaceae into a distinct family Paeoniaceae. If the group represents paraphyly an attempt should be made to identify the remaining members and include these within the group to make it monophyletic. Sosef (1997) compares the existent hierarchial models of classification. He argues that a phylogentic tree can be subdivided according to a monophyletic hierarchial model, in which only monophyletic units figure or, according to a 'Linnean' hierarchial model, in which both mono- and paraphyletic units occur.

Most present day phylogenists try to fit the monophyletic model within the set of nomenclatural conventions that fit the Linnean model. However, the two models are intrinsically incongruent. The monophyletic model requires a system of classification of its own, at variance with currently accepted conventions. Since, however, the monophyletic model is unable to cope with reticulate evolutionary relationships, it is unsuited for the classification of nature, the Linnean model is to be preferred. This renders the acceptance of paraphyletic supraspecific taxa inevitable.

Ancestors of Angiosperms

The identity of the ancestors of the flowering plants is a most difficult problem which is as yet far from being solved. Several groups of plants have been considered as ancestral stock for angiosperms.

Bennettitalean Ancestry

Since Saporta (Saporta and Marion, 1885) Bennettitales have often been proposed as possible ancestors of angiosperms, and in this connection the resemblance in structure between the strobili of the Mesozoic genus *Cycadeodea* (Fig. 5.1) and the flower of *Magnolia* has often been pointed out. But this resemblance is wholly superficial, they are alike only in that both are bisexual and both consist of an elongated axis on which are arranged successively and in the same order, protective bracts (perianth members in *Magnolia*), microsporophylls and megasporophylls. But along with these few similarities there are profound differences. The microsporophylls (stamens) of *Magnolia* (as in other primitive angiosperms) are free and arranged spirally on the axis, but in Bennettitales they are whorled and mostly connate. The megasporophylls of the Bennettitales are very reduced, simplified stalk-like structures, sometimes very abbreviated, each bearing at its apex a solitary erect ovule. Between these stalk-like megasporophylls and alternating with them, are sterile organs (interseminal scales) which appear to be modified sterilized megasporophylls. These sterile scales, with their tightly packed expanded apices, form a kind of protective armour round the ovules. Protection of the ovules is achieved, therefore, in a very different way from that found in the Angiosperms. Another special feature of the Bennettitales is the presence in the ovule of a distinct micropylar tube, formed by the integument and serving for the reception of the microspores. In angiosperms there is no such micropylar tube; the microspores are caught by the stigma, not by the ovule. The Bennettitalean seed differs from that of the primitive Angiosperms in being exalbuminous, the embryo itself filling almost all the seed cavity, and nutritive tissue being entirely absent or very scanty. All these show that the Bennettitales cannot have been the ancestors of the Angiosperms.

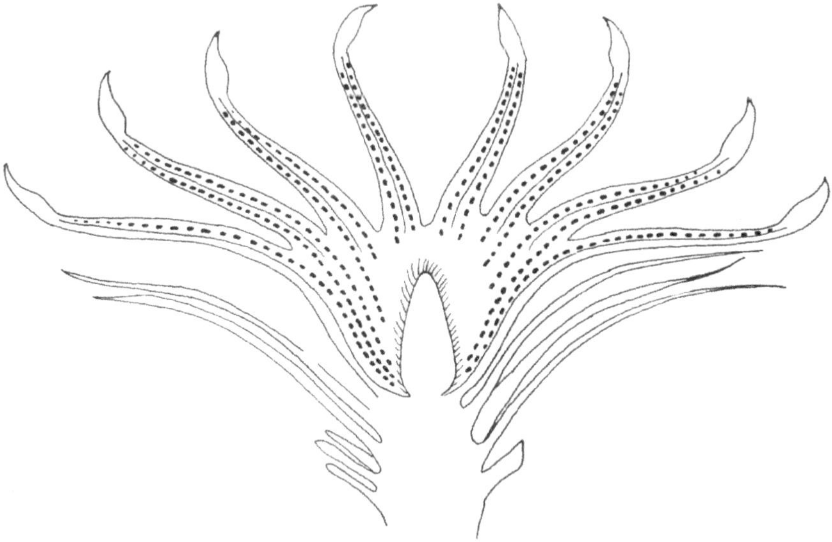

Fig. 5.1: Bisexual strobilus of Cycadeoidea.

Though the Bennettitales cannot be the ancestors of the Angiosperms, it is quite possible that they are connected through common ancestry like seed-ferns (see Arber and Parkin, 1907). Bennettitalean theory and its variations were supported by Arber and Parkin (1907), Takhtajan (1969), Becker and Theissen (2003), Baum and Hileman (2006) and Theissen and Melzer (2007).

Gnetalean Ancestry

This theory was proposed by Wettstein (1901) and supported by Markgraf (1930), Fagerlind (1947) and Donoghue and Doyle (2000). The Gnetales resemble Angiosperms in many respects. Both of them have two cotyledons, unlike all other Gymnosperms have vessels in their secondary wood, two integuments and net veined leaves.

Gnetales, in many respects, have attained an evolutionary level higher than that of some of the primitive Angiosperms. For instance, the living genera *Ephedra, Welwitschia* and *Gnetum* have vessels in the secondary wood, yet some Angiosperm families such as Winteraceae (woody) and Nymphaeaceae (herbaceous) have xylem completely devoid of them. Further more, special investigations (Thompson, 1918) have shown that the vessels of *Welwitschia, Ephedra* and *Gnetum* originated in an entirely different way from those of Angiosperms.

Isoetalean Ancestry of Monocotyledons

Proposed by Campbell (1928), this theory is essentially based on marked similarity between *Isoetes* and Monocots especially *Najas flexilis. Isoetes* is predominantly herbaceous and geophilous. It is also either aquatic or amphibious. Like most of the Filicineae, it is found in humid tropics, a habitat, which is also characteristic of a large number of monocotyledons. Campbell also pointed out the marked similarity in habit and resemblances in the embryo and anatomy of the older sporophyte between Isoetales and some other lower aquatic monocotyledons.

Engler and his associates have also postulated the probable origin of Monocotyledons from various groups of Pteridophytes, through intermediate and hypothetical group, protangiosperms. According to Engler this group is not represented in the fossils because many of them were herbaceous. On the basis of similarities in their vascular bundles Engler also postulate direct derivation of Monocotyledons from Ophioglossaceae.

This theory of Pteridophytes ancestry is highly unacceptable because the monocotyledons are now considered as most advanced group of Angiosperms and derived from Dicotyledons.

Coniferalean Ancestry of Amentiferae

This theory has been proposed by Eichler (1875), Engler (1882, 1892), Engler and Prantl (1924), Rendle (1904, 1930), Hagerup (1934, 1936) and Doyle (1945). These authors have pointed out the resemblances of Angiosperms to conifers and considered that coniferales might have given rise to primitive (hypothetical) group of Angiosperms known as Amentiferae. The inflorescence of the amentifers like

Casuarinaceae, Salicaceae and Fagaceae with their simple and naked flowers were compared with those of conifers.

This theory is not acceptable because the Amentiferae group is now considered as an advanced group of Angiosperms. The advanced features in Amentiferae include anemophilous flowers, unisexual flowers, naked flowers etc. The wood anatomical data suggest that Amentiferae are highly advanced. Hence coniferales as the ancestors of Angiosperms and Amentiferae as primitive group of Angiosperms is ruled out.

Pteridospermean Ancestry

Pteridosperms (Seed ferns) were considered as ancestors of Angiosperms by Long (1966). Features like reticulate venation, monopodial branching, presence of cambium, presence of microsporophylls and megasporophylls on the same plant. Origin of sepals from leaves and petals from sepals and stamens, development of triploid endosperm as extreme reduction of female gametophyte, similarities of seed structure and existence of one or several ovules subtended by a cupule together with information provided by *Glossopteris* reproductive structures are enough to bring seed ferns (Pteridosperms) closest to Angiosperms. Pteridospermalean theory was supported by Melville (1969), Retallack and Dilcher (1981), Zhang (1995), Sun *et al.* (2001) and Frohlich (2003)

But Pteridosperm ancestry of Angiosperms appears to be untenable because of two serious objections. The scalariform xylem elements never occur in Pteridosperms while they are common in Angiosperms. No satisfactory explanation is given for cupule as an equivalent of carpellary wall in case of multiovulate cupules.

Pentoxylalean Theory

Pentoxylales have many similarities with *Pandanus*. These include: erect habit, dioecious nature, tuft of leaves, axillary inflorescence, peduncle with tracheids and spiral thickening and bordered pitting, two layered seed coat of which the inner one is usually tanniferous, seeds enclosed in a fleshy layer sarcotesta, endospermic seeds and minute embryo.

Based on these similarities, Meeuse (1961) considered Pentoxylales as ancestors of Angiosperms. In spite of resemblances, authors like Pant and Kidwai (1971) hold that they could equally be the result of parallel evolution.

Caytonialean Ancestry

Caytonialean ancestry of Angiosperms was proposed by Thomas (1925, 1936) and supported by Stebbins (1974). Caytoniales are fossil plants of middle Jurassic period. Caytoniales had Angiosperm-like anthers produced in groups or single on branching pinnate structures, which may be described as sporophylls. Such structures are comparable with the branched stamens found in such plants as *Ricinus, Hypericum* and *Calothamnus*. Caytoniales are characterized by a curved cup-like structure called cupule, in which the ovules are enclosed. Proponents of Caytoniales as the ancestors of Angiosperms point to the almost sealed cupule as suggestive of the way a carpel might have evolved. The caytoniales also exhibited leaves with a net-like pattern of venation and tubes leading to the micropyle of the ovule. Thomas suggested that the carpel wall of Angiosperm may represent, a pair of concrescent cupules and that the possible origin of stigma should be considered in the light of these ancient forms.

Caytoniales, however, are now shown to have relationships with Pteridosperms and they are now classified as Mesozoic remnants of that group. The ancestors of Angiosperms must have had open megasporophylls with exposed ovules, as is evident from the morphological studies on the nature of their carpel. Thus the ovary-like pouches of Caytoniales cannot be taken to be the fore-runners of Angiosperm carpel.

Further the presence of pollination drop, direct pollination, fleshy canals in the inner side of the curved cupules, highly cuticularised winged pollen and absence of connective like structure in tetralocular synangia of caytoniales negate the theory of Caytonialean ancestry of Angiosperms.

Durian Theory of Origin of Angiosperms

Durio zibethinus, a member of Bombacaceae from Burmese and Malayan forests was considered by Corner (1949), as a surviving model of primitive angiosperms. It is a cauliflorous tree and bears large, coloured, loculicidal spiny capsules with fleshy arillate seeds. There are about 45 Angiospermous families, which show arillate genera mostly distributed in the tropics. According to this theory the primitive Angiosperms are mesophytic, tropical in distribution, with cycad like tree habit, compound leaves, probably monocarpic and producing a large terminal cluster of arillate follicles.

The theory has been criticized by a number of morphologists including Pijl (1952), Parkin (1953), Metcalfe (1954) and Eames (1961).

Other theories on origin of Angiosperms include Chloranthoid theory proposed by Leroy (1983) and Stuessy (2004); Gonophyll theory proposed by Melville (1969) and supported by Stewart and Rothwell (1993); multiple origin hypothesis proposed by Anderson (1934), Wu *et al.* (2002) and Nair (1979) and Paleoherb theory proposed by Taylor and Hickey and supported by Burger (1981).

Although some groups have been eliminated as likely ancestors of Angiosperms, Darwin's abominable mystery is not completely solved. However, it appears likely that plants having some Angiosperm features could have evolved from the seed fern line late in the Jurassic or very early Cretaceous period. Certainly a vast gene pool was available at the time to give rise to the first Angiosperms, which might be called *Proangiosperms*.

Time of Origin and Diversification

Some botanists suggested that the Angiosperms originated in the early Mesozoic or even the late Paleozoic and that they underwent extensive diversification by the Aptian-Albian stages of the lower Cretaceous period. The best available evidence from the fossil record indicates that the Angiosperms originated during the early cretaceous period around 130 to 135 million years ago (see Wolfe *et al.*, 1975). Taylor (1981) suggests that pollen and megafossil data indicate the origin of the Angiosperms during the very early Cretaceous or the later stages of the Jurassic period.

In the older cretaceous sediments, the Angiosperm fossil records are overshadowed by those of ferns and Gymnosperms and it is not until the late part of the Cretaceous that the Angiosperms became dominant. The first monosulcate Angiosperm pollen grains, characteristic of the primitive Dicotyledons and the Monocotyledons, appear in the Barremian age of the lower cretaceous. Tricolpate pollen grains which are found in the more advanced Dicotyledons are first reported from slightly younger Aptian rocks. Pollen sequences have clearly shown a pattern of increasing diversity in the cretaceous. The progression of pollen from primitive types in older strata to more derived types in younger sediments indicates to paleobotanists that Angiosperms underwent much diversification during cretaceous (Raven and Axelrod, 1974; Wolfe *et al.*, 1975).

Only a few records are available of dicotyledonous woods of early cretaceous age. They represent primitive types of xylem and lack of the features of advanced Angiosperms. An investigation of fossil leaves reveal a pattern of increasing diversity during cretaceous period. The ancestors of flowering plants diverged around 245-202 million years ago. These diversified enormously and got widely dispersed around 120 million years ago but replaced conifers as the dominant taxa only around 60-100 million years ago.

Phylogeny of Angiosperms

In the last two to three decades, molecular phylogenetic studies (see full literature in Friedman *et al.*, 2004; Soltis *et al.*, 1999, 2000; Zanis *et al.*, 2003) have indicated that extant gymnosperms may be sister to the flowering plants. Hence, for the time being, establishing angiosperm outgroups will continue to be critical to assessing character state polarities and homologies of the most important angiosperm features. There is also conflict regarding which lineages constitute the earliest divergent angiosperms. Previous views focused on Magnoliaceae and its close relatives, but very recent analyses have indicated that the monotypic *Amborella* is sister to all other angiosperms (see full literature in Friedman *et al.*, 2004) and that Nymphaeales is sister to all other angiosperms exclusive of *Amborella*; further Illiciaceae, Schisandraceae, Trimeniaceae and Austrobaileyaceae (Austrobaileyales) make up a clade that is sister to the remaining angiosperms.

The homologies of carpels, tetrasporangiate anthers, bitegmic ovules, three-celled male gametophytes, eight-nucleate female gametophyte etc., which are believed to be synapomorphies of angiosperms, with those of nonflowering seed plants (i.e. gymnosperms), are totally unknown. It is generally assumed that first flowers of angiosperms were hermaphrodite because the female flowers of *Amborella* produce staminodes, thus demonstrating a basically bisexual nature to its flower (it is also worth to indicate that unisexual flowers were common among the earliest angiosperm macrofossils). *Amborella* and many members of Austrobaileyales have helically arranged perianth lobes but not distinct sepals and petals which are characteristics of eudicots. Molecular genetics of floral organogenesis has provided some clues to the diversification of floral morphology; particularly the role of the homeotic transcription factors, the MADS-box genes (like MIKC type), is very significant in this connection. Also important is the role of floral organ identity genes (the ABC + DE genes), as their gene functions are mainly conserved across a breadth of taxonomic groups and hence can be used to identify widely conserved organ identity programmes (Friedman *et al.*, 2004). It has been suggested that the alterations in expression patterns of floral development genes gave rise to the angiosperm flower and all its variations. Investigations on the floral transcription factor LEAAFY and the B-class MADS-box genes in gymnosperms, for example, have led to the proposal of the mostly male theory of floral origin (Frohlich, 2003), which suggests that the male and female reproductive units were combined in angiosperm flower by emergence of ectopic ovules on microsporophylls in the originally male cones of a gymnosperm ancestor. Another theory of flower origin suggests a homeotic transformation of reproductive organ to either male or female from unisexual gymnospermous cones, resulting in the origin as an ancestral bisexual flower (Theissen *et al.*, 2002; Theissen and Meizer, 2007). In this hypothesis, changes in expression of the homologous to the B-class floral homeotic genes along the reproductive axis of the cone cause the required developmental transformation to form a flower.

References

Bahadur, B., T. Pullaiah and K.V.Krishnamurthy. 2015. Angiosperms: An overview. In: Bahadur, B. *et al. Plant Biology and Biotechnology.* Springer, New Delhi.

Naik, V.N. 1988. *Taxonomy of Angiosperms*. Tata-McGraw Hill, New Delhi.

Pullaiah, T. 2013. *Text book on Biosystematics*. Astral, New Delhi.

Radhakrishnaiah, M. 1996. *Essentials of Plant Taxonomy*. Hyderabad,

Takhtajan, A.L. 1969. *Flowering Plants — Origin and Dispersal*. Oliver & Boyd. Edinburgh.

6

Plant Nomenclature

For communicating about the plants around him man has given names to plants. Naturally he named plants around him in his own language. Since there are several languages, even in the same country, for the plants different names were given in different languages in different countries. When a person wants to communicate about a plant to a person speaking different language it becomes difficult or impossible. Further for the same plant, even in the same language, several names have been given. Similarly same name has been given for different plants, even by scientists. It was found necessary to have a common name throughout the world. Hence an International Code of Nomenclature for algae, fungi and plants has been devised to bring an order for naming a plant. The nomenclature to taxonomy is similar to the constitution to a country.

It was already emphasized earlier that plant nomenclature is one of the important aspects of plant taxonomy.

What Language should be Used for Plant Names?

Plant names should be given in such a way that they should be used internationally. But in what language? Is it English, German, Sanskrit or a language used in most of the countries? Or the oldest language? Which language gives clear cut meaning? Which language has no changes? Which language is not used now? It was decided that the names should be in such a language in which most of the earliest names were given. Latin was such a language. Other reasons that weighed for selecting Latin are:

1. In the past European scholars learned their subjects in Latin.

2. All the terms used for description of living organisms are available in Latin language.

3. Europe's earlier literature was mostly in Greek and Latin. Besides these, the language selected should be International. Most of the European languages were derived from Latin. The usage of Latin is not only for the names but also for description. In this connection the students should consult the book *Botanical Latin* (Stearn, 1973).

Plant Nomenclature is one of the main aspects of Plant Taxonomy.

History of Plant Nomenclature

Until the first half of eighteenth century polynomial system was used in naming of the plants. Polynomial means a plant name consisting of several words in a series revealing the diagnostic characters of a plant. For example, *Sida cordifolia* L. (1753) was named as '*Althaea maderaspatana subrotundo folio molli & hirsuto multipilis*' by Plukenet (1692). *Strychnos nux-vomica* L. was then known as *Strychnos folis ovatis caule enesme*. Such polynomials are difficult to remember and for indexing plant names.

Polynomial system was superseded by the binomial system. Binomial means a name of species in a combination of two words. The first denotes the name of a genus and second the specific epithet. The first application of a binomial system to the plant kingdom as a whole was done by Linnaeus in his *Species Plantarum* (1753) wherein each species is given the usual diagnostic phrase together with trivial name (specific epithet) written in the margin of the text against each species.

Linnaeus in his *Fundamenta botanica* (1736) gave various aphorisms for naming plants. Linnaeus elaborated these aphorisms and pronouncements in his *Critica botanica* (1737). This was written primarily to explain the taxonomic and nomenclatural principles employed by Linnaeus in his *Genera Plantarum* and *Hortus Cliffortianus*. The aphorisms constituted his principles of nomenclature and many of them are paralleled closely by contemporary rules of nomenclature. Later in his *Philosophia Botanica* (1751), Linnaeus amplified the views expressed in his Aphorisms and principles and established the beginnings of sound nomenclature for plants.

In the seventeenth and eighteenth centuries there was no generally accepted rules for naming plants, except the Linnaen principles and it was difficult to realize the confusion which existed over plant names; every botanist followed his own way in naming the plants. According to the Linnaean principles, no two genera could have the same generic name and no two species within a genus could have the same specific epithet. It was accepted also that when a genus was divided into two or more genera, the original generic name must be retained for one of them. In a general way, priority of publication was given recognition; most authors after the time of Linnaeus endeavoured to avoid assigning different names to identical species or identical names to different species. However, the number of species and genera new to science increased so rapidly during the early part of the nineteenth century that confusion in nomenclature did exist. In 1813 Augustin de Candolle published *Theorie elementaire de la botanique* containing instructions on various nomenclatural procedures mostly followed by Linnaeus.

For many years there was no index from which botanists of one country or district might learn names that had been employed elsewhere by other botanists. The first important post-Linnaean index to the names of flowering plants was Steudel's *Nomenclator botanicus* (1821). This work comprised a list of the Latin names of all flowering plants then known, together with their synonyms.

At present the index widely used is *Index Kewensis*. This was published in two parts and 14 supplements. In this genera and species of phanerogams published after 1753 were indexed. The names published between 1753-1885 were published in two volumes. Later at the interval of every 10 years a supplement is being published. *Index Londonensis* is another index where illustrations of flowering plants and ferns were listed. Web site *Plant List* (www.theplantlist.org) is the latest which gives the details of all plant species names and their synonyms.

Paris Code (1867)

The first organized efforts toward standardization and legislation of nomenclatural practices were the first International Botanical Congress held in Paris in August 1867. Alphonse de Candolle proposed his "*Lois de la nomenclature botanique*" (Laws of botanical nomenclature) mainly based on his father's (Augustine de Candolle) *Theoire elementaire de la botanique* (1813).

The Paris Code' was printed and published in 1867. It consisted of only three chapters with a total of 68 articles. The rule of priority was treated as fundamental. Considerable attention was given to the matter of author citation and to the terms to be applied to the categories of plants. The requirements for valid publication and for acceptance and rejection of names were established. The rules of the Paris code represented an excellent beginning in the right direction, but their application revealed numerous inherent deficiencies.

Rochester Code (1892)

A set of rules of plant nomenclature prepared by N.L. Britton and a group of botanists was discussed at the meeting of the American Association for the advancement of science held in Rochester, New York in 1892. In this code it was established that each species must be based on a specified herbarium specimen to be known as nomenclatural type. It advocated the strict application of the rule of priority. The code accepted all binomials, even tautonyms (the specific epithet repeating the generic name). Page priority or the precedence of a name within a publication was also taken into consideration in the application of the rule of priority.

Vienna Code (1905)

The third International Botanical Congress, which met in Vienna in June 1905, gave extensive consideration to Nomenclatural matters. The rules were based on those of Paris code but were modified and amplified considerably. The important changes made in Vienna code are: Establishment of Linnaeus' '*Species Plantarum*' (1753) as the starting point of the nomenclature of Vascular plants and the acceptance of descriptions of genera as established by Linnaeus in the fifth edition of his *Genera Plantarum* (1754), to validate the publication of generic names appearing in his *Species Plantarum* (1753), the establishment of a list of generic names to be conserved (*nomina generic conservanda*); the requirement that names of new groups to be accompanied by Latin diagnosis; the ruling that tautonymy was not admissible, and that the date of publication alone determined the priority of a name.

American Code (1907)

The proposals of the Rochester code formed the basis for the American code and agreed in some respects with Vienna code. It did not accept the *nomina generica conservanda* or the requirements of Latin diagnosis for new taxa.

Brussels Code (1912)

The Fourth International botanical congress met in 1910 in Brussels. It's only significant action was the establishment of different starting points for priority of names of non vascular plants, the recognition of value of type concept and classification of phraseology of the Vienna rules.

Cambridge Code (1935)

The fifth International Botanical Congress was held at Cambridge (England) in 1930. At the Cambridge congress, every effort was made to harmonize the basic differences between Vienna rules and American code. This was the first code of nomenclature came into use by the botanists all over the world. The code adopted the type concept.

Amsterdam Code (1947)

The sixth International Botanical Congress was held at Amsterdam in 1935. The rules of Nomenclature were divided into chapters and sections and these were subdivided into Articles and Recommendations.

Stockholm Code (1952)

The seventh International Botanical Congress was held in Stockholm in 1950 and the rules of nomenclature were the subject of lengthy and considerable discussion. Minor changes and a few additions were legislated. The word *taxon* (Pl. *taxa*) was introduced for the first time to designate any taxonomic group or entity.

Paris Code (1956)

The eighth International Botanical Congress met in Paris in July 1954. The nomenclature section decided that the code should be published in English, French and German languages. One of the main new features of this code was the separation of the Preamble and Principles from Rules and Recommendations. Appendix II (*Nomina Generica conservanda et rejicienda*) was amended and supplemented. Appendix IV (Determination of types) was reprinted verbatim.

Montreal Code (1961)

The ninth International Botanical Congress was held in August 1959 at Montreal. In the Montreal code *Nomina familiarum Conservanda* for Angiospermae as Appendix II was introduced. It was made clear that the naming of fossil plants should follow the same lines as that of recent ones.

Edinburgh Code (1966)

The tenth International Botanical congress was held in August 1964 at Edinburgh. The Edinburgh congress set up a committee to draw up a glossary of the main technical terms occurring in the code and that resulted in An annotated glossary of botanical nomenclature (McVaugh *et al.*, 1968).

Seattle Code (1972)

The eleventh International Botanical Congress met in Seattle, U.S.A. in August 1969. Most of the proposals made here are concerned with refinements and some amendments. The word autonym (automatically established name) has been introduced in this code.

Leningrad Code (1978)

The twelfth International Botanical Congress was held in July 1975 in Leningrad (Russia). In this code the Principle of automatic typification is extended to those names of taxa above the rank of family that are ultimately based on generic names. Application of the principle of priority is recommended when, selecting among names thus typified. The code makes it clear that it does not apply to names of organisms treated as Bacteria.

Sydney Code (1983)

The Thirteenth International Botanical Congress was held at Sydney (Australia) in August 1981 and the code was published in 1983.

Berlin Code (1988)

The fourteenth International Botanical Congress was held in August 1987 at Berlin (Germany). For the first time *Nomina specifica conservanda* was introduced here. Names of two species have been conserved against Rule of priority, since these names were widely used and any change may result in confusion. These are *Lycopersicon esculentum* P. Miller and *Triticum aestivum* Linn. Article 66 and 67 were removed.

Tokyo Code (1994)

Fifteenth International Botanical Congress was held at Tokyo in August 1993 and the code was published in 1994. Registration of new names has been introduced.

St. Louis Code (2000)

Sixteenth International Botanical congress was held in 1999 and the code was published in 2000.

Vienna Code (2006)

Seventeenth International Botanical Congress was held in 2005 and the code was published in 2006.

Melbourne Code (2012)

Eighteenth International Botanical Congress was held in 2011 and the code was published in 2012.

The following are some of the major decisions taken at Melbourne:

1. Name of the Code has been changed to *International Code of Nomenclature for algae, fungi and plants* and abbreviated as *ICN.*
2. Acceptance of certain forms of Electronic publication – 1st January 2012 as starting point for Electronic publication,
3. Option of using English as an alternative to Latin for the descriptions or diagnoses of new taxa of non-fossil organisms,
4. Registration of new names of Fungi,
5. Abolition of the provision for separate names for fungi with pleomorphic life history,
6. Abandonment of the morphotaxon concept in the nomenclature of fossils.

Shenzhen Code (2017)

Nineteenth International Botanical Congress was held at Shenzhen, China during July23- 29, 2017.

Broad Outline of Melbourne Code

Preamble

Division I Principles (1-6).

Division II Rules and Recommendations

Chapter I Ranks of taxa and the terms denoting them (Art. 1-5)

Chapter II Names of taxa (general provisions) (Art. 6-15).

Chapter III Nomenclature of taxa according to their ranks (Art. 16-28).

Chapter IV Effective and valid publication (Art. 29-50).

Chapter V Retention, choice and rejection of names and epithets (Art. 51-72).

Chapter VI Orthography of names and epithets and gender of generic names (Art. 73-76).

Division III Provisions for the modification of the code.

Appendix I Names of hybrids.

Appendix II *Nomina familiarum conservanda.*

Appendix III A. *Nomina generica conservanda et rejicienda.*

III B. *Nomina specifica conservanda.*

Principles

1. The nomenclature of algae, fungi and plants is independent of zoological and bacteriological nomenclature. This code applies equally to names of taxonomic groups treated as algae, fungi and plants whether or not these groups were originally so treated.
2. The application of names of taxonomic groups is determined by means of nomenclatural types.
3. The nomenclature of a taxonomic group is based up on priority of publication
4. Each taxonomic group with a particular circumscription, position and rank
5. can bear only one correct name, the earliest that is in accordance with the rules, except in specified cases.
6. Scientific names of taxonomic groups are treated as Latin regardless of their
7. derivation.
8. The rules of nomenclature are retroactive unless expressly limited.

Rules

Ranks of Taxa and the Terms Denoting them (Art. 1-6)

Taxonomic groups of any rank will be referred to as *taxa* (singular: *taxon*). Every individual organism is treated as belonging to an indefinite number of taxa of consecutively subordinate rank among which the rank of species is basic. The principal ranks of taxa in descending sequence are: kingdom (regnum), division or phylum, class, order, family, genus and species. The secondary ranks of taxa in descending sequence are tribe between family and genus, section and series between genus and species and variety and forma below species. If greater number of taxa is desired, the terms for these are made by adding the prefix sub to the terms denoting the ranks. An organism may thus be assigned to taxa of the following ranks of the plant kingdom: Regnum, subregnum, Divisio, subdivisio, classis, subclassis, ordo, subordo, Familia, subfamilia, Tribus, subtribus, Genus, subgenus, Sectio, subsectio, series, subseries, species, subspecies, varietas, subvarietas, forma, subforma.

Table 6.1: Ranks of Taxa and endings provided by ICBN

Ranks	Endings	Example
Regnum	-ile	Vegetabile
Division	-phyta	Magnoliophyta
	-mycota	Eumycota
Subdivision	-phytina	Pterophytina
	-mycotina (Fungi)	Eumycotina
Class	-opsida	Magnoliopsida
	-phyceae (Algae)	Chlorophyceae
	-mycetes	Ascomycetes

Ranks	Endings	Example
Subclass	-opsidae	Pteropsidae
	-idea (seed plants)	Asteridae
	-physidae (Algae)	Cyanophysidae
	-mycetidae (Fungi)	Basidiomycetidae
Order	-ales	Magnoliales
Suborder	-ineae	Rosineae
Family	-aceae	Magnoliaceae
Subfamily	-oideae	Mimosoideae
Tribe	-eae	Heliantheae
Subtribe	-inae	Rosinae
Genus	-us, -um, -is, -a, -on	*Pyrus, Solanum*
		Arabis, Passiflora,
		Heteropogon
Subgenus		*Hypericum* subg. *Hypericum*
Section		*Hypericum* section *Hypericum*
Subsection		*Scrophularia* subsection *Vernales*
Series		*Scrophularia* series *Lateriflorae*
Species		*Tridax procumbens*
Subspecies		*Acacia nilotica* subsp. *indica*
Varietas		*Brassica oleracea* var. *botrytis*
Forma		*Tectona grandis* f.*punctata*

Typification (Art. 7-10)

The application of names of taxa of the rank of family or below is determined by means of nomenclatural types. The code lays great emphasis on typification of various taxa in order to active stabilization of names. A nomenclatural type is that constituent element of a taxon to which the name of the taxon is permanently attached, whether as a correct name or as a synonym. In article 7.2 it is clearly stated that this nomenclatural type is not necessarily the most typical or representative element of a taxon; it is merely that element with which the name is permanently associated. Article 9 gives the details of various kinds of types. These are explained below.

Holotype

A holotype is the one specimen or illustration used by the author or designated by the author as the nomenclatural type (Art. 9.1). As long as holotype is extant, it automatically fixes the application of the name concerned. The holotype is chosen by the original author from a single gathering made by a collector at one time, and expressed definitely at the time of original publication

e.g. *Dimorphocalyx kurnoolensis* Venkataraju & Pullaiah

Holotype Venkataraju 6661 A (MH).

Isotype

An isotype is any duplicate (part of a single gathering made by a collector at a time) of the holotype. For instance, if a new species is based on a single collection (consisting of 5 specimens and 5 herbarium sheets made out of them) bearing the same field number, the original author will designate one among them as holotype and the rest of the 4 specimens will be isotypes.

e.g. *Dimorphocalyx kurnoolensis* Venkataraju & Pullaiah

Isotype Venkataraju 6661 B-D (SKU)

Paratype

A paratype is any specimen cited in the protologue other than the holotype, isotype(s) or syntypes. If a new taxon is described based on more than one gathering made at different times, the author will designate one gathering as holotype and isotype(s). The rest of the gathering bearing different field numbers form the paratypes. For instance, if the new taxon is collected with flowers in one season and in fruit in other season, these two gatherings will bear two different field numbers. While describing the new taxon, the author(s) will choose one gathering for holotype and isotypes, the next gathering cited in the protologues forms the paratype.

e.g. *Andrographis nallamalayana* Ellis

Paratype: Ellis 16709 A-E

Syntype

A syntype is any specimen cited by the author when there is no holotype, or any one of two or more specimens simultaneously designated as types (Art. 9.5).

Lectotype

A lectotype is a specimen or illustration designated from the original material as the nomenclatural type when no holotype was indicated at the time of publication or if the holotype is missing. When two or more specimens have been designated as types by the author of a specific or intraspecific name, the lectotype must be chosen from among them. Designation of lectotype should be undertaken only in the light of an understanding of the group concerned. In choosing a lectotype, all aspects of the protologue should be considered as a basic guide. If a holotype is lost or destroyed, the lectotype is chosen from the isotype(s). If no holotype was designated and if syntypes exist, one of them must be chosen as lectotype. If neither an isotype, nor a syntype is extant, a paratype if such exists, may be chosen as lectotype.

e.g. 1. Rangachari and Tadulingam (in *J. Bombay nat. Hist. Soc*. 24: 846. 1916) while describing the new species *Cynodon barberi* cited several specimens in the protologue but failed to indicate one as the type or holotype. Hence all the specimens cited became syntypes. Jain (in *Bull. Bot. Surv. India* 8: 204. 1967) after a critical study, correctly selected one of the specimens "Madras Herbarium, South India Flora No. 13715" deposited in CAL, as the lectotype

e.g. 2. In 1826, Blume published the new genus *Mastixia* (Bijdr. 654. 1826) with the species *M. pentandra* Bl. and *M. trichotoma* Bl. The type species of the genus however was not indicated at the time of original publication. Hutchinson (Gen. Fl. 2: 45. 1967) chose *M. pentandra* Bl. as the lectotype of the genus and his choice should be retained.

Neotype

A Neotype is a specimen or illustration selected to serve as nomenclatural type if no original material is extant or as long as it is missing (Art. 9.7). A neotype may be designated only when all the originally cited material is believed lost or destroyed. In selecting a neotype more care and critical knowledge are essential, as the reviewer usually has no guide except his own judgment as to what best fits the protologue.

e.g. The new species *Farsetia macrantha* was described by Blatter and Hallberb (in *J. Bombay Nat. Hist. Soc.* 26: 220. 1918). In the protologue they cited 3 numbers viz. Blatter 5785, 7302, 7305. These specimens presumably were deposited in the Blatter Herbarium, Bombay (BLAT), where the entire collection of Blatter, including the 'types' have been preserved. Bhandari (*J. Bombay Nat. Hist. Soc.* 72: 606. 1975) while working on this species could not trace the type specimens in the Blatter Herbarium and concluding that all the original material has, presumably been lost or destroyed, designated Bhandari 507, as Neotype, exemplifying the true application of the name *F. macrantha* Blatt. & Hallb. and to ascertain with certainity the sense in which the same must be used. The specimen comes from the type locality and is in complete agreement with the original diagnosis.

Epitype

A specimen or illustration selected to serve as an interpretative type when the holotype, lectotype or previously designated, or all original material associated with a validly published name, is demonstrably ambiguous and cannot be critically identified for purposes of the precise application of the name to a taxon. When an epitype is designated, the holotype, lectotype or neotype that the epitype supports must be explicitly cited.

Topotype is often the name given to a specimen collected from the me locality from which the holotype was originally collected.

Rule of Priority (Art.11-12)

There are many more names published for plant taxa than the actual number of taxa. This obviously means that several names have frequently been attributed to one and the same taxon, and it is essential to have a way of deciding which of the several names should be selected for the taxon. The code has certain provisions for this.

Principle III and IV deal with the principle of priority and limitation of the principle of priority. Each family or taxon of lower rank with a particular circumscription, position and rank can bear only one correct name, special exceptions being made for nine families and one subfamily for which alternative names are permitted.

For any taxon from family to genus, inclusive, the correct name is the earliest legitimate one with the same rank except in cases of limitation of priority by conservation. For any taxon below the rank of genus, the correct name is the combination of the final epithet of the earliest legitimate name of the taxon in the same rank with the correct name of the genus or species to which it is assigned. The principle of priority does not apply for names of taxa above the rank of family.

Limitations of Priority (Art 13-15)

The rule of priority however has certain limitations. It can not be applied to all names published. Instead, valid publication of names of different groups of plants is treated as beginning on various dates as provided in the code. In the case of Spermatophytes and Pteridophytes it is 1st May 1753, the date, of publication of Linnaeus *Species Plantarum* (ed. 1). For other groups of plants various other dates are prescribed, e.g., for Mosses it is 1st January, 1801, for Fungi 31st December 1801, for Algae 1st May 1753, for Hepatics and Lichens it is 1st May 1753 and so on. Name published prior to this date is not considered while applying Rule of priority. The two volumes of Linnaeus' *Species Plantarum* ed. 1 (1753), which appeared in May and August respectively, are treated as having been published simultaneously on the former date (1 May, 1753).

By adhering strictly to the Principle of Priority, several changes were brought about in the current names of the plants. In order to avoid disadvantageous changes in the nomenclature of species, genera, families and intertaxa, the code provides lists of names that are conserved. These names are conserved due to their very long usage in botanical publications and in the interest of nomenclatural stability.

Appendix II of the code consists of list of conserved families (*Nomina familiarum conservanda*). Appendix III lists the generic names conserved over all their earlier nomenclature synonyms (based on the same type) and taxonomic synonyms (based on different types) listed against them (*Nomina generica conservanda*). One such example is *Desmodium* Desv. (1813) a leguminous genus. There are two more names to this genus, *Meibomia* Heist ex Fabricius (1759) and *Pleurolobus* J. St. Hilare (1812). All these three genera have the same circumscription. According to the Principle of Priority, *Meibomia* should have been accepted as a correct name but since *Desmodium* is better known and widely used, it has been conserved. The Berlin code (Greuter *et al.*, 1988) has added one more appendix (Appendix IV) i.e. *Nomina Specifica conservanda et rejicienda*.

Nomenclature of Taxa According to their Rank

The name of a family is a plural adjective used as a noun, it is formed from the genitive singular of a name of an included genus with the termination 'aceae'.

The following names, of long usage are treated as validly published. Palmae (Arecaceae; type, *Areca* L.); Cruciferae (Brassicaceae; type, *Brassica* L.); Leguminosae (Fabaceae; type *Faba* Mill. (=*Vicia* L.); Papiolionaceae (Fabaceae; type *Faba* Mill.); Guttiferae (Clusiaceae; type *Clusia* L.); Umbelliferae (Apiaceae; type, *Apium* L.); Labiatae (Lamiaceae; type *Lamium* L.); Compositae (Asteraceae; type, *Aster* L.). Gramineae (Poaceae; type, *Poa* L.). Botanists are authorized to use the above alternative names.

The name of a subfamily is formed by adding the termination 'oideae' to the stem of the genus-, e.g., Amaranthoideae (from *Amaranthus*), Rosoideae (from *Rosa*); tribes with the termination 'eae' e.g., Amarantheae (from *Amaranthus*); Roseae (from *Rosa*); and subtribe with the termination 'inae' e.g., Amaranthinae (from *Amaranthus*).

The name of a genus is a noun in the nominative singular, or a word treated as such, and is written with an initial capital letter. It may be taken from any source whatever, and may even be composed arbitrarily e.g., *Impatiens, Rosa.* The generic names may be based on any source, but the common sources are:

(a) Commemoration of a person such as *Janakia* for E.K.Janaki Ammal, *ia* for Pierre Magnol, *Caesalpinia* for Caesalpino, *Bauhinia* for Bauhin.

(b) Based on place such as *Araucaria* after Arauco, a province of Chile, *Arabis* for Arabia, *Sibirica* for Siberia.

(c) Based on important character such as *Echinops* for echinate character, *Acanthospermum* for spiny fruit, *Hygrophila* for marshy habitat.

(d) Aboriginal name taken directly from a language other than latin without alteration of ending, e.g., *Vanda* Sanskrit name, *Sasa* Japanese, *Narcissus* Greek name for daffodils, *Ginkgo* a Chinese word.

The generic name of a tree, whatever be the ending, takes a feminine form since trees are generally feminine in classic Latin. *Pinus, Quercus* and *Prunus* are thus all feminine genera.

The name of a subdivision of a genus is a combination of a generic name and a subdivisional epithet connected by Wa term (subgenus, section, series, etc.) denoting its rank, e.g., *Costus* subg. *Metacostus*; *Euphorbia* sect. *Tithymalus*; *Euphorbia* subsect. *Tenellae.*

The name of a species is a binary combination consisting of the name of the genus followed by a single specific epithet. If an epithet consists of two or more words, these are to be united or hyphened.

e.g. *Andrographis nallamalayana* Ellis

Strychnos nux-vomica Linn.

The generic names and specific epithets when printed must be printed in Italics. When these names are written or typed they must be underlined. The first letter of the generic name should be a capital letter while the rest should be small letters. All the letters of the species should be in small letters (lower cap).

The names of the species may indicate the continents. e.g., *Crinum asiaticum.* Sometimes they may indicate the countries, e.g., *Mangifera indica* Linn., *Azadirachta indica* A. Juss., *Ocimum americanum* Linn., *Flaveria australasica* Hook, *Pavonia zeylanica* (Linn.) Cav., *Albizzia chinensis* (Osbeck) Merril. Species may indicate places, e.g., *Pimpinella tirupatensis* Bal. & Subr., *Parahyparhenia bellarensis* (Hook. f.) Clayton, *Phyllanthus maderaspatanus* Linn., *Asystasia gangetica* (Linn.) T. Anders., *Commelina benghalensis* Linn., *Crotalaria mysorensis* Roth, *Liagora visakhapatnamensis*, *Dimorphocalyx kurnoolensis* Venkataraju & Pullaiah.

Species are sometimes named in memory or commemoration of great scientists, e.g., *Ceropegia pullaiahii* Kullais. *et al.*, *Indigofera linnaei* Ali, *Rhynchosia beddomei* Baker, *Euphorbia santapaui* Henry, *Habenaria rammayyana* Ramachandrachary & Wood, *Iseilema venkateswarlui* Murthy & Sathyavathi. Specific epithets can also indicate the characters, e.g., *Gynandropsis pentaphylla* (Linn.) DC, *Sida cordata* (Burm. f.) Borssum. *Rhizophora apiculata* Blume, *Cleome viscosa* Linn., *Leucaena leucocephala* (Lamk.) de Wit, *Acanthospermum hispidum* DC.

The name of an infraspecific taxon is a combination of the name of a species and an infraspecific epithet connected by a term denoting its rank. e.g. *Crotalaria madurensis* Wight var. *kurnoolica* Ellis & Swaminathan. *Leucas flaccida* R. Br. var. *sebastiana* (Subbarao & Kumari) Moulali & Pullaiah.

Effective and Valid Publication (Art. 29-45)

The name must be published in a medium appropriate to the code and the names should be published according to the provisions of the code. These are known as Effective publication and Valid publication respectively.

Publication is effected, under this code, by distribution of printed matter (through sale, exchange or gift) to the general public or at least to scientific institutions with generally accessible libraries. Publication is also effected by distribution on or after 1 January 2012 of electronic material in Portable Document Format (PDF) in an online publication with an International Standard Serial Number (ISSN) or an International Standard Book Number (ISBN). It is not effected by communication of new names at a public meeting or by placing of names in collections or gardens open to public or by the issue of microfilm made from manuscripts, type-scripts or other unpublished material.

The name of a taxon when published must fulfill a number of conditions for valid publication. The name of a taxon must be accompanied by a description or diagnosis of the taxon or by a reference to a previously and effectively published description or diagnosis of it. A name of a new taxon of plants published on or after 1-1-1935 must be accompanied by Latin description or diagnosis or by a reference to a previously and effectively published Latin description or diagnosis of the taxon. Option of using English as an alternative to Latin for the descriptions or diagnoses of new taxa of non-fossil organisms has been permitted since Melbourne Code (1st January 2012).

Publication of the names of a new taxon of the rank of family or below on or after 1-1-1958 is valid only when the nomenclatural type is indicated. The place where the type specimen is permanently conserved also should be indicated. New names or combinations published on or after 1-1-1953 are not validly published without a clear indication of the rank of the taxon. It should be clearly indicated whether it is a new genus (*gen.nov.*), new species (*sp.nov.*) or a new combination (*comb. nov.*) etc.

Registration of Names: New names of plants and Fungi will have to be registered in order to be validly published after 1st January 2000. After 1st January 2000 when one or more of the other conditions for valid publication have not been met prior to registration, the name must be resubmitted for registration after these conditions have been met. The coordinating centre will be the Secretariat of IAPT, currently at the Botanischer Garden and Botanical Museum, Berlin-Dahlem, Germany. The International Mycological Institute in Egham, U.K. will act as associate registration centre for the whole fungi.

Citation of Author's Names

For the indication of the name of a taxon to be accurate and complete, and in order that the date may be readily verified, it is necessary to cite the name of the author(s) who first validly published the name concerned

Annonaceae Juss.

Annona L.

Annona squamosa L.

Authors' name put after the names of plants may be abbreviated, unless they are very short e.g., L. for Linnaeus, DC. for de Candolle, H.B.K. for Humboldt, Bonpland and Kunth, Hook. or Hk. for William Jackson Hooker.

If son's name is to be abbreviated letter 'f' (for filial) is added after abbreviation e.g. Hook.f. or Hk.f. for Joseph Dalton Hooker, L.f. for son Linnaeus.

When a name is published jointly by two authors, the names of both are connected by *et* or by an ampersand (&). It is wrong to use 'and' between the authors names since the name of the plant is in Latin and this includes the authors names also.

e.g. *Dimorphocalyx kurnoolensis* Venkataraju et Pullaiah (or Venkataraju & Pullaiah)

Kalanchoe cherukondensis Subbarao & Kumari

When a name is published jointly by more than two authors, the citation should be restricted to that of the first one followed by *et al.* (i.e. and others).

e.g. *Ceropegia pullaiahii* Kullayis., Sandhyarani & Karuppusamy. This should be cited as Ceropegia pullaiahii Kullayis. *et al.*

Retention, Choice and Rejection of Names and Epthets

Retention of Names

An alteration of the diagnostic characters or of the circumscription of a taxon does not warrant a change in its name. When a genus is divided into two or more genera, the generic name must be retained for one of them. Similarly when a species is divided into two or more species, the specific epithet must be retained for one of them. When a subdivision of a genus is transferred to another genus without change of rank, its epithet must be retained.

e.g. *Saponaria* sect. *Vaccaria* DC. when transferred to *Gypsophila* is *Gypsophila* sect. *Vaccaria* (DC.) Godr.

When a species is divided into two or more species, the specific epithet must be retained for one of them or, if it has not been retained, must be reinstated. When a species is transferred from one genus to another genus or placed under another generic name for the same genus without change of rank, the specific epithet must be retained or if it has not been retained, must be reinstated unless one of the following obstacles exists:

(1) The resulting binary is a later homonym or a tautonym;

(2) An earlier legitimate specific epithet is available.

e.g. *Uvaria longifolia* Sonner when transferred to *Polyalthia* it becomes *Polyalthia longifolia* (Sonner) Thwaites.

Linnaeus in 1767 gave the name *Hedysarum prostratum* to a Leguminous genus. Later he transferred it to *Indigofera*. He should have named it as *Hedysarum prostratum* but instead he gave it the name *Indigofera enneaphylla*. Ali (1958) has pointed out this mistake but he could not restore the name because in the meanwhile *Indigofera prostrata* was given to another species. Hence Ali (1958) named this species as *Indigofera linnaei* Ali.

Choice of Names

When two or more taxa of the same rank are united, the oldest legitimate name or (for taxa below the rank of genus) the oldest legitimate epithet is retained.

e.g. K.Schumann (in Engler & Prantl, Pflanzenfam. III. 6:5. 1890) uniting three genera *Sloanea* L. (1753), *Echinocarpus* Blume (1825) and *Phoenicosperma* Miq. (1865) rightly adopted the oldest of these three generic names, *Sloanea* L. for the resulting genus.

When the rank of a genus or infrageneric taxon is changed, the correct name or epithet is the earliest legitimate one available in the new rank.

Rejection of Names

A legitimate name must not be rejected merely because it, or its epithet, is inappropriate or disagreeable or because another is preferable or better known or because it has lost its meaning. e.g. The name *Scilla peruviana* L. is not to be rejected merely because the species does not grow in Peru.

A name is illegitimate and must be rejected if it was nomenclaturally superfluous when published. For example, the generic name *Elaeodandron* Jacq. (1782) is illegitimate because it was a superfluous name for *Cassine* L. (1753); the two genera had precisely the same circumscription. Hence the name *Elaeodendron* Jacq. published at a later date is to be rejected

A name is illegitimate and must be rejected if it is a later homonym i.e. it is spelled exactly like a name previously and validly published for a taxon of the same rank based on a different type. For example, the name *Banalia* Moq. (1849) given to a genus of Amaranthaceae is a later homonym of *Banalia* Rafin. (1840), a name previously and validly published for a genus of Euphorbiaceae; *Banalia* Moq. is therefore rejected, as was done by A.N. Henry and Bela Roy (in *Bull. Bot. Surv.* 10: 274. 1908) who renamed it *Indobanalia.*

A specific or infraspecific epithet is illegitimate and is to be rejected if it was published in contravention of Arts. 51, 53, 55, 56 or 60.

Nomen Nudum (abbreviated *nom.nud.*)

Names which are contrary to Art. 38 and published without an accompanying description are called *Nomina nuda*. These names are illegitimate and are to be rejected unless validated by a subsequent description. Thus many names published by Roxburgh in his *'Hortus Bengalensis'* and by Wallich in *'Wallich's Catalogue'* are *nomina nuda.*

e.g. *Odina wodier* Roxb.

Erigeron asteroides Roxb.

Centranthera humifusa Wall.

Many such names have been validated later by the provision of validation description.

Nomina Specifica Conservanda

The Berlin Congress for the first time introduced *Nomina specifica conservanda,* in which names of species are conserved against Rule of priority. These names are conserved because they have been used for quite a long time and any change in the name by the strict application of Rule of priority will lead to confusion.

Linnaeus described two species, *Triticum aestivum* and *T.hybernum* in his *Species Plantarum,* both bearing same date of publication, 1753. According to ICN when two species with the same date of publication are united, the author who unites them first has the choice of selecting the correct binomial. For a long time it was known that the first person to unite the two species were Fiori and Paoletti in 1896 who selected *T. aestivum* Linn. as the correct name. It was pointed by Kergnelen (1980), however, that the first person to unite these two species was actually Merat (1821) and he had selected *T. hybernum* Linn. and not *T. aestivum*. This discovery led to the danger of *T. aestivum* Linn. being dropped in favour of *T. hybernum* Linn. A proposal for conserving the name *T. aestivum* Linn. was thus made by Hanelt and Schultze-Motel (1983), and being an important economic plant this was accepted at the Berlin Congress.

P. Miller in 1768 proposed a new name, *Lycopersicon esculentum,* for the tomato, a species described earlier by Linnaeus (1753) as *Solanum lycopersicum*. Karsten (1882) made the name change *Lycopersicum lycopersicum* (Linn.) Karst. Retaining the epithet used by Linnaeus, but since the name became a tautonym it was not considered the correct name for tomato. Nicolson (1974) suggested an orthographic correction

Lycopersicon lycopersicum (Linn.) Karst. suggesting that *Lycopersicon* and *lycopersicum* are orthographic variants. Since the name *Lycopersicon lycopersicum* was no longer a tautonym, it was accepted as the correct name. But since *Lycopersicon esculentum* Mill., 1768 was a more widely known name, a proposal for its conservation was made by Terrel (1983) and accepted at the Berlin Congress.

Terms in Code

We came across several terms in the code and these are explained below:

Synonym

Synonyms are the different names used for the same taxonomic group or taxon. A synonym, as commonly used is meant for a later synonym or a synonymous name that is to be rejected. There are two kinds of synonyms.

Nomenclatural synonyms and taxonomic synonyms.

For example *Miliusa tomentosa* (Roxb.) Sinclair, *Saccopetalum tomentosum* (Roxb.) Hook.f. & Thoms. and *Uvaria tomentosa* Roxb. are names based on the same type and hence they are nomenclatural synonyms. There is no question of their identity.

Taxonomic synonyms are the names based on different types considered to belong to the same taxon; often called as heterotypic synonyms.

For example, Saha and Krishnamurthy (1962) described a new species *Boerhavia punarnava*. Earlier Linnaeus (Sp. Pl. 13. 1753) described another species *B. erecta*. But Nair (1967), Rao and Hemadri (1970) who verified the two types found them to be identical. Hence the later taxonomic synonym i.e. *B.punarnava* Saha & Krishn. is rejected.

Homonym

A name spelled exactly like a validly published name for a taxon of the same rank based on a different type.

e.g. *Jatropha heterophylla* Steudel, Nom. Bot. ed. 2.1: 799. 1840, and

Jatropha heterophylla Heyne ex Hook.f. Fl. Brit. India 5: 382. 1887.

The later homonym i.e. *J.heterphylla* Heyne ex Hook.f. is illegitimate and is replaced by the new name *J. heynei* Bal. (in *Bull. Bot. Surv. India* 3: 41. 1961).

Tautonym

A binary name in which the specific epithet exactly repeats the generic name

e.g. *Malus malus*

Sophia sophia (L.) Britton

Linaria linaria

Justicia adhatoda L. when transferred to the genus *Adhatoda*, becomes *Adhatoda adhatoda* a tautonym, so Nees (1882) published it under a new epithet *Adhatoda vasica* Nees.

But names such as *Cajanus cajan*, *Sesbania sesban* and *Bambusa bambos* are not considered as tautonyms as the specific epithets do not repeat exactly the generic names.

Tautonyms are not valid in botanical nomenclature.

Autonym

An automatically created tautonym for infrageneric or infraspecific taxa.

e.g. *Hypericum* subgenus *Hypericum* section *Hypericum*, *Hypericum perforatum* L. subsp. *perforatum* var. *perforatum*, *Crotalaria madurensis* Wight var. *madurensis*.

Basionym

A specific or infraspecific epithet that has priority and is retained when transferred to a new or different taxon.

e.g. 1. *Delonix elata* (L.) Gamble Fl. Pres. Madr. 396. 1919. *Poinciana elata* L. Cent. Pl. 2.16. 1756

e.g. 2. *Desmodium triflorum* (L.) DC. Prodr. 2:334. 1825. *Hedysarum triflorum* L.Sp. Pl. 749. 1753

Abbreviations

ICN : International Code of Nomenclature for algae, fungi and plants

IAPT : International Association for Plant Taxonomy

IAAT : Indian Association for Angiosperm taxonomy

sp. nov. : *species nova*, a species new to science.

comb. nov. : *combinatio nova*, new combination, a name change involving the epithet of the basionym, name of the original author being kept within the parenthesis.

nom. nov. : *nomen novum*, new name, when the original name is replaced and its epithet cannot be used in the new name.

The abbreviations *sp. nov.* and *nom. nov.* are used only when first published. In future references these are replaced by the name of the publication, page number and the year of publication for full citation.

References

McNeill, J. *et al.* (eds.). 2012. *International Code of Nomenclature for algae, fungi and plants*. Konigstein

Henry, A.N. and M. Chandrabose. 1980. *An Aid to the International Code of Botanical Nomenclature.* New Delhi.

Stearn, W.T. 1966. *Botanical Latin*. Great Britain.

7

Plant Identification

Taxonomic Literature

Various forms of literature incorporating description, illustrations and identification keys are useful for proper identification of unknown plants. The library is therefore as important in taxonomic work as a herbarium and a knowledge of taxonomic literature is vital to the practicing taxonomist. The literature of taxonomy is one of the oldest and most complicated literatures of science. Several bibliographic references, indexes, guides and databases are available to help taxonomists to locate relevant literature concerning a taxonomic group or a geographical region. The major forms of literature helpful in identification are described below.

Floras

A Flora is the systematic enumeration of plants of a given region. It may be a small area, a district, a state, a country or a continent. Floras, especially modern Floras, give complete citation of the species according to ICN, description, distribution, keys to the genera, keys to species and often illustrations.

e.g., Continent: *Flora Europaea* by T.G. Tutin *et al.* (1964-80),
Flora Australiensis by G. Bentham (1863-78).

Country: *Flora of British India* by J.D. Hooker (1872-97),
Flora of India by Botanical Survey of India (1994-),
Flora Malesiana by C.G. Steenis (1948),
Flora USSR by V.L. Komarov and B.K. Shiskin (1934-64).

State: *Flora of Andhra Pradesh* by Pullaiah *et al.* (1997),

Flora of Karnataka by C.J. Saldanha (1986),

Flora of Orissa by Saxena and Brahmam (1994).

District: *Flora of Anantapur district* by Pullaiah and Yesoda (1989),

Flora of Medak District by Pullaiah *et al.* (1998),

Flora of Hassan Disctrict by Saldanha and Nicolson (1978).

Manuals

A manual is a more exhaustive treatment than a Flora, always having keys for identification, description and glossary. Manuals do not necessarily indicate place of deposition of plant specimens included. The main emphasis is placed on providing suitable keys and diagnostic descriptions.

e.g., *Manual of cultivated plants* by L.H. Bailey (1949),

Manual of Aquatic Plants by N.C. Fassett (1957),

Manual of Indian Forest Botany by Bor (1953)

Monographs

A Monograph is a comprehensive taxonomic treatment of taxonomic group, generally a genus or a family. It gives significant information of a morphological and taxonomic nature concerning taxon. Strictly speaking it should cover the taxon as it exists throughout the world. Monograph gives information on anatomy, cytology, chemistry, genetics, geography, ecology etc.

e.g., The genus *Pinus* by N.T. Mirov (1967),

The genus *Crepis* by Babcock (1947),

The genus *Datura* by Blackslee *et al.* (1959).

Monograph on Indian Subtribe Cassiinae (Caesalpiniaceae) by V. Singh (2001).

Revision

Revision is a taxonomic treatment of a genus or family. It covers only morphological aspects.

Journals

Several Botanical societies, Botanical gardens, Herbaria and Publishing companies are publishing Journals of Taxonomic research. Some of them are given below.

Rheedia Published by Indian Association for Angiosperm Taxonomy,

Nelumbo (formerly *Bulletin Botanical Survey of India*) published by Botanical Survey of India,

Journal of Economic and taxonomic Botany published by Scientific Publishers, Jodhpur,

Taxon published by International Association of Plant Taxonomy, Berlin.

Kew Bulletin published by Royal Botanic Gardens, Kew, England,

Systematic Botany from New York

Annals of Missouri Botanic Garden from U.S.A.

Plant Systematics and Evolution published by Springer-Verlag, New York.

General Taxonomic Indexes

From time to time several indexes have been prepared for Angiosperms. Indices covering other groups of plants have also been published: *Index Filicum* for Pteridophytes, and *Index Muscorum* for Bryophytes. Indices serve as an aid to locating quickly the source of original publication of a name.

Index Kewensis Plantarum Phanerogamarum

This is a comprehensive index of scientific names of seed plants. It was published in 1893-1895 in two volumes under the direction of J.D. Hooker and B.D. Jackson. It consists of an alphabetical list of genera published from the time of Linnaeus to the year 1885. Under each generic name was given, in alphabetical sequence, every species epithet known to have been published followed by the name of the author, the place of publication and an indication of the native country of that plant. Later on supplements were published regularly at the interval of 5 years. So far 19 supplements have been published by Kew Botanical Gardens. The index Kewensis and supplements are now available on CD-ROM and it includes 968,000 records.

Gray Herbarium Card Index, Cambridge, Mass.

A card index is a data base of 287,225 records of New World Vascular plant taxa at the level of species and below. Only Gray Index has nomenclatural synonyms cross-reference to basionyms. The information is now accessible over the internet via keyword searches from the Email Data Server and through the Biodiversity and Biological collection Gopher.

Index Holmiensis

Published from Swedish Museum of Natural History it gives Plant distribution maps of vascular plants. 8 Volumes have been published up to 1995.

Index Londonensis, Oxford 1920-1941.

This is an alphabetical index of illustrations of flowering plants, ferns and fern allies appearing from 1753 to 1995.

Flowering Plant Index of Illustrations and Information

It was compiled by R.T. Saacson (1979). It is a 2-volume index of all illustrations of flowering plants.

Index Nominum Genericorum

It is a 3-volume work published in 1979 and gives a list of all generic names. The first supplement appeared in 1986. It is also available on the internet.

Hu Card Index

The Hu Card Index is a file of 158,844 cards for Chinese plant names now housed in the Harvard University Herbaria building where it is available for use in person. The Index was produced by Dr. Hu Shiu-Ying (Arnold Arboretum of Harvard University) and his staff. The Hu Card index was prepared in the early 1950s when Arnold Arbooretum undertook the project to prepare Flora of China.

International Plant Names Index

Royal Botanic Gardens Kew, The Harvard University Herbarium and the Australian National Herbarium, under the collaborative project, have developed International Plant Names Index (IPNI), a single web data base which combines citation data for seed plants from Index Kewensis, The Gray Herbarium Card Index and the Australian Plant Names Index (APNI). It provides information on names and associated basic bibliographic information about plant names. The data are freely available and are gradually being standardized and checked. IPNI is intended to be a dynamic source, depending on direct contributions by all members of the botanical community.

Dictionaries

Numerous valuable Dictionaries have been published but by far the most useful is Dictionary of Flowering Plants and Ferns by J.C.Willis. The 8th edition revised by Airy Shaw appeared in 1973. The book contains valuable information concerning genera and families followed by name of the author, distribution, family and the number of species in the genus.

Taxonomix Keys

Keys are the devices useful in identifying an unknown. They represent one type of taxonomic literature. The first modern type keys (dichotomous) clearly designed for identification were those of Lamarck in his *Flore Francaise* in 1778 (see Voss, 1952, for an interesting account of the history of keys and phylogenetic trees in systematic biology). Flora usually incorporate diagnostic key by the use of which the reader may identify an unknown plant with one included in the work concerned. Keys do not offer descriptions of the plants concerned, but state only the essential diagnostic characters by means of which the taxa can be identified. Ideally they use the most conspicuous and clear cut characters without special regard to these considered taxonomically the most important. For this reason the sequence of taxa is often quite artificial and such keys are frequently termed artificial keys.

A synopsis is a devise, usually in the format of a key presenting graphically the technical characters which in general or in the aggregate differentiate taxa. Such keys are known as synoptic keys. They were used in nineteenth century floras and monographs. These are usually very unsuitable for the purpose of identification.

Single Access Key - Sequential Keys

A sequential key is an artificial analytical devise or arrangement whereby a choice is provided between two contradictory propositions resulting in the acceptance of one and the rejection of the other. A sequential key may be short and limited to a single pair of contradictory propositions (a couplet), or it may be composed of an extensive series of these.

The currently conventional and most acceptable type of key is the dichotomous key, a type usually of one of two formats. The Dichotomous key was first used by R. Morrison in his *Plantarum Umbelliferarum Distributio nova* (1672). In the Dichotomous key each statement of couplet are arranged in Yokes, and each lead is identified by a letter or figure. Each successive subordinate Yoke is indented under the one preceding it. An example of the dichotomous key with indented or Yoked leads is given below.

This example has been taken form *Flora of Anantapur District* by Pullaiah and Yesoda (1989).

Families of Gamopetalae

Indented Key

1. Stamens more than petals:
    2. Flowers usually unisexual; stamens inserted on the receptacle .... EBENACEAE
    2. Flowers bisexual, stamens inserted on the corolla .... SAPOTACEAE
1. Stamens as many as petals or fewer:
    3. Ovary inferior:
        4. Anthers united around style (syngenesious):
            5. Flowers in involucrate heads, ovary unilocular............................ ....ASTERACEAE
            5. Flowers not in heads, ovary 2-more locular ...... LOBELIACEAE
        4. Anthers free ....RUBIACEAE
    3. Ovary superior:
        6. Ovary unilocular:
            7. Ovule solitary .... PLUMBAGINACEAE
            7. Ovules 2-many .... GENTIANACEAE
        6. Ovary 2-many locular:

8. Plants carnivorous, with small insectivorous bladders ....LENTIBULARIACEAE

8. Plants not carnivorous:

9. Leafless parasitic herbs, without chlorophyll .... OROBANCHACEAE

9. Leafy plants with chlorophyll:

10. Anthers forming a column with style and stigma (Gynostegium), pollen agglutinated into wax like masses or pollinia... ASCLEPIADACEAE

10. Anthers free, pollen not agglutinated like masses .... APOCYNACEAE

Bracketed Key

1. Stamens more than corolla lobes 2
1. Stamens as many as petals or fewer 3
2. Flowers usually unisexual, stamens inserted on the receptacle .... EBENACEAE
2. Flowers bisexual, stamens inserted on the corolla .... SAPOTACEAE
3. Ovary inferior 4
3. Ovary superior 6
4. Anthers united around style (Syngenesious) 5
4. Anthers free .... RUBIACEAE
5. Flowers in involucrate heads, ovary unilocular .... ASTERACEAE
5. Flowers not in heads, ovary 2-more locular .... LOBELIACEAE
6. Ovary unilocular 7
6. Ovary 2-many-locular 8
7. Ovule solitary .... PLUMBACINACEAE
7. Ovules 2-many .... GENTIANACEAE
8. Plants carnivorous, with small insectivorous bladders .... LENTIBULARIACEAE
8. Plants not carnivorous 9
9. Leafless parasitic herbs without chlorophyll .... OROBANCHACEAE
9. Leafy plants with chlorophyll 10
10. Anthers forming a column with style and stigma (Gynostegium), pollen agglutinated into wax like masses or pollinia .... ASCLEPIADACEAE
10. Anthers free, pollen not agglutinated like masses .... APOCYNACEAE

Example No. 2

Genera in Caesalpiniaceae (from Flora of Andhra Pradesh by Pullaiah and Chennaiah 1997).

Indented Key

1a. Plants spiny:

    2a. Leaflets reduced to scales or minute .... *Parkinsonia*

    2b. Leaflets not reduced:

        3a. Inflorescence corymb .... *Poinciana*

        3b. Inflorescence not corymb:

            4a. Pod winged at apex.. *Pterolobium*

            4b. Pod not winged at apex .... *Caesalpinia*

1b. Plants not spiny:

    5a. Leaves bilobed:

        6a. Leaflets separated; fruits compressed .... *Hardwickia*

        6b. Leaflets jointed up to half of the length; fruits not compressed .... *Bauhinia*

    5b. Leaves compound:

    7a. Leaves 2-pinnate:

        8a. Inflorescence long raceme .... *Peltophorum*

        8b. Inflorescence corymbose clusters .... *Delonix*

    7b. Leaves 1-pinnate:

        9a. Petals 5 .... *Cassia*

        9b. Petals 3 perfect, 2 sterile .... *Tamarindus*

Bracketed Key

1a. Plants spiny 2

1b. Plants not spiny 5

2a. Leaflets reduced to scales or minute .... *Parkinsonia*

2b. Leaflets not reduced 3

3a. Inflorescence corymb .... *Poinciana*

3b. Inflorescence not a corymb 4

4a. Pod winged at apex .... *Pterolobium*

4b. Pod not winged at apex .... *Caesalpinia*

5a. Leaves bilobed 6

5b. Leaves compound 7

6a. Leaflets separated, fruits compressed .... *Hardwickia*

6b. Leaflets jointed up to half of the length, fruits not compressed. *Bauhinia*

7a. Leaves 2-pinnate 8

7b. Leaves 1-pinnate 9

8a. Inflorescence long raceme .... *Peltophorum*

8b. Inflorescence corymbose clusters .... *Delonix*

9a. Petals 5 .... *Cassia*

9b. Petals 3 perfect, 2 sterile .... *Tamarindus*

Example No. 3

Indented Key

Species of *Cleome*

A. Gynandrophore present .... *C. gynandra*

AA. Gynandrophore absent:

B. Plants viscous with stalked gland including ovary and capsule.

C. Stamens 6 .... *C. monophylla*

CC. Stamens 10 or more:

D. Leaves simple .... *C. simplicifolia*

DD.Leaves compound, palmately lobed .... *C.viscosa*

BB. Plants not viscous, ovary and capsule glabrous:

E. Leaves 3-foliolate, stamens 6:

F. Fruits cylindric .... *C. aspera*

FF. Fruits compressed, flat .... *C. felina*

EE. Leaves 5-9-foliolate, stamens numerous .... *C. chelidonii*

When this key is changed into a bracketed key, then:

A. Gynandrophore present .... *C. gynandra*

AA. Gynandrophore absent .... B

B. Plants viscous with stalked gland including ovary and capsule C

BB. Plants not viscous, ovary and capsule glabrous .... E

C. Stamens 6 .... *C. monophylla*

CC. Stamens 10 or more .... D

D. Leaves simple .... *C. simplicifolia*

DD. Leaves compound, palmately lobed .... *C.viscosa*

E. Leaves 3-foliolate, stamens 6 .... F

EE. Leaves 5-9-foliolate, stamens numerous .... *C. chelidonii*

F. Fruits cylindric .... *C. aspera*

FF. Fruits compressed, flat .... *C. felina*

There are advantages and disadvantages to each of this type of dichotomous keys. The yoked type has the advantage of grouping similar elements in such a manner that they can be grasped visually as groups. However in extended keys of this type there is a sloping and shortening of lines to the right with a result of loss of economy of page space. In the parallel or bracketed type, the advantages of the indented format are lost and conversely the disadvantages are offset, because there is no opportunity in the bracketed type to group blocks of leads visually with elements having one or more characters in common. In this type of format, however, all leads are of approximately the same line length and produce maximum efficiency of page space.

Suggestions for the Use of Keys

1. Select appropriate keys for the materials to be identified. The keys may be in a flora, manual, guide, handbook, monograph or revision. If the locality of unknown plant is known, select a flora, guide or manual treating the plants of that geographic area. If the family or a genus is recognized, one may choose to use a monograph or revision. If locality is unknown select one of the manuals treating such plants since most floras do not include cultivated plants unless naturalized.
2. Read both leads of a couplet before making a choice. Even though the first lead may seem to describe the unknown material, the second lead may be even more appropriate.
3. Use a glossary to check the meaning of terms you do not understand.
4. Verify your results by reading a description, comparing the with an illustration or an authentically named herbarium specimen.

Suggestions for Construction of Keys

1. Identify all groups to be included in a key.
2. Prepare a description of each taxon.
3. The key characters should be contrasting and not overlapping. Use macroscopic morphological characters and constant character states when possible. Avoid characters and constant character states when possible. Avoid characters that can only be seen in the field.
4. Construct strictly dichotomous keys.
5. Use at least two characters per lead when possible.
6. Start both leads of a couplet with the same word if at all possible and successive leads with different words.

7. Mention the name of the plant part before descriptive phrases, e.g., leaves alternate not alternate leaves or flower blue not blue flowers.
8. Place those groups with numerous variable character states in a key several times when necessary.
9. Construct separate keys for dioecious, for flowering or fruiting materials and for vegetative materials when pertinent.

Multi-access Keys

Where as the sequential keys have a single commencing point i.e. a fixed sequence, multiaccess keys can be commenced at any position. Multiaccess keys are usually produced not on pages in a book, but on separate punched cards. They were apparently first used by A.T.J. Bianchi in 1931 and by S.H. Clarke in 1936. There are two types of punched card key. In the edge-punched key there is one card for each combination of attributes (i.e. one for each taxon) and each attribute is represented by one of the holes punched around the perimeter of the card (Fig. 7.1). If a taxon possesses a particular attribute that position is clipped out to form an open notch instead of a circular hole. All the cards are stacked up with their corresponding holes aligned, and an attribute possessed by the specimen to be identified is chosen. A thin rod or knitting needle is then pushed through the appropriate position and the rod is lifted horizontally and gently shaken. All those cards (taxa) possessing that attribute will fall away from the stack leaving the cards without that attribute on the rod. The latter are put aside and the others are gathered up and the process repeated with them using further characters, until only one card falls out of the stack.

In the body-punched key (polyclave) the holes are punched in rows in the main body of the card. In this case each card represents an attribute and each taxon occupies a standard position on the card, if the taxon possesses that attribute its position is punched out. To identify a specimen some of its attributes are listed and the appropriate cards are selected. These are aligned and taxa which possess all the attributes being tested will show a hole right through the stack when held up to the light. Cards (attributes) are added to the stack until only one hole (taxon) remains. A specimen card of punched cards key is given in Fig. 7.1. This was prepared by Saldanha and Rao (1974) for the Dicot families of South India.

Other punched card keys include *A polyclave to the monocotyledonous families of the world — a computer generated identification key* by C.K. Rao and J. Pankhurst (1986) and *British grasses: a punched-card key* by R.J. Pankhurst and J.M. Allison (1985).

Computers in Identification

Computers have also been found use in plant identification, whereby we no longer need trained botanists for this task. The following main approaches are used in computer identification. Both the above keys have been published by British Museum of Natural History.

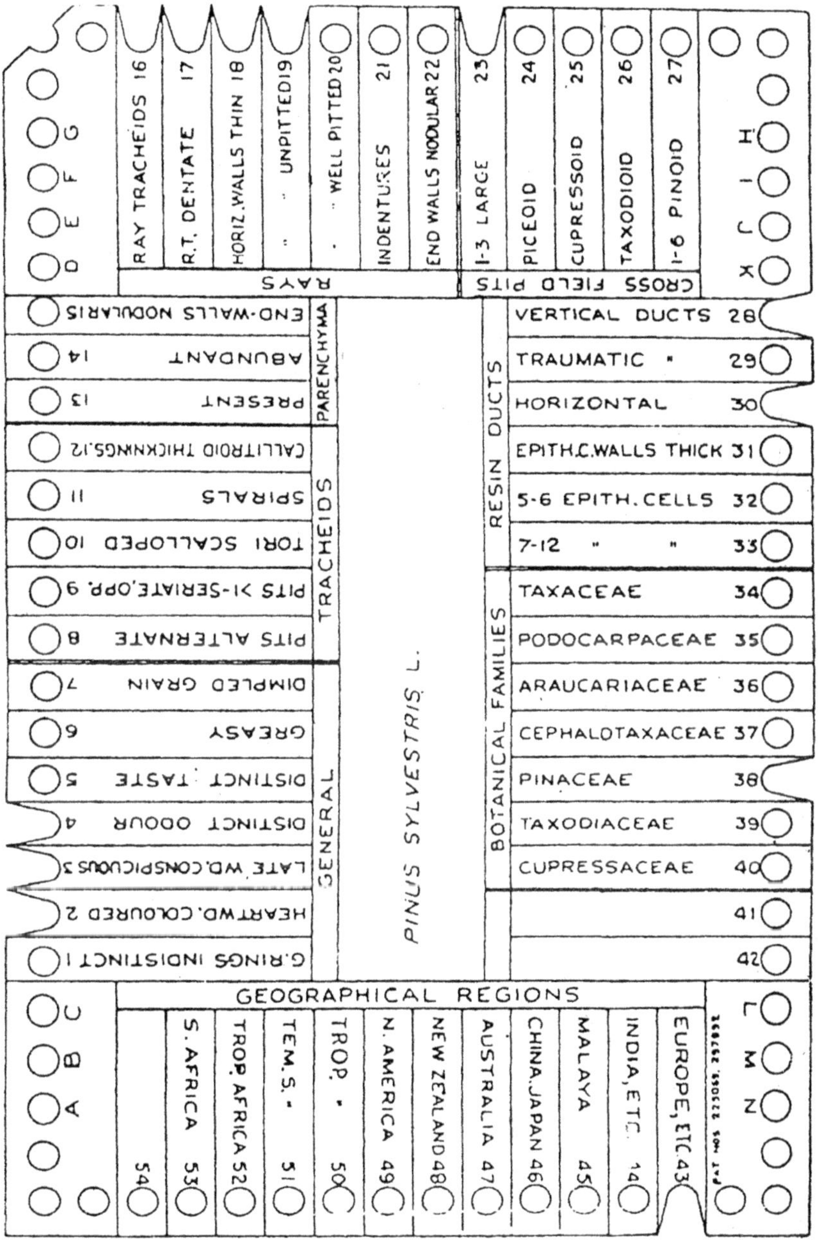

Fig. 7.1: Edge punched card key for the identification of soft wood timbers. The card shown is that for *Pinus sylvestris* and has the appropriate character holes clipped out.

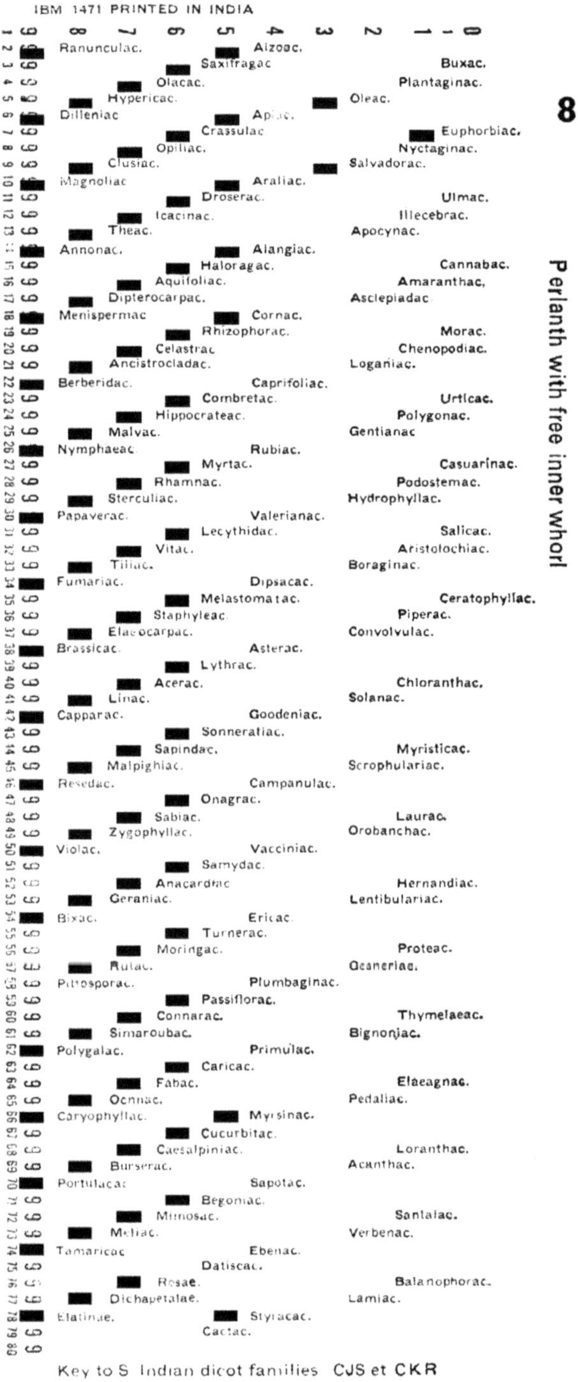

Fig. 7.2: A specimen of the card of the punched card key.

Computer Stored Keys

Dichotomous keys are constructed in the usual manner, fed into computer and run using an appropriate programme which may appropriately designed for step wise processing of the key through a dialogue between the user and computer. The computer programme starts with the first couplet of the key, enquires about the attribute in the unknown plant and on the information provided, and handles the key asking relevant questions until finally the actual identification is achieved.

Computer Constructed Keys

Appropriate programmes may be developed which can construct a taxonomic key based on the information about the taxa, in the same way and based on the same logic which is used by man to construct keys manually. Such keys permanently stored in a computer can be handled as above for the step-wise process of identification.

Simultaneous Character-Set Identification

Taxonomic keys are an aid to rapid identification and always provide only a provisional identification, confirmation being achieved only after comparison with a detailed description of the specific taxon. This comparison with detailed description is not done in the first place, as comparing the description of all taxa of the group or the area would be laborious, time consuming and often impossible. Such a comparison can be achieved through a computer in a matter of seconds. With such an approach, the whole set of characters of the unknown plant may be fed into the computer simultaneously and a computer programme used to compare the description with the specific group and to suggest the taxon with which the description matches. In case computer programme is not available, the computer programme may be able to suggest possible alternate identification.

Autonomous Pattern Recongnition Methods

Computer technology has now developed to the extent that fully automated identification can be achieved. The computer fitted with optical scanners can observe and record features, compare the same with those already known and make important conclusions.

Interactive Keys

Last two decades have seen the development of sophisticated computer based programmes designed to collect, integrate and use it for organizing descriptions and associated taxon data through user friendly interfaces.

DELTA System

The DELTA system is an integrated set of programmes based on the DELTA format (Description Language for Taxonomy) which is a flexible and powerful method for recording taxonomic descriptions for processing by a computer. DELTA, a shared programme, has been adopted as a standard for data exchange by the International Taxonomic Data base Working Group. It enables the generation and typesetting of descriptions and conventional keys, conversion of DELTA data for use by classification programmes, and the construction of Intkey packages for interactive identification and information retrieval. The system developed in Natural Resources and Biodiversity Programme of the CSIRO (Australia) is in use world-wide for diverse kinds of organisms including fungi, plants and wood. The programmes are continually refined and enhanced in response to feed back from users.

The DELTA programme key generates conventional identification keys. Characters are selected by the programme for inclusion in the key based on how well the characters divide the remaining taxa. This information is then balanced against subjective determined weights, which specify the ease of use and reliability of the characters.

DELTA data can be readily converted to the forms required by programmes for phylogenetic analysis, e.g. PAUP, Henning 86 and MacClade. The characters and taxa for these analyses can be selected from the full data set. Numeric characters are converted into multistate characters, as numeric characters cannot be handled by these programmes. Printed descriptions can be generated to facilitate checking of the data.

8

Field and Herbarium Methods

In the first chapter we discussed the necessity of Herbarium in teaching and Research in Plant Taxonomy. Let us know how to establish, maintain and improve a Herbarium.

A herbarium is a collection of plant specimens that usually have been dried and pressed, arranged in the sequence of an accepted classification, and are available for scientific study (Lawrence, 1951). According to Jain and Rao (1977) Hebarium is a store house of plant specimens collected from far and wide, mounted on appropriate sheets, arranged according to some known system of classification and kept in pigeon holes of steel or wooden cup boards, usually specially prepared for the purpose. Tournefort in about 1700 used the term Herbarium, as an equivalent to *Hortus siccus* (Steam, 1957), and this use was taken up by Linnaeus who also adopted it as a substitute for *Hortus siccus, Hortus mortus* and others.

The term herbarium, used in the strictest sense today, is simply a collection of dried specimens.

Luca Ghinni (1490-1556) seems to have been the sole initiator of the art of herbarium-making and this art was spread by his students. Gherards Caibo, a pupil of Ghinni, began collecting and preserving specimens as early as 1532 and his herbarium is extant today. The first herbarium was founded in University of Padua, Italy.

Uses of Herbarium

Herbarium as a Conservatory of Material and Data

As the Herbarium is a storehouse of plant material, it provides valuable material for study of flora and vegetation of different regions. By studying the specimens in the herbarium one can know the flora of distant regions without actually visiting it. Floras of foreign countries can also be studied in the herbarium.

Herbarium also provides information about the flora of a region in the past. If a particular region is submerged by the construction of a dam or if there is a natural disaster like earthquake or volcanoes, one can know its past flora by studying the Herbarium specimens collected from that place and stored in a Herbarium. Thus Herbarium is a conservatory of plant material.

Herbarium in Teaching and Research

Graduate and Post-Graduate students while learning are unable to see the live plants because the plants are distributed in different geographical regions. A Herbarium in such a situation becomes handy in learning about those plants by studying the herbarium specimens. A teacher while teaching about a particular family can show the herbarium specimen and explain about the details.

Herbarium is very much useful to research. The specimens in a herbarium are identified correctly and hence are very useful in palynological, anatomical and chemotaxonomic research. Palynologists usually study the pollen grains from herbarium specimens and thus avoid visiting different geographical regions. Herbarium also provides specimens for biosystematic and numerical taxonomic studies. Embryological and Cytological studies must always be supported by deposition of voucher specimens in a standard Herbarium. This will help later workers to verify whether the author has worked on the exact species.

Many plant collectors collect plants and give valuable data on the herbarium labels. They may not always publish this information. Later workers who study the herbarium specimens may bring out to light such hidden information.

Scientists working in medical industry have to collect plants form forests. In such cases these scientists visit the herbaria to see the herbarium specimens so that they can identify these species in the field. In the Herbarium they note down the characters of the species and location of occurrence and go to field for collection.

Herbaria also provide an opportunity for various scientists to meet and exchange their views.

Kinds of Herbaria

Depending on the nature of work Herbaria are of different types.

1. Herbaria of the organizations like Botanical Survey of India have specimens of all types of plants collected from all localities.
2. Herbaria of the institutions interested in drugs and medicines usually have specimens of medicinal plants.
3. Herbaria of universities and colleges have specimens, which are mostly used for teaching.
4. Herbaria of Agricultural institutions usually have specimens of wild relatives of cultivated plants and weeds of cultivated fields.

Important Herbaria in the World

Herbaria range from small personal collections to large collections of Universities and Research Institutions. Shelter (1969) estimates that the herbarium resources of the world may include as many as 250 million specimens although the total number actually reported by the various institutions was 148 million. Of this estimate about 78 million are in European collections and 36 millions are in North America. *Index Herbariorum* (Holmgren *et al.*, 1990) is an extremely useful index, which includes location and curators of herbaria of the world, size and type of collections, type of institution, publications, and important historical collections. Table 8.1 lists the largest herbaria in the world and other details about them, while table 8.2 gives herbaria in India.

Table 8.1: Largest Herbaria of the World.

Sl. N.	Herbarium	Place/Country	No. of specimens	Year of founding	Abbreviation
1.	Museum National d' Histoire Naturelle	Paris/France	9,500,000	1635	P
2.	NewYork Botanical Garden	New York/ USA	7,800,000	1891	NY
3.	V.L.Komarov Botanical Institute	St. Petersbourg, Russia	7,160,000	1823	L
4.	RoyalBotanic Gardens, Kew	Kew, England	7,000,000	1853	K
5.	Missouri Botanical Garden	St. Louis/ USA	6,600,000	1859	MO
6.	Conservetoire et Jardin Botaniques	Geneva/ Switzerland	6,000,000	1817	G
7.	US National Museum	Washington D.C./ USA	5,000,000	1868	US
8.	British Museum (Natural History)	London/ England	5,000,000	1753	BM
9.	Natur Historisch Museum	Vienna/ Austria	5,000,000	1748	W
10.	Swedish Museum of Natural History	Stockholm, Sweden	4,400,000		S

Table 8.2: Largest Herbaria in India

Sl. N.	Herbarium	Place	No. of specimens	Year of founding	Abbreviation
1.	Central National Herbarium	Howrah	2,500,000	1793	CAL
2.	Forest Research Institute	Dehradun	330,000	1816	DD
3.	Botanical Survey of India Southern circle	Coimbatore	260,000	1956	MH

Sl. N.	Herbarium	Place	No. of specimens	Year of founding	Abbreviation
4.	Botanical Survey of India Eastern circle	Shillong	260,000	1956	ASSAM
5.	Blatter Herbarium St. Xavier College	Bombay	200,000	1906-07	BLAT
6.	Botanical Survey of India Western circle	Pune	170,000	1956	BSI
7.	National Botanic Research Institute Herbarium	Lucknow	120,000	1948	NBG
8.	Botanical Survey of India Northern circle	Dehra Dun	84,575	1956	BSD
9.	Botanical Survey of India Central circle	Allahabad	63,000	1955	BSA
10.	Botanical Survey of India Industrial Section	Kolkata	50,000	1887	BSIS

Besides the above there are herbaria in several universities at Bangalore, Jaipur, Kolkata, Delhi, Poona, Hyderabad, Anantapur etc.

Field Methods

Purpose of Collection

Herbarium specimens are collected for different purposes. They are:

1. For establishing new herbaria and for enriching old herbaria,
2. For writing a flora,
3. For class work,
4. For musea,
5. For plant introduction and hybridization,
6. For medical herbarium,
7. For ethnobotanical studies,
8. For revisionary studies and monographic work, and
9. For biogeological studies

Collection Methods

Before going for plant collection proper planning and related arrangements have to be made. These include collecting details about the place to be visited, writing to the forest officials for arrangements, arrangements for stay and transport etc.

Necessary equipment and implements have to be taken with the team. These are: Vasculum, polythene bags, cutter, collecting pick, knife, tree prunner, field note book, field shoes, camera, bags, first aid box, pencils, scale, ball point pen, pocket lens, old news papers, blotting papers, plant pressers, ropes, identity cards, torch with batteries, binoculars, altimeter (to note the altitude) and GPS (to note the coordinates).

Vasculum is a metallic case of tin or aluminium and provided with tight lid (Fig. 8.1). The vasculum is used for keeping the plants immediately after collection. If the weather is too dry and hot, some water is sprinkled on the shoots before placing in the vasculum. In the absence of vasculum, polythene bags can serve the purpose, these are very light and are easy to carry to the field. In the market different sizes of polythene bags are available. Even empty polythene fertilizer bags can also be used. The author used only polythene bags for more than three decades with successful result.

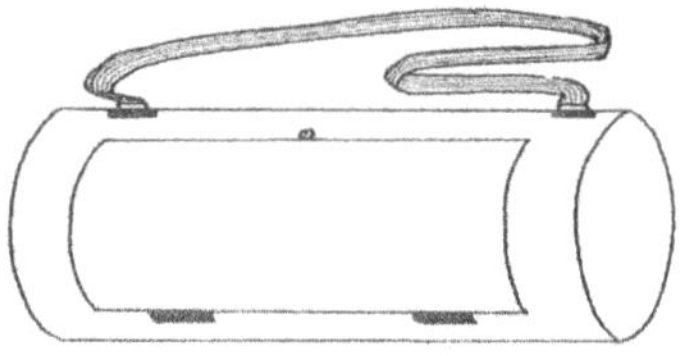

Fig. 8.1: A vasculum.

For cutting small twigs from trees and shrubs plant cutter or a knife is used. A pruning shear is used to cut the twigs of tall trees. The collecting pick is essential for digging up rhizomes, deep-seated bulbs or corms and the roots of most herbaceous plants.

Field notebooks are specially prepared notebooks for labeling the plants and for recording notes about the plants in the field. The pages of these note- books are specially printed (Fig. 8.2), punched and perforated according to an almost uniform pattern. One book has hundred leaves, the pages are serially numbered, first book having 1-100, second book 101-200 and so on. The six (or four) tags on each page have same number, these are detachable on lines of perforation, and can be tied to the specimens with the thread provided in the punched hole of each tag. Thus, each page has same number marked at five places one on the page itself and four on the tags or tickets.

Detailed notes should be entered in the field note book at the time of collection in the field itself. The details to be filled in the field note book are, date, name of the plant (if known), family, place of collection, altitude, habit, vegetation, local name, local use, description and collector's name. Regarding description importance should be given to such of the characters, which later can not be deduced from herbarium specimens. These include whether the plant is armed or unarmed, latex present or absent, if present its colour, smell of the flower, colour of the flower, if stem is angled how many angles are there, the height of the plant etc. Field notes should be filled preferably with ballpoint pen because the writing will not get spoiled during rain.

The size of the card sheet on which the specimens will finally be mounted is 28×42cm, and this limits the size of the collecting plants or twigs. In case of small herbs, the whole plant with roots or underground parts should be collected. If the herb is long it can be bent in V, N, or W shapes

HERBARIUM, DEPT. OF BOTANY
SRI KRISHNADEVARAYA UNIVERSITY
ANANTAPUR – 515 003 (A.P.)

Flora of Dist. ____________ State ____________
Serial No. 16071 ____________ Date ____________
Name ____________
Family ____________
Local Name ____________
Habit ____________ Height ____________ Fl ___ Fr ___
Locality ____________ Alt ____________
Soil ____________
Vegetation type ____________
Associated plants ____________
Distribution ____________ Abundance ____________
Uses ____________
Significant Notes ____________
Photograph ____________ Collector ____________
Identified by ____________

F.A.P. 16071
F.A.P. 16071
F.A.P. 16071
F.A.P. 16071

Fig. 8.2: A page in the field note book.

.When plants are collected they should contain flowers or fruits, or if available both. In the case of grasses and sedges the whole plant including underground part should be collected. If the plants are unisexual care should be taken to collect both sexes.

During collection sufficient precautions have to be taken. In the Western Ghats ground leaches are plenty and they will climb the body. Plant collectors should wear pants (including girls) and shoes. Stinging plants like *Tragia, Mucuna* etc. should be either avoided or collected carefully and kept separately, other- wise they will give itching for the whole day.

When the plant specimens are collected, the number tags should be detached from the field notebook and tied to the specimens. These numbers should be tied in the field itself and never postpone to the camp house, lest one will forget the details of the specimens.

Pressing and Drying Specimens

Plant pressers are of different types. These pressers can be made with wooden pieces or small iron rods (Fig. 8.3). Best results can be obtained by using Iron pressers. Between these two pressers blotting papers are placed and tied tightly with a rope made from cotton threads. Nylon threads are of no use because they are elastic.

For pressing plant specimens old newspapers can be used. But blotting papers give good results. Heavy and good absorbent papers are manufactured by Hand Made Paper Institute, Agriculture College Compound, K.B. Joshi Road, Shivaji Nagar, Pune, Maharashtra, 411 005.

The specimens collected in the field are poisoned before placing them between the plant pressers. The poisoning is generally done by dipping the whole plant in a saturated solution of mercuric chloride in ethyl alcohol.

Mercuric chloride is corrosive for metals and so enamel trays should be used. The solution is poured in a tray and the specimen is dipped in it with the help of pincers. Dipping fingers in the solution should be avoided and rubber gloves should be used while poisoning. After dipping in the mercuric chloride solution the specimens are placed in between the blotting papers. Spreading should be done only next day when alcohol solution has dried up.

Mercuric chloride is a deadly poisonous chemical and its effect on human beings is cumulative, hence much care should be taken while poisoning. Poisoning work should be avoided by persons having cuts in their hands.

Poisoning kills the plant and thereby the formation of abscission layer is prevented. The above method of dipping the plant in mercuric chloride solution is called dry method. Plant specimens can also be poisoned by dipping in Formalin and this method is called wet method.

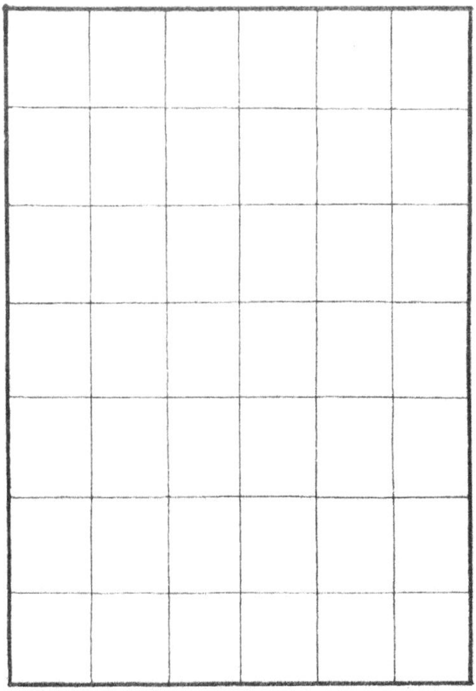

Fig. 8.3: Plant Presser.

Pressing is the process of placing specimens between blotting papers under heavy pressure. All plant parts such as leaves and flowers are spread out neatly. Some leaves are placed facing up and others facing down to show the characters on both the surfaces. This is specially important in case of ferns, which have sori on the abaxial side. In case of gamopetalous flowers one flower should be split open to show the androecium and gynoecium. If the specimens are longer than the sheet, they can be folded like 'V', 'N', 'M' or 'W' (Fig. 8.4). If there are too many leaves and branches, a few are removed, so that there is as little overlapping as possible and all parts are easily visible. The extra branches or leaves are cut a little above the base to show their arrangement.

While pressing, the specimens should be placed in such a way that there is almost uniform thickness of the bundle in the middle and on sides. This will provide uniform pressure in the press. For this some specimens are placed in centre and some along the sides.

The main objective of pressing is to flatten and dry the specimens. This is done by changing the blotters every day for 6-10 days depending on the weather. The plants gradually lose their moisture and finally become completely dry.

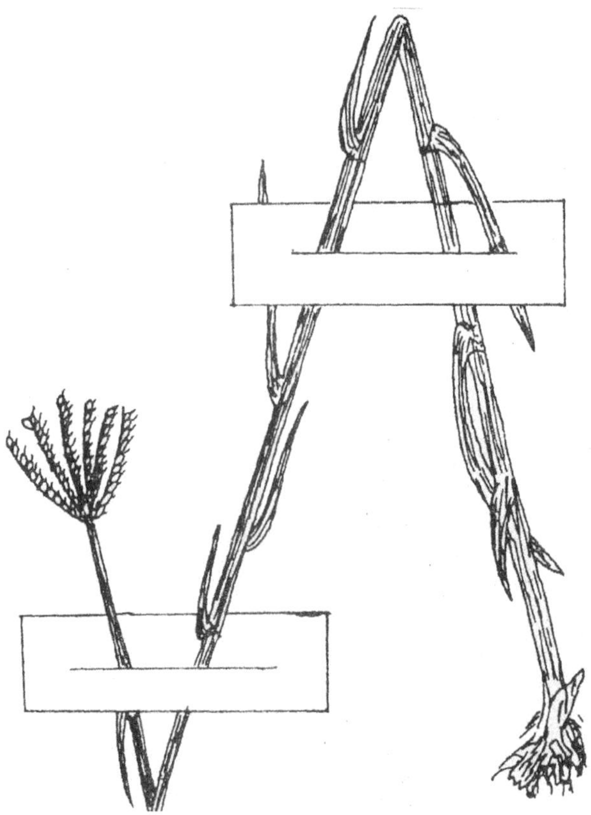

Fig. 8.4: Method of pressing a grass specimen with slit slips.

Collection of Special Kinds/Groups of Plants

The collection methods described in the earlier pages are for collection of terrestrial plants of average habit and form. Certain groups like succulents, aquatic plants, aroids, large bamboos require special methods and precautions, which are described below.

Collection of Succulents

Succulent plants like Cacti, Euphorbias, members of Crassulaceae etc. present unusual difficulties in making herbarium specimens. Their thick, succulent tissues take very long time to dry and so they require special attention. Unless dried by artificial heating or frequent changing of dryers, they are prone to rot and catch fungal infection. Hence the tissue should be killed by dipping the plant in boiling water. Cacti may be split at the joints, killing the tissue and then drying. While recording the characters of the succulents, details like the shape, size and the arrangement of spines and joints should be noted.

Collection of Minute Plants

Some plants like *Lemna, Wolfia* and *Spirodela* are microscopic and cannot be processed for the herbarium in the usual way. These plants should be collected in mass with collection number, notes, etc. sun dried and put in a packet (Fig. 8.5) and the packet pasted on the mounting board. These can also be preserved in F.A.A. whose composition is as follows:

Ethyl alcohol 95%	70cc
Glacial acetic acid	5cc
Formalin	5cc
Water	20cc

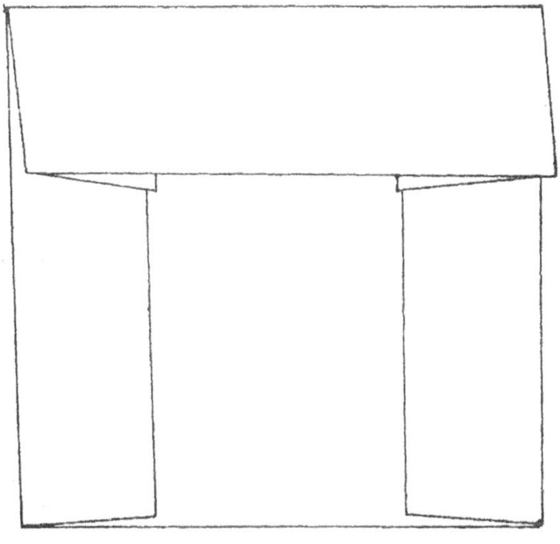

Fig. 8.5: Paper packet.

The collection number, place, date, etc. should be written on a slip in lead pencil and put inside the jar or pasted on it.

Collection of Ferns

Ferns should be collected with their basal portion, because the shape of the rhizome, the hairs and scales of rhizome are important taxonomic characters. If the frond is too big, a portion of the frond with basal portion must be collected and the exact size noted in the field notebook along with other details. Fronds without sori are of little value.

Collection of Slender Aquatic Plants

After collecting the slender aquatic plants, like *Utricularia*, in a tray containing water and are spread out, then a wire press or sieve plate with white paper or muslin cloth is inserted below the specimen and taken out, the paper or cloth is lifted slowly with both hands and placed between dryers. Some extra dryers are provided below and above. The plants should be changed along with the paper or muslin cloth.

Collection of Very Large Plants like Bamboos, Palms, Bananas, etc.

Some plants like Bamboos, Palms and bananas and some ferns like *Cyathea* and *Angiopteris* are very large in size. They require special methods. The habit of the plant and approximate size of the culm (or pseudostem) leaves and inflorescence must be noted in the field notebook. The whole plant and its main parts should be photographed with a scale on the spot. Portion of leaves should be collected noting the actual size. In case of bamboos, few ligules must be collected, noting the serial number of the nodes from the base, from which they are taken.

Collection of Aroids

Aroids frequently have virtually impenetrable epidermis and unless killed properly, they may continue to grow even in the press. Some bulbous orchids have also similar growth. These plants are killed by heating them at high temperatures. The tubers of the Aroids are often essential for correct identification, but their preservation is tedious. Sometimes whole tubers are collected and kept in museum or notes are taken on their shape and size and their small sections are kept in herbarium specimens.

Specimen Preparation

After the plant specimen is poisoned, pressed and dried it is affixed on a mounting sheet along with a label. The mounting sheets are made from heavy long lasting white card sheet whose standard size is 28×42cm. The aim of mounting is that the specimens should be neatly and uniformly spread and fixed on the sheet, and all parts of the plant should be easily visible for study.

Specimens are often attached to their sheets by gluing. The glue paste is made by adding flakes of glue to boiling water gradually and in small quantities, till it makes a syrupy paste. In Recent years synthetic glue Fevicol is also being used for pasting the specimens. Authors are using Fevicol and it is giving good results. The specimen is laid on an old newspaper lower side facing up. Then with the help of a brush the glue is applied to all parts of the specimen. The specimen is then lifted, using forceps, and placed on the mounting board. The specimen then firmly pressed by placing a blotting sheet on its top. This blotter removes the excess of glue that might come off on the sides of leaves, etc. The newspaper on which the specimen is placed for applying glue is discarded after few applications, as the excess glue on it tends to stick on the upper side of the specimens. While applying glue on the specimens with brush, some parts may get detached from the main twig or plant. In such case care should be taken to place these parts on to the specimens in their natural position. Small fruits, seeds and fragments may be placed in small packets and pasted on mounting board. The specimen is then covered with a blotting paper. The mounting sheets with specimens glued on them are kept in press for one day for proper sticking and drying. Next day the bundle is opened and the intermediate blotters removed. Some of the stiff parts are stitched with the help of a needle and thread. Stitches should be small and independent and thread should not be carried from one stitch to another on the lower side of the mounting sheet (Figs. 8.6, 8.7). On each side of the stem/twig a hole is made and a thread is inserted. A knot is put at the back and thread is cut after each knot.

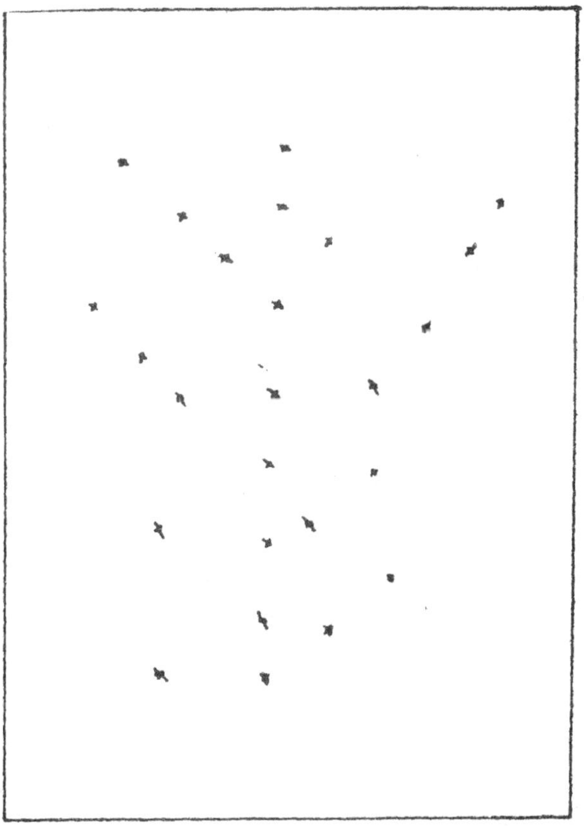

Fig. 8.6: Right way of making stitches.

The gummed cloth, or gum tape method is said to be largely employed in some western countries. In this method the specimen is affixed to the sheet by small straps/strips of gummed cloth. This method is quicker and easier than gluing, but in tropical climate, often the tapes soon lose their sticking property and come away.

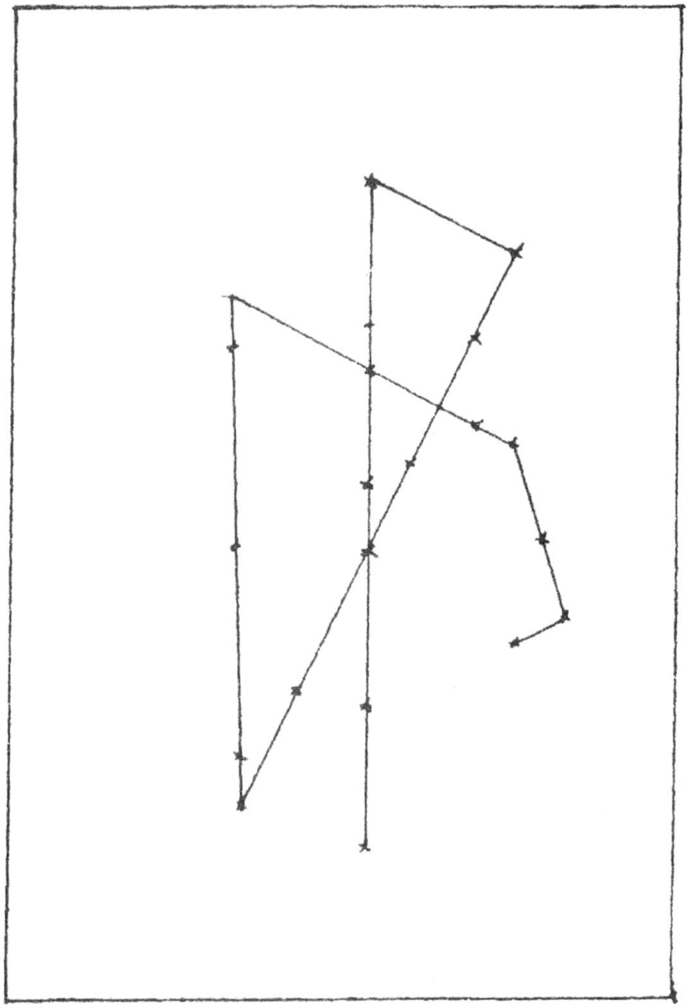

Fig. 8.7: Wrong way of making stitches.

Herbarium Label

Mounting of specimens is followed by pasting of Herbarium labels. The international standard size of herbarium label is 8 × 12 cm. The purpose of the label is to provide the person using the specimen with pertinent data not apparent from the material. The label should contain the following data (Fig. 8.8).

HERBARIUM

Department of Botany

SRI KRISHNADEVARAYA UNIVERSITY
ANANTAPUR 515 003
ANDHRA PRADESH (INDIA)

Flora ________________________________

No ________________________________

Name ________________________________

____

____

Family ________________________________

Locality ________________________________

____

Alt ________________ Date ________________

Notes ________________________________

____

____

____

Collector ________________________________

Identified by

Fig. 8.8: Herbarium label.

- Name of genus and species,
- Name of the family,
- Locality of collection, its altitude, latitude and longitude
- Date of collection,
- Description/notes,
- Collector's name and number,
- Local name and local use, if any.

Herbarium label is fixed on the bottom right hand corner of the mounting sheet.

After gluing the specimens and pasting the labels, the data on the labels should be entered clearly with some permanent ink.

Sorting and Filling of Specimens

Once the specimens have been glued, sewn and allowed to dry, they should be sorted for filing. First they are identified and given accession number. These sheets are then sorted out family, genus and species-wise. All the sheets of the same species are filed in lighter covers called the "species cover"; and all the species (in species covers) belonging to one genus are placed in one or more heavy folders of heavy paper, called the 'genus cover'.

The specimens are usually arranged in the herbarium according to some recognized system of classification. In many Indian herbaria, the order of families and genera is according to Bentham and Hooker's *Genera plantarum.*

Bundles are then kept in pigeonhole. In this pigeonhole, where the bundle of a new family start is marked by a fixed label or by hinged flap-board separator cardboard. The name of the family is printed or written on this in bold letters.

Special Collection

In addition to the general collection, many herbaria maintain various special collections.

Type Collection

Type specimens such as holotypes and isotypes are often housed separately from the general collection, where unnecessary handling and risk of damage can be greatly reduced. These are commonly filed in the same sequence as the general collection or alphabetically.

Synoptic Collection (Teaching or Identification)

Special collection for routine identifications and student use are sometimes separated from general collections. The synoptic collection can greatly reduce time spent consulting specimens in a herbarium as well as specimen wear. The collections are usually small and may be arranged and rearranged (particularly for teaching purposes) according to recent schemes of classification, e.g., Cronquist, Takhtajan, APG IV.

Fumigation

Protection of mounted or unmounted specimens from pests is done through fumigation. For this purpose volatile poisonous liquids like methyl bromide, carbon disulphide or carbon tetrachloride is used. These liquids are placed in small petri dishes in each herbarium case and these are kept closed for about a week. Paradichlorobenzene is sometimes used; small cloth bags of this chemical are placed in many pigeonholes. Fumigation should be done at regular intervals. A mixture of ethylene dichloride (3 parts) and carbon tetrachloride (volume 1) gives satisfactory results.

Exchange

Growth of herbarium takes place in different ways (1) Scientists in herbarium collect specimens (2) Specimens are received as gift, on purchase, or exchange (3) Specimens received for identification are retained. Specimens brought from outside are fumigated before they are brought in. Type specimens should be sent through insured post.

General Instructions for Herbarium Users

1. Keep sheets flat; do not shuffle or leaf through a folder like a book. Specimens are brittle and easily damaged.
2. Store properly, preferably, specimens should be stored in herbarium cases or at least on shelves. Do not crowd too many specimens into a box or pigeonhole. Keep materials in covers when not in use.
3. Do not lay books or other heavy objects on specimens.
4. Place loose fragments in packets or envelopes if the specimen to which the fragment belongs can be ascertained.
5. No flower or fruit is to be taken for dissection from Herbarium sheets or mounted specimens. Flowers and fruits can be taken from unmounted specimens.
6. Those specimens, which need repair, should be brought to the notice of the curator.
7. Full bundles should be taken out; pulling out one or more sheets will surely damage the specimens and break flowers.
8. Putting bundles back into pigeonholes should be done with extreme care.
9. Specimens should not be kept on tables for longer periods. They may catch infection or moisture.
10. Almirahs should be closed after taking out required bundles.
11. Open flames and smoking are prohibited in the herbarium .
12. All specimens before entering the herbarium must be properly fumigated or poisoned.

References

Holmgren, P.K, N.H. Holmgren and L.C. Barnett. 1990. *Herbariorum. Part I. The Herbaria of the World.* 8th Ed. Reg. Veg. 120. New York.

Jain, S.K. and R.R. Rao. 1977. *Hand Book of Field and Herbarium Methods.* Today and Tomorrows. New Delhi.

Santapau, H. 1955. *Botanical collector's Manual.* Botanical Survey of India, Calcutta.

Shah, N.C. J. Singh, Vinay Ramjan and S. Kumar. 1996. Herbaria: A repository of plant knowledge. *J. Med. Aromatic Plant Sci.* **18**:63-70.

9

Botanical Gardens

Botanical gardens are those where trees, shrubs, herbs, climbers and other plants are collected from different places and different countries and grown in a garden in a scientific and systematic manner. These botanical gardens are different from recreation parks in that, parks are for recreation, while botanic gardens are mainly for scientific studies, learning and plant conservation. Botanical Gardens contain green houses, glass houses, palm houses, lakes, flower gardens, etc. Many a times Herbaria, laboratories and libraries are attached to these botanic gardens. Both Botanic gardens and Botanical gardens can be used, grammatically both of them are correct.

Role of Botanic Gardens

1. Maintenance of the unique and diversified collections of living plants in the garden helps the botanist to study the living plant of its morphology and evolution.
2. Botanic gardens have served as acclimatization centres for various economic plants like coffee, tea, rubber, cotton, tea, rubber, cotton, hemp, vanilla, cinchona and others, which are natives of one place and were introduced in other countries.
3. Botanical gardens maintain germplasm, commercial plants and flower plants and these are useful for hybridization. Many better varieties of fruit, vegetable and flower plants have been introduced from these gardens commercially.
4. Gardens are a source of living material for academic studies. Specialists like cytologists, embryologists, phytochemists and anatomists can collect these living materials from the botanical gardens. In nature these plants may be distributed in different geographical regions.

5. Green houses are useful for studying the plants from a different climatic region.
6. Botanic gardens supply seeds and important plants on exchange basis to encourage the introduction of some useful and economic exotic plants.
7. Dissemination of information on plant based industries and research departments.
8. In glass houses and green houses endangered plants are being conserved. In this way world flora can be preserved.
9. Botanical gardens are useful for training people in growing horticultural plants, landscape and gardening.
10. Botanical gardens are places for recreation.

Types of Botanical Gardens

1. General Botanical Gardens
2. Arboretum: Gardens where only trees are grown.

   e.g., Arnold Arboretum, USA.
3. Orchidarium: Only Orchids are grown.

   e.g., National Orchidarium, Shillong and Yercaud.
4. Pineatum: Conifers are grown.
5. Bambusatum: Bamboos are maintained

   e.g., Indian Botanic Garden, Howrah has a good Bambusatum.

History of Botanical Gardens

Man has been growing plants in gardens from times immemorial. In India plants useful in food and medicine were grown as early as during 4000 to 2000 B.C. Emperors and Kings established gardens around their palaces. World famous hanging gardens of Babylonia are such gardens. Romans brought plants from the places they conquered and grew them in Italy. Similarly, Chinese, Aztecs and Persians have shown interest in gardens. They grew flower plants and perfume plants.

Between A.D. 600-1600, when botany has not developed as an independent science people used to grow flowers and vegetables plants in their compounds. Medicine has developed tremendously during 17th Century. Garcia de Orta in 1565 has written about the medicinal plants of India and this has been translated into Latin. Since then many Europeans were attracted to India. During 16th Century herbalists have grown medicinal plants collected from different parts of the world in their gardens. Many Universities have established botanical gardens during this period. By 18th Century, many famous botanical gardens have been established. Hyams and McQuitty (1969) in their book Botanical Gardens of the World have given a detailed account of these gardens. Holttum (1970) has mentioned that there are about 525 large Botanical Gardens.

Padua Botanic Garden, Padua, Italy

Orto Botanico is generally regarded as the oldest Botanic Garden and was founded in 1545. This garden was mainly established for demonstration purpose of medicinal plants. This garden was established in an area of about 20,000 square metres. Padua garden is rectangular with Central circle of 84 metres in diameter and surrounded by rectangular beds. Each bed is planted with single plant species. Plants were labeled. Collections of grasses, alliums, *Iris, Paeonia,* Mediterranean endemics in rock garden, tiny garden of medicinal plants, aquatics in small pond and cacti, aroids, orchids and tropical exotics are worth seeing. Old trees in arboretum include *Ginkgo biloba, Platanus orientalis* and *Cedrus deodora.*

Many plants have been introduced into Italy and Europe through this garden. These include *Cedrus deodora* (1828), *Mesembryanthemum* (1713), *Robinia pseudoacacia* (1662), *Pelargonium cucullatum* (1801) and *Jasminum nudiflorum* (1590).

Pisa Botanical Garden, Italy

Luca Ghinni in 1545 established Pisa Botanical Garden. It was shifted in 1562, which is seen even today. Mattioli and Andrea Caesalpino contributed to the growth of this garden. The flora Tuscan was also published for the first time by Georgio Santi (1782-1814). Later Giovanni Arcangelli (1881-1995) listed Italian plants in *Compedio della Flora Italiana.* Paleontological studies have been carried out first in this garden. Pisa garden is famous for *Aesculus hippocastanum, Magnolia grandiflora, Liriodendron tulipifera, Ailanthus glandulosa* and *Cinnamonum camphora.*

Palermo Botanic Garden, Italy

Established in the year 1795 by the King Ferdinand. The garden is famous for *Ficus magnolioides* and large *Dracaena draco.* The garden has exotic plants, xerophytic plants, citrus garden, winter garden, fibre plant garden, medicinal plants garden and essential oil plants.

Leiden Botanic Garden, Netherlands

First planting in this garden began in 1594, seven years after its foundation. This garden was established to serve as demonstration garden to the medical faculty of University of Leiden. Famous herbalists who contributed to the growth of this garden include Caroline Clusius and Boerhaave. The first green house was established in this garden in 1599 for the protection of plants introduced from Cape of Good Hope. Many of plants collected by East India Company were grown in this garden. *Vanilla, Cinchona* and Camphor have been introduced through this garden.

Royal Botanic Garden, Edinburgh, Scotland

This garden, established in 1670, is famous for Rododendrons and Azaleas. It has a demonstration garden, glass house, herbarium and library. The Rock garden is finest in the world and remarkable for its Himalayan alpine plants. *Araucaria, Rhododendron, Primula, Acontium, Adenophora, Arisaema, Clematis, Delphiniums, Buddleia, Crotoneaster,* Jasmine species are grown in this garden.

Royal Botanic Garden, Kew, England

Originally Kew Garden belonged to Kew house owned by Richard Bennet. Later on it was taken over by the Government and became Royal Botanic Garden. William Aiton took over as Superintendent and established a Botanical Garden. Sir William Jackson Hooker took over as Director of the Garden in 1841. He established a green house and started collecting plants and exchanging them with other gardens. A palm house was built during this period and herbarium was also established. He was succeeded by Joseph Dalton Hooker. J.D. Hooker added *Rhododendron* to the garden. The next Director, Thistleton Dyer also enlarged herbarium and added Alpine house, Rose Garden, Bamboo Garden and a Lily pond.

Special gardens at Kew comprise Rhododendron Dell, Bamboo Garden, Chalk Garden, Rock Garden, Shrub Garden, Health Garden and Rosary. The plant layout at Kew includes Willows, Oak, elm, Cedar, Lilacs, Orangery, Pine, Sweet Chestnut, Tulip, Azaleas, Daffodils, Poplar, ferns, Thuja, Cypress, Japanese Ceder, Red wood etc. The living collections at Kew are most diverse with 351 families, 5465 genera and over 28,000 species growing successfully.

New York Botanical Garden, USA

This garden was established as a private garden in 1801 by Dr. David Hosack. It was taken over by the State of New York in 1891. John Torrey, Nathaniel Lord Britton and Dr. Arthur Cronquist are some of the Botanists who contributed to the growth of this garden. The garden today covers 100 ha in the heart of New York city along the Bronx river. In addition 778 ha Mary Flager Cary Arboretum at Millbrook have been added to the jurisdiction of the garden. There are 15,000 species distributed in demonstration gardens, Montgomery conifer collection, Stout day lily garden, Havemeyer lilac collection, Rhododendron and Azalea collection, Everett rock garden, herb garden, rose garden, arboretum and conservatory complex. The garden has a systematic arrangement of trees and shrubs that make it a place of interest for general public as well as botanists and horticulturists. The garden plays a major role in conservation of rare and endangered species.

Arnold Arboretum, Michigan, USA

One of the best arboreta in the world is Arnold Arboretum. This Arboretum has a rich treasure of *Rhododendron, Forsythia, Prunus, Sargentina, Magnolia campbelli, Paeonias, Malus* etc. It was established in 1872.

Missouri Botanical Garden, St. Louis, Missouri, USA

Founded in 1859 Missouri Botanical Garden is one of the Oldest botanical institution in United States of America. The Garden is a centre for botanical research and science education. It spread over 32 hectares and includesClimatron (Geodesic dome conservatory), Japanese strolling garden, Tower Grove House and Linnean House. It has living collections of plants from all over the world. In the climatron tropical plants are being grown. The garden has several interesting plants like *Amrphophallus titanum* which flowered in August 2017.

10

Modern Methods in Plant Taxonomy

Most systems of classification of plants have been evolved on the basis of exomorphic characters placing particular reliance on the characters of the flower, the flower being considered more conservative than the vegetative organs. Even so the goal of reaching a truly natural system of classification has remained unfulfilled.

From the beginning of the twentieth century, with the development of technology, newer evidences became available to taxonomists. These newer branches like Anatomy, embryology, palynology, cytology, phytochemistry and ultrastructure have been utilized in deducing the relationships of plants. Taxonomists in recent times like Cronquist, Takhtajan, Dahlgren and APG have utilized these newer evidences in classifying angiosperms. In the following pages the utility of these evidences in solving taxonomical problems is discussed.

Anatomy in Relation to Taxonomy

The first attempt to use anatomy in plant classification was made by Bureau (1864). He used the anatomical characters for delimitation of taxa of various levels in the family Bignoniaceae. Solereder (1899) in *Systematische Anatomie der Dicotyledonae* emphasized the importance of xylem structure in plant classification. Bailey (1951) and Metcalf and Chalk (1950) have listed numerous anatomical features that can be made use in taxonomy. The *Anatomy of Dicotyledons* (Metcalfe and Chalk, 1950) and *Anatomy of Monocotyledons* (Metcalfe, 1960 onwards) are some of the books in this direction.

Epidermal Characters

Stomata

Leaf characters especially epidermal characters and stomatal characters are very much useful in taxonomy. The importance of epidermal characters was emphasized by Stace (1965) while the micromorphology (study of morphological characters with scanning electron microscope) has been emphasized by Dehgan (1980) and Barthlott (1981).

Van Cotthem (1970) divided the stomata into 8 types. They are:

Anomocytic (Ranunculaceous Type)

Stoma surrounded by a limited number of cells that are indistinguishable in size, shape or form those of remainder of epidermis (Fig. 10.1a). e.g., Ranunculaceae, Malvaceae, Papaveraceae, *Tridax procumbens, Boerhavia diffusa, Cleome* spp.

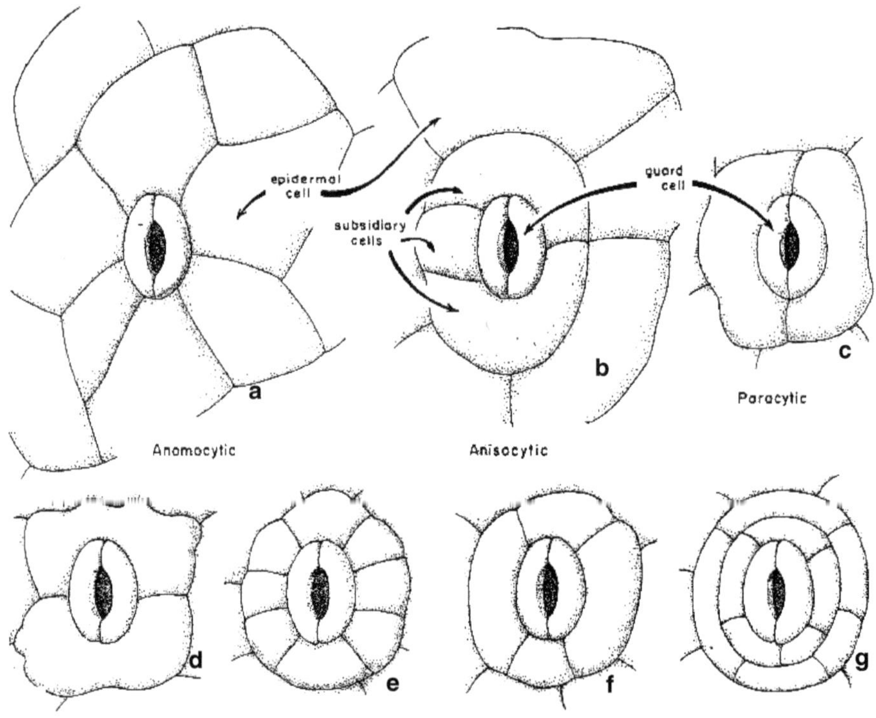

Figure 10.1: Types of Stomata

a: Anomocytic; b: Anisocytic; c: Paracytic,; d: Diacytic; e: Actinocytic; f: Tetracytic; g: Cyclocytic

Anisocytic (Cruciferous Type)

Stoma surrounded by three cells of which one is distinctly smaller than the other two (Fig. 10.1 b), e.g., Cruciferae, Solanaceae.

Paracytic (Rubiaceous Type)

Stoma accompanied on either side by one or more subsidiary cells parallel to the long axis of the pore and guard cells (Fig. 10.1c). e.g., Rubiaceae, Magnoliaceae.

Diacytic (Caryophyllaceous Type)

Stoma enclosed by a pair of subsidiary cells whose common wall is at right angles to the guard cells (Fig. 10.1d), e.g., Caryophyllaceae, Acanthaceae.

Tetracytic

Four subsidiary cells are present, two lateral and terminal (Fig. 10.1f), e.g., Most families of Monocotyledons.

Actinocytic

Stoma surrounded by a circle of radially elongate subsidiary cells (Fig. 10.1e), e.g., Commelinaceae, Musaceae.

Cyclocytic

Stoma surrounded by four or more subsidiary cells which form a ring around each stomata (Fig. 10.1g), e.g., Tribe Laguncularieae of the family Combretaceae.

Hexacytic

Stoma accompanied by six subsidiary cells consisting of two lateral pairs parallel to the long axis of the pore and two polar (terminal) cells; the second lateral pair as long as the stomatal complex.

Besides the above, several other characters can be used in taxonomy. Presence of stomata on the upper side (epistomatic) or lower side (hypostomatic) or on both sides (amphistomatic) are also useful in taxonomy. For example the presence or absence of stomata on the upper surface of a leaf is a good diagnostic feature in the separation of *Alnus subcordata* and *A. orientalis* of Betulaceae, which are otherwise difficult to distinguish. In the former epistomatic condition prevails while in the latter, the hypostomatic. Similarly shape of the guard cells are also useful in plant classification. For example stoma with bean- shaped guard cells are met with Dicotyledons while stoma with dumbell-shaped guard cells are seen in Poaceae In Cyperaceae rectangular guard cells are seen.

Development of stomata has been divided by Pant (1965) into three types.

Perigenous

Stoma in which the subsidiary cells do not have a common origin with the guard cells, but are formed by cells lying around the meristemoid that divides to form the guard cells.

Mesogenous

Stoma in which the subsidiary cells are produced from the same meristemoid (initial) as the guard cells.

Mesoperigenous

Stoma in which at least one of the subsidiary cells has a common origin with the guard cells, but others do not.

Development of stomata is useful in plant classification. Ramayya (1977) has emphasized the importance of stomatal characters in taxonomy. Ramayya and Rajagopal (1968, 1971) gave a key to genera of Portulacaceae and Aizoaceae based on epidermal characters.

In the family Acanthaceae the stomata are diacytic, whereas in the closely related Scrophulariaceae the stomata are anomocytic. In the family Combretaceae the stomata are paracytic in the subfamily Strephonematoideae and anomocytic in the subfamily Combretoideae except the tribe Lagunculaireae where they are cyclocytic. Stomatal characters are not, however always so reliable and the different types have obviously each originated on many different occasions and they often exist together on one plant. In the genera *Streptocarpus* and *Saintpaulia* of the Gesneriaceae; for example, the cotyledons bear aomocytic stomata and the mature organs anisocytic stomata (Saharsrabudhe and Stace, 1974). In *Lippia nodiflora* of Verbenaceae anomocytic, anisocytic, diacytic and paracytic stomata occur on the same leaf (Pant and Kidwai, 1964). In *Alternanathera pungens* of family Amaranthaceae also all the above four types of stomata are found.

Inamdar (1968) who studied the stomata of *Nyctanthus* expressed the opinion that this genus should be kept in Oleaceae. In *Nyctanthus* and Oleaceae the stomata are Tetra-Mesoperigenous (sometimes aperigenous), paracytic while Verbenaceae shows Dia-Mesogenous stomata. So Inamdar (1968) said the *Nyctanthus* should be kept in Oleaceae instead of Verbenaceae.

Bremekamp (1953) said that the genus *Elytraria* should be removed from Acanthaceae and should be included in Scrophulariaceae. Paliwal (1966) studied the phylogeny of stomata to ascertain this. He found that in *Elytraria* and Acanthaceae the stomata are diacytic and development is Syndetocheilic, but in Scrophulariaceae stomata are anomocytic and development is Haplocheilic. Hence Paliwal (1966) retained the genus *Elytraria* in the family Acanthaceae.

Trichomes

Trichomes are epidermal outgrowths. They exhibit great variation in structure and function and thus are of much taxonomic value. Glandular and non-glandular hairs, papillae, setae and prickles are all considered as hairs. These trichomes are useful in plant classification and understanding the relationship between groups. Schulz (1936) used the type of hair as the major criterion in the subdivision of the family Cruciferae into tribes and genera. Ramayya (1962, 1969) studied the trichomes of Compositae and gave key to the genera based on trichome characters. Based on trichome characters he considered Heliantheae as the primitive tribe. Rollins (1944, 1945, 1946, 1949) studied the trichomes of *Parthenium argentatum* and *P. incanum* and their natural hybrids. He has shown that in *P. argentatum* the trichomes are T-shaped while in *P.incanum* they are whip-like with a long thread. Intermediate types of trichomes are observed in the hybrids between these species.

In the family Combretaceae Stace (1965, 1969a, b, 1973) found that trichome anatomy is of immense significance in classification at all levels, from the circumscription of the family down to the separation of species and even varieties. In particular, it has led to an improved tribal classification within the family and an improved subgeneric and sectional classification within the largest genus *Combretum*.

Epidermal Cells

The epidermal cells, other than those modified by their relationship to trichomes, stomata, the venation system or other special structures, provide many characters of taxonomic value. These include their size, shape, orientation, anticlinal wall undulation and periclinal wall curvature. The epidermal cell modification occurs also on the leaf margin and when silicified, calcified, suberized, gelatinized or containing various types of crystals. Silicified and suberized epidermal cells are of major taxonomic importance in the Poaceae.

Other Epidermal features are also useful in plant classification. The genus *Morus* is divided into species based on Idioblasts in epidermis. Tomlinson (1956, 1959a, b) and Metcalfe (1968) said that silica bodies in the leaves are useful in classification of the families Zingiberaceae, Musaceae, Arecaceae and Rosaceae up to genera and even species.

Internal Structure

Carlquist (1961) said that leaves in Angiosperms show great diversity and this character is useful in classification. The nature of epidermis, its thickness, type of mesophyll, sclerenchyma patterns, sclereids, crystals, venation patterns, etc, may be useful in classification. The importance of foliar anatomy in the systematics of grasses was realized quite early by Duval Jouve (1875). According to Avdulov (1931) Poaceae have two types of internal structure in leaves. In the Festucoid (Pooideaea) grasses the mestome around the stele is layered and mesophyll cells are straight while in Panicoid grasses mestome may or may not be present and the mesophyll cells are arranged radially.

Similarly Ayensu (1974) found that sclerenchyma patterns in the Velloziaceae can be used to separate the genera *Vellozia* and *Barbasenia*. The genera in Cyperaceae can be segregated based on internal structure of the leaves (Govindarajulu, 1969).

Crystals, especially calcium oxalate crystals, are found to be useful in the classification of Angiosperms. Foliar sclereids have, of late, been demonstrated to be of taxonomic importance in many taxa.

Laticifers are cells or a series of fused cells containing latex and are common features of many succulent plants and other plants of arid regions. They vary widely in their structure and latex in their composition. Their presence or absence and when present, their structure has been of some taxonomic value. In Aroideae (De Bary, 1884), for example, certain species lack laticifers or any related structures while others have longitudinal rows of elongated, cylindrical sac-like cells. Some authors consider that Fumariaceae, which lack laticifers (Sperlich, 1939), should be considered distinct from the related family Papaveraceae having a well developed latex system. In the former family laticifers are replaced by idioblasts. It is one of reasons for considering these two families as distinct from the rest of the families in Engler's Rhoedales.

The taxonomic usefulness of petiole vascularization patterns has become increasingly apparent in recent years (Radford *et al.*, 1974). Sinnot (1914) gave a classification of nodal types in megaphyllous plants. In general there are three major types of node — unilacunar, trilacunar and multilacunar (Fig. 10.2). Trilacunar nodes occur in the majority of Dicotyledons, mutilacunar nodes in the primitive orders such as Magnoliales, Piperales, Trichodendrales and advanced orders such as Umbellales and Asterales and unilacunar nodes in Laurales, Caryophyllales, Ericales, Diapensiales, Ebenales, Primulales, Myrtales and a majority of families of Asteridae (Paliwal and Anand, 1978).

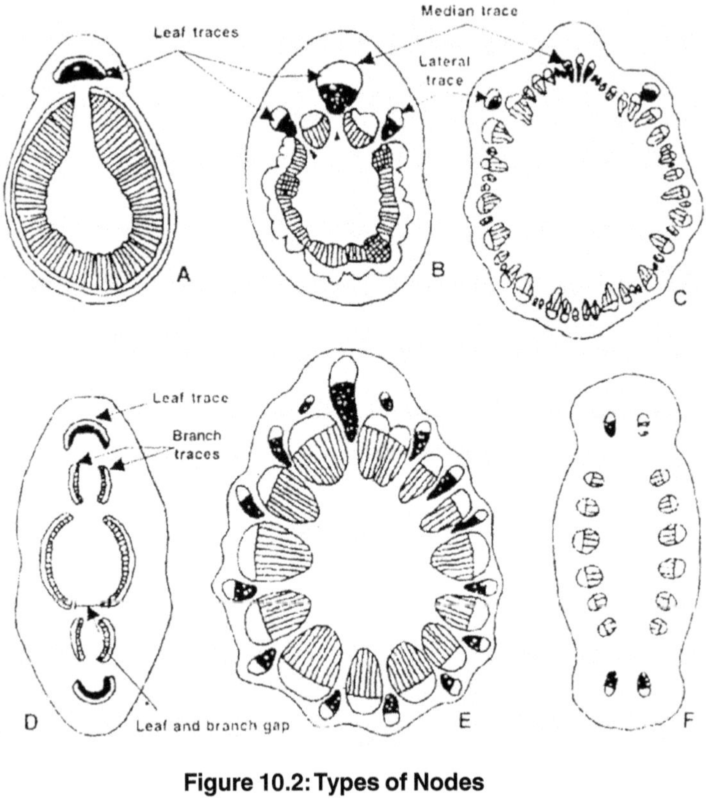

Figure 10.2: Types of Nodes

A, D and F: Unilacunar; B and C: Trilacunar; E: Multilacunar

Stem anatomy has been studied in the herbaceous members where it has proved to be of some diagnostic value. In the stem of *Ranunculus repens* there is relatively little strengthening tissue and is correlated with the creeping habit of the plant, while in the closely related, species, *R. acris*, interfascicular tissue is sclerified and therefore an erect habit is possible.

Stace (1970) has shown that in the subgenus *Genuini* of *Juncus*, anatomical characters of stem and foliar bract can be used to distinguish most British species. Further, it is also possible to identify parents of several hybrids on anatomical grounds.

There is considerable variety in the anatomy of stems of *Dioscorea* and can be successfully employed in the delimitation of species that are not easily separable on exomorphic grounds. Ayensu (1970) found anatomy useful in separating *Dioscorea rotunda* and *D. cayennensis*, which are closely related and difficult to distinguish.

The classical case of anatomy in relation to taxonomy is the distribution of vascular bundles. In Dicotyledons vascular bundles are arranged in a ring while in Monocotyledons vascular bundles are scattered.

Bentham and Hooker (1862-83) included Basellaceae as a subfamily in Chenopodiaceae. Anatomically Basellaceae are characterized by a single ring vascular strands with intraxylary phloem in the larger bundles. On the other hand, Chenopodiaceae are characterised by the development of polycyclic, anomalous secondary vascular bundles.

Bentham and Hooker (1873-1876) split the Gentianaceae into two sub- families — Menyanthoideae and Gentianoideae. But Hutchinson (1969), Cronquist (1981) and Takhtajan (1980) raised them to independent families. Vascular bundles in Menyanthaceae are collateral while Gentianaceae show bicollateral vascular bundles.

The most important anatomical features that have been used in taxonomy and phylogeny are those of the secondary wood. This is because of their conservative nature. Wood anatomy can be used at all taxonomic levels. At and above the family level it has been successfully used along with other lines of evidence to establish the systematic position of the primitive vesselless Angiospermic families Winteraceae (all seven genera-*Drimys, Pseudowintera, Bubbia, Belliolum, Exospermum, Zygogynum* and *Tetrathalamus*), Trochodendraceae, Tetracentraceae, Amborellaceae and Chloranthaceae. Similarly it has been agreed by all phylogenists that the Englerian group of primitive Angiosperms, namely Amentiferae can not be considered primitive since they have specialized wood (Tippo, 1938). Anatomical evidence also supports the case for considering the remarkable genera *Paeonia* and *Austrobaileya* as the type of independent families (Bailey and Swamy, 1949).

At and below the generic level, the most useful contribution of wood anatomy is to provide evidence for placing of taxa of uncertain affinity. In *Quercus* (Fagaceae), different subgenera and sections can be delimited with the help of anatomical evidence. The placement of the genus *Myristica* and the related genera close to Lauraceae is supported by the evidence of wood structure (Eames, 1961), while on the same basis Myristicaceae are considered unrelated to Annonaceae and Eupomaticeae. Similarly, placing of Calycanthaceae in or near Rosales or Myrtales is not supported by wood structure.

Sieve cells also provide taxonomical evidence. For example, unspecialized sieve cells occur in lower vascular plants — Pteridophytes and Gymnosperms. Only one Angiosperm, *Austrobaileya scandens*, shows unspecialized sieve cells in phloem.

Floral Anatomy and Taxonomy

Floral anatomy has a wide application in determination of status of high ranking groups such as genera and species. On the basis of similarity of floral anatomical pattern affinities of families Papaveraceae, Capparaceae, Brassiaceae and Magnoliaceae could be established. These families have a common parietal placentation and inverted placental bundles with regard to floral axis on inner side of secondary marginal strands with normal orientation. This fact also supports placement of Moringaceae in Rhoedales. Puri (1947, 1949) suggested affinities between Passifloraceae, Moringaceae and Cucurbitaceae on floral anatomical basis.

Burtt (1965) opined that *Cyrtandromoea* has no close affinity with other members of Gesneriaceae and advocated its transfer to Scrophulariaceae with which it exhibits more resemblance. Singh and Jain (1978) have supported this treatment on the basis of floral anatomy. Kale and Pai (1979) have supported the removal of the genus *Trichopus*, on the basis of floral anatomy, from the family Dioscoreaceae of Bentham and Hooker and erection of a unigeneric family Trichopodiaceae as has previously been suggested on anatomical evidence by Ayensu (1966, 1972).

Fruit Wall and Seed Coat Anatomy

Fruit wall and seed-coat anatomy have also come to the help of taxonomists in many cases. The distinction between the two genera *Ammannia* and *Rotala* (Lythraceae) has been puzzling taxonomists for quite some time. The diagnostic feature was in the fruit-irregularly rupturing fruits in *Ammannia* and valvular capsules in *Rotala*. But this difference was very difficult to appreciate, especially in herbarium specimens. Van Leemjen (1971) found the fruit-wall character to distinguish the two taxa. The orientation of the cells of the two main layers of pericarp are different in*Ammannia* the cells of both layers are very similar, whereas in *Rotala* the cells of the inner layer are linear, strongly lignified and transverse to those of the outer.

Two different views were expressed regarding the systematic position of the Asteraceous genus *Lagascea*. Cassini (1815), Lessing (1832), DeCandolle (1836) and Merxmuller (1954) included the genus *Lagascea* in the tribe Vernonieae. Bentham (1873), Hoffmann (1894) and others place *Lagascea* in the tribe Heliantheae. Stuessy (1976) on the basis of morphological and chromosomal studies, supports the assignment of the genus *Lagascea* to Heliantheae. Pullaiah (1981) on the basis of pericarp structure favours retention of *Lagascea* in the tribe Heliantheae. In Heliantheae three different zones are distinguished in the pericarp. Between the middle and outer zones characteristic schizogenous cavities develop which get filled with a brown resinous and hardening substance. Such schizogenous cavities are absent in Vernonieae.

Seed coat morphology has yielded valuable taxonomic information. Seed coat anatomy has also been used for systematic purposes.

Ultrastructure in Relation to Taxonomy

Morphology and taxonomy have benefitted greatly in recent years from information obtained by using advanced technology in microscopy, particularly Scanning Electron Microscopy (SEM) and Transmission Election Microscopy (TEM).

However, only a few ultrastructural characters are of wide application in plant classification. 'Dilated Cisternae' (DC) in the endoplasmic reticulum were first noted and described by Bonnet and Newcomb (1965) in the root cells of *Raphanus sativus* (radish) and later by Favali and Gerola (1968) in the phloem parenchyma of foliar veins in *Brassica chinensis*. Jorgensen *et al*. (1977) detected them in foliar veins in *Capparis cynophallophora*. Iverson and Flood (1970) found that such structures are common occurrence in the Cruciferae. According to Behnke (1977) these organelles are typical of Brassicaceae and Capparaceae.

Behnke (1972, 1977, 1981) in a series of observations since about three decades ago, has demonstrated ultrastructural details of sieve element plastids, which can be used in plant taxonomy. Behnke studied more than 2500 species from 380 families with regard to this character. There are broadly two types of such plastid elements, one accumulating starch (S-type) and the other accumulating protein (P-type). There are, however, certain groups of plants-Rafflesiaceae, Crassulaceae and some species of Ulmaceae, Moraceae and Urticaceae in which the plastids have neither starch nor protein accumulations, and they are called S_0-type. S-type sieve element plastids with starch accumulation in grains of different sizes are the commonest type in flowering plants. About 65 per cent of these have such plastids in the sieve tube elements (Behnke, 1977). The subclasses Ranunculidae, Hamamelididae and Dilleniidae have without exception S-type sieve element plastids, while in the Rosidae a great majority of the orders have this type. Half the members of the Magnoliidae have S-type and half P-type elements. In Caryophyllidae majority of the members show P-type sieve element plastids (Fig. 10.3).

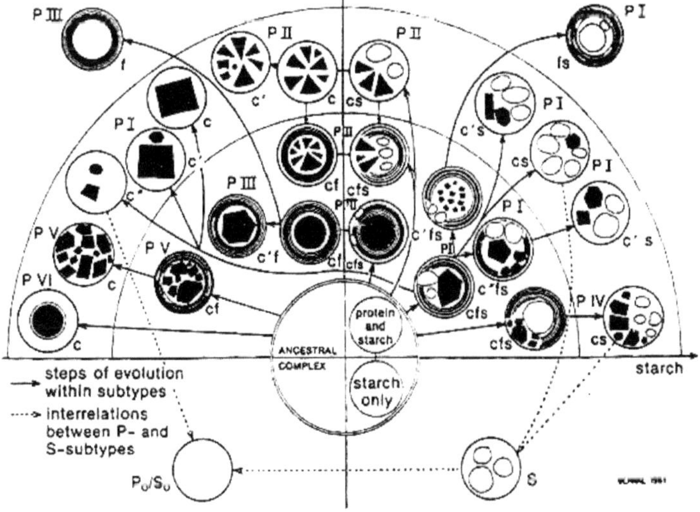

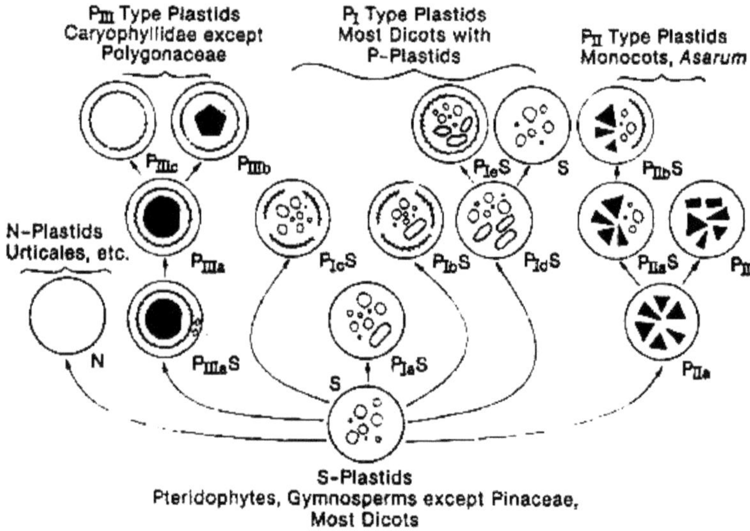

Figure 10.3: Sieve Element Plastids in Angiosperms

P-type plastids are further differentiated on the number and shape of the crystalloids and the nature of the filaments surrounding them.

(i) P I subtype: The plastids contain single crystalloids of different sizes and shapes and/or irregularly arranged filaments. PI subtype sieve element plastids are present in Magnoliales, Laurales and Aristolochiales.

(ii) P II subtype: The plastids contain several cuneate crystalloids oriented towards the centre of the plastid. PII subtype sieve element plastids are present in Monocotyledons.

(iii) PIII subtype: The plastids contain a ring-shaped bundle of filaments. These plastids are seen only in Centrospermae

(iv) P IV subtype: The plastids contain a few polygonal crystalloids of various sizes. This type of plastids are seen only in Fabales.

(v) PV subtype: The plastids contain many crystalloids of different sizes and shapes. It is seen in the order Ericales and family Rhizophoraceae.

(vi) PVI subtype: The plastid contains a single circular crystalloid. It is found in family Buxaceae.

Behnke (1974) investigated the order Rhamnales and found that the family Rhamanaceae had S-type while Vitaceae and Leeaceae had P-type sieve element plastids. The heterogeneity of the Magnoliidae as to its plastid types is attributed to its basal position in the evolution of flowering plants (Behnke, 1975). P-type are considered to be ancestral and S-type has been thought to be derived by loss of protein. P I subtype is held to be the most primitive among P-subtypes of flowering plants.

The best use of plastids has been made in elucidating the interrelationships and circumscribing the order Caryophyllales (Centrospermae) (Behnke, 1976). All the families kept in this order show P III subtype sieve element plastids. These families are : Aizoaceae, Nyctaginaceae, Phytolaccaceae, Portulacaceae, Caryophyllaceae and Molluginaceae. In earlier classification other families like Polygonaceae, Vivianiaceae, Bataceae, Gyrostemonaceae, Plumbaginaceae, Fouquieriaceae, Frankeniaceae, Rhabdodendraceae and Theligoniaceae have been kept in Centrospermae but in recent classifications these have been removed from Centrospermae because they show S-type sieve element plastids.

Embryology in Relation to Taxonomy

The role of embryology in taxonomy was first brought to prominence by the German embryologist Schnarf in 1931. Subsequently, enormous work has been done on this aspect and the data collected has proved to be of considerable taxonomic significance. Often a single embryological character can mark out the family, e.g., Pseudoembryo sac in the Podostemaceae, composite endosperm in the Loranthaceae and formation of a single pollen grain from a pollen mother cell in the Cyperaceae. The importance of embryology in taxonomy has been summarized by many embryologists like Maheshwari (1950), Johri (1963), Davis (1966), Poddubnaja-Arnoldi (1976), Herr (1984), Batygina and Yakovlev (1980-1990) and Johri *et al.* (1992).

Embryological Features of Taxonomic Importance

1. Anther: Number of vertical rows of cells of archesporium, type of wall development (Basic, Monocotyledonous, Dicotyledonous and Reduced), the tapetum glandular or amoeboid, number and ploidy of nuclei in tapetal cells, successive or simultaneous cytokinesis, types of tetrad.
2. Pollen grain: Number and position of germ pores, number of cells at the shedding stage.
3. Ovule: Type of ovule (orthotropous, anatropous, campylotropous, amphitropous or circinotropous), number of integuments, micropyle formed by one (inner or outer) or two integuments, nucellus crassinucellate, pseudocrassinucellate or tenuinucellate, presence or absence of endothelium, epistase, hypostase, aril, caruncle, obturator etc.
4. Archesporium: Number or archesporial cells, extent of parietal tissue.
5. Megasporogenesis and development of embryo sac: Shape of the megaspore tetrad, position of functional megaspore in the tetrad; embryo sac development mono, bi- or tetrasporic, number of nuclear divisions.
6. Mature embryo sac: Number and organization of cells in the embryo sac.
7. Fertilisation.
8. Endosperm: Nuclear, Cellular or Helobial, presence or absence of haustoria, seeds albuminous or exalbuminous.
9. Embryo: Development type (Asterad, Onagrad, Chenopodiad, Solanad and Caryophyllad), presence or absence of suspensor haustoria.

10. Seed coat: Number of layers and their derivation (from one or two integuments).
11. Special features: Apomixis, polyembryony.

Application of Embryology to Problems of Taxonomy

Embryological characters are useful in delimitation of plant groups at all levels. At the highest level Angiosperms are characterized by double fertilization (syngamy and triple fusion) which is not seen in any other division. Similarly Angiosperms differ from Gymnosperms in that the nutritive tissue for development of embryo in Angiosperms is a post fertilization phenomenon while in Gymnosperms it is a pre-fertilization character. The Dicotyledons and Monocotyledons differ in the cytokinesis of pollen mother cells after meiosis. The Dicotyledons are characterized by the simultaneous cytokinesis while Monocotyledons are characterized by the successive cytokinesis.

Within Dicotyledons and Monocotyledons the subclasses are also differentiated on embryological grounds. The Asteridae are characterized by unitegmic and tenuinucellate ovules while the Caryophyllidae are characterized by bitegmic and crassinucellate ovules.

Embryological data supplement other evidence in many cases. Polygonaceae has been associated by various taxonmists with the Centrospermae group. Polygonaceae differs embryologically from other Centrospermae group. Polygonaceae are characterized by the orthotropous ovules while Centrospermae have either Campylotropous or amphitropous ovules.

Some Families Marked out by their Embryological Features

Podostemaceae

An exclusively tropical family, Podostemaceae includes perennial aquatic herbs, which grow attached to stones in running water. The family shows interesting embryological features. A unique feature is the formation of pseudoembryo sac due to the disintegration of nucellar cells below the embryo sac (Fig. 10.4B). A combination of the following characters further mark out this family.

1. Occurrence of pollen grains in pairs (Fig. 10.4A),
2. Anatropous, bitegmic and tenuinucellate ovule with micropyle formed by the outer integument (10.4B),
3. Bisporic embryo sac,
4. Solanad type of embryogeny,
5. Prominent suspensor haustoria (Fig. 10.4C),
6. Absence of antipodal cells (except in *Dicraea*),
7. Absence of triple fusion and consequently endosperm,
8. Presence of pseudoembryo sac.

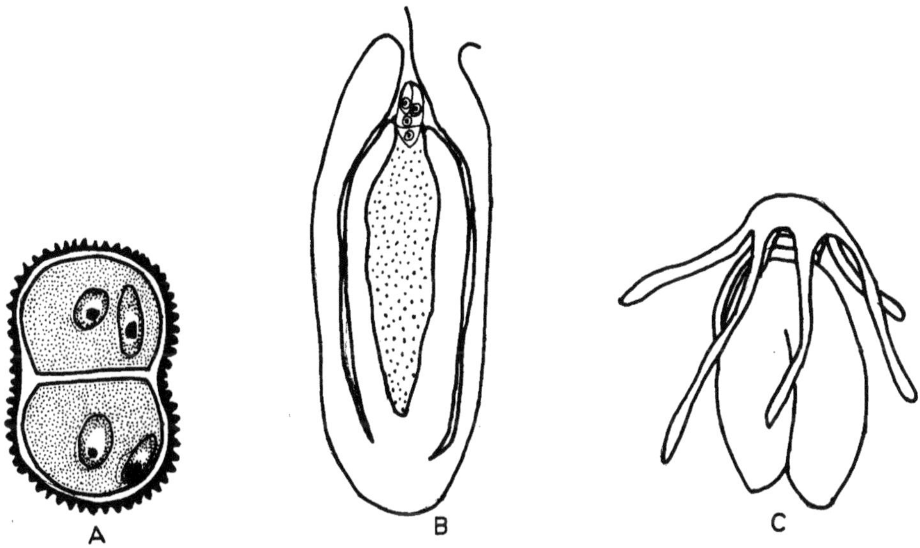

Fig. 10.4: Podostemaceae. A. Paired pollen grains of *Podostemon subulatus.* B. Ovule of *Indotristicha ramosissima.* C. Mature embryo of *Dicraea elongata* (A, C after Magnus, 1913; B. after Mukkada, 1962).

Onagraceae

The family Onagraceae is characterized by the presence of Oenothera type of embryo sac development (Fig. 10.5). Oenothera type of embryo sac development is seen in all the members of the family Onagraceae and is not found in any other family. Further of the four megaspores formed after meiosis, the chalazal three megaspores degenerate, while the micropylar megaspore develops into embryo sac. The family is also characterised by the absence of antipodals. Onagraceae are further characterized by a diploid endosperm.

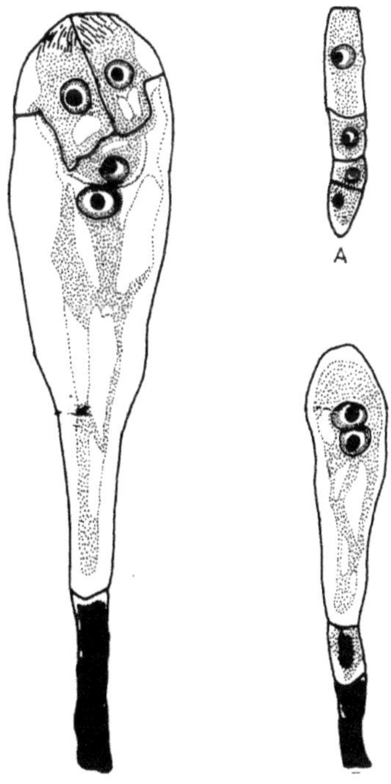

Figure 10.5: Embyro Sac Development in Onagraceae

Cyperaceae

The development of microspore from the pollen mother cell in the family Cyperaceae is quite unique. In all other Angiosperms the pollen mother cell divides by meiosis, resulting in four haploid nuclei, each of which forms a functional microspore. The members of the Cyperaceae exhibit a special mode of pollen development. Of the four nuclei formed after meiosis, only one functions. The functional nucleus remains in the centre and the three non-functional nuclei are cut off on one side of the cell, and they do not form pollen grains. The functional nucleus divides to form a vegetative cell and a generative cell. The non-functional nuclei eventually degenerate (Fig. 10.6).

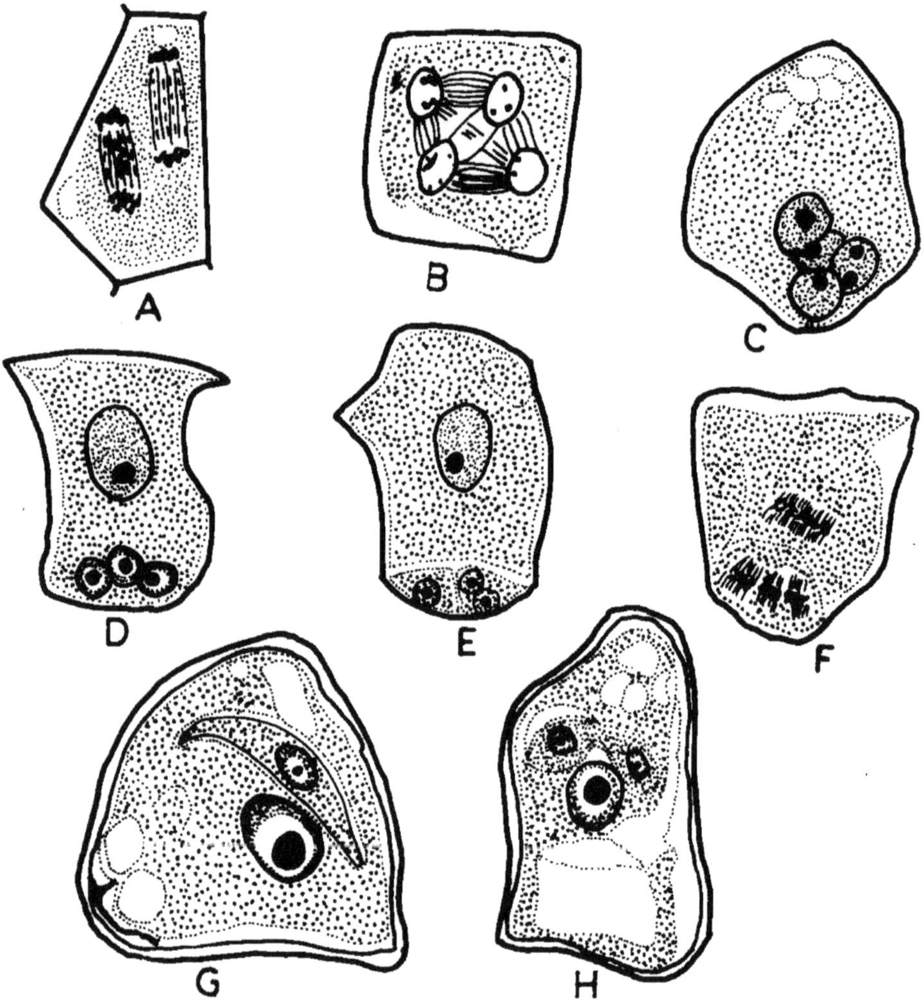

Figure 10.6: Pollen Development in Cyperaceae

Loranthaceae

Members of the Loranthaceae show some embryological features, which are not seen in any other Angiosperms. In this family members do not show any differentiation in ovules and placenta. Here no ovules are present as such and the placental ovular complex is known as a mamelon. The integuments have disappeared and there is a gradual reduction in the lobing and differentiation of mamelon. Finally in some genera even the mamelon is absent and the archesporial cells differentiate at the base of the ovarian cavity. The embryo sacs ascend to various heights in the style. The family is also characterized by the composite endosperm formed by several embryo sacs.

Same other examples of the value of embryology in Taxonomy.

Paeonia

In most of the classical systems of classification the genus *Paeonia* is kept in the monogeneric tribe *Paeonieae* of the Ranunculaceae. The vasuclar anatomy and certain characters of leaves and fruits led Worsdell (1908) to conclude that *Paeonia* constitutes a natural and independent group and should be assigned to a separate family Paeoniaceae which is closer to Magnoliaceae than Ranunculaceae. Kumazawa (1935) studied its wood anatomy and confirmed this conclusion. In the structure of exine Wodehouse (1936) found *Paeonia* to differ from the rest of the Ranunculaceae. Gregory (1941) noticed that the basic chromosome number of *Paeonia* is five in contrast to seven, eight or nine in the Ranunculaceae. Eames (1951) also pointed out that the floral anatomy of *Paeonia* is quite different.

Embryological data also support the view of removing *Paeonia* from the Ranunculaceae to a separate family Paeoniaceae. The most peculiar embryological feature of this genus is its embryogeny (Yakovlev and Yoffe, 1957; Murgai, 1959, 1962). According to Yakovlev and Yoffe (1957) the zygote undergoes repeated nuclear divisions forming coenocytic structure (Fig. 10.7A-C). Later, the nuclei become lodged in the peripheral layer of cytoplasm around a large central vacuole (Fig. 10.7D). Now, wall formation occurs, and the coenocyte becomes cellular, except in the central part (Fig. 10.7E). Some of the peripheral cells give rise to embryo initials (Fig. 10.7F) of which one matures into adult embryo.

Other embryological features which justify the separation of *Paeonia* from Ranunculaceae to a separate family are given in Table 10.1.

Table 10.1: Embryological Differences Between *Paeonia* and Ranunculaceae

Feature	*Paeonia*	Ranunculaceae
Stamens	Spirally arranged, centrifugal	Spirally arranged, centripetal
Anther	Multilayered endothecium, mostly 2-layered tapetum	One-layered endothecium and one- layered tapetum
Pollen	Reticulately pitted exine, large and elongate generative cell	Granular, papillate or smooth exine, small lenticular generative cell.
Female archesporium	Multicelled, many megaspore mother cells function	Uni- or multicelled, one cell functions
Antipodal cells	Persistent, not polyploid	Persistent (ephemeral in *Adonis*), nuclei one or more than one, polyploid.
Embryogeny	Unique, Free nuclear	Onagrad or rarely Solanad type
Seed	Arillate	Non arillate

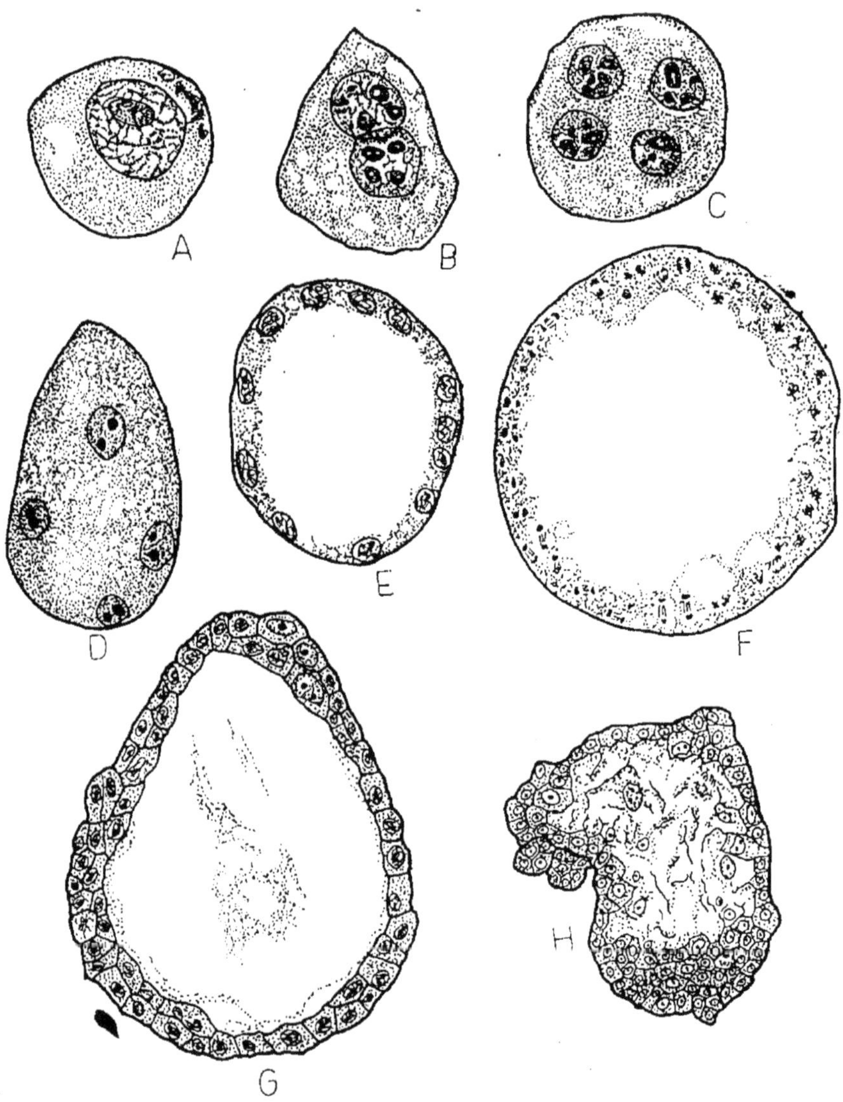

Figure 10.7: Embryo Development in Paeoniaceae

Trapa

Bentham and Hooker (1883) placed the genus *Trapa* in the family Onagraceae, while Raimann (1898), Engler and Diels (1936) and Hutchinson (1959) included it in the unigeneric family Hydrocaryaceae now called Trapaceae. Eames (1953) remarked that on the basis of anatomical evidence *Trapa* does not belong to Onagraceae. The embryological data presented in Table 10.2 also support the view that *Trapa* should be removed from Onagraceae to Trapaceae.

Table 10.2: Embryological Differences Between *Trapa* and Onagraceae

Feature	*Trapa*	Onagraceae
Pollen grain	Pyramidal with 3 much folded meridional crests	Bluntly triangular and basin shaped
Ovary	Semi-inferior, bilocular with a single pendulous ovule in each chamber	Inferior, mostly trilocular with many ovules per chamber on an axile placenta
Embryo sac	Polygonum type	Oenothera type
Endosperm	Absent	Present, nuclear
Embryo	Solanad type	Onagrad type
Suspensor	Well developed suspensor haustorium	Short and inconspicuous
Cotyledons	One cotyledon extremely reduced	Cotyledons equal
Fruit	Large, one-seeded drupe with prominent spines	Loculicidal capsule

Basellaceae

Several divergent views have been expressed about the systematic position of Basellaceae, a small family of herbaceous twining vines that occur chiefly in tropical America and Asia. Eichler (1878), Volkens (1893) and Lawrence (1951) include the family in Centrospermae close to Portulacaceae. A similar assignment is offered by Bessey (1915), Gunderson (1950), Takhtajan (1997) and Cronquist (1989), although these systematists have adopted the alternative ordinal name, Caryophyllales. Hutchinson (1973) included the family among group of advanced Caryophyllaceous families which he elevated to a separate order, Chenopodiales. Hooker (1880) included it in Chenopodiaceae as the subfamily Baselleae and Franz (1908) lowered it to a tribal rank, placing it in subfamily Montioideae of Portulacaceae.

Maheswari Devi and Pullaiah (1975) investigated the life history of *Basella rubra* and summarised the embryological features of the family Basellaceae. They also offered a comparison of those families included in Centrospermae in Table 10.3.

Basellaceae and Chenopodiaceae differ markedly embryologically, and the inclusion of Basellaceae as a subfamily of Chenopodiaceae is untenable. This conclusion is also supported by anatomical features summarized by Metcalfe and Chalk (1950). For example, stems in Basellaceae contain a single ring of vascular strands with intraxylary phloem in the larger bundles. Chenopodiaceae on the other hand, are characterized by the development of polycyclic, anomalous secondary vascular bundles.

Basellaceae and Portulacaceae share many embryological features (Table 10.3) supporting a close relationship, as also suggested by Takhtajan (1997). However, to include Basellaceae as a tribe in Portulacaceae does not appear justified in view of certain differences.

Since the Basellaceae show all the embryological features which predominate in Caryophyllales, the retention of Basellaceae as one of the independent families in Caryophyllales seems to be fully justified on embryological grounds.

Gentianaceae and Menyanthaceae

Benthem and Hooker (1873-1876) and Rendle (1952) split the Gentianaceae into two subfamilies: (a) Menyanthoideae including the marshy plants, and (b) Gentianoideae including all the others. Wettstein (1935) raised Menyanthoideae to a family rank, which was accepted by Don (1938) and Lindsey (1936). This was supported by both vegetative (Perrot, 1897) and floral anatomy (Lindsay, 1938) and embryology (Soueges, 1943; Maheswari Devi, 1962).

Morphological and embryological dissimilarities of the subfamilies Menyanthoideae and Gentianoideae are tabulated in Table 10.4.

Table 10.4: Differences Between Menyanthoideae and Gentianoideae

Feature	Menyanthoideae	Gentianoideae
Habit	Aquatic or marshy	Tetrrestrial
Morphology	Leaves alternate, corolla valvate or imbricate	Leaves opposite and decussate, corolla twisted
Anatomy	Vascular bundles collateral, sieve tubes absent in phloem	Vascular bundles bicollateral or amphi-cribal, sieve tubes present in phloem
Anther tapetum	Secretory type, cells bi- or multi-nucleate, nuclear fusion common	Secretory (*Exacum*) or amoeboid (*Canscora*), cells remain uninucleate
Placenta	Suppressed parietal placenta	Well developed parietal placenta, some times the placenta fuse at the centre to form a bilocular ovary
Endothelium	Present	Absent
Endosperm	Cellular	Nuclear
Embryogeny	Asterad type	Solanad type

Maheswari Devi (1962) is of the opinion that these differences fully justify a family rank for the Menyanthoideae which would be better designated as Menyanthaceae. The name Gentianaceae should be retained for Gentianoideae.

Palynology and Taxonomy

Pollen morphology offers reliable data and evidence of taxonomic relationships. Palynological evidence may be used to place taxa of uncertain affinities, to suggest rearrangements, withdrawals and separations as well as corraborating other lines of evidence. Sometimes, the study of fine structural details in fossil pollen opens up possibility of new sources of phylogenetic evidence.

Palynological evidence may be used to place taxa of uncertain affinities, to suggest rearrangements, withdrawals and separations as well as corraborating other lines of evidence. Sometimes, the study of fine structural details in fossil pollen opens up possibility of new sources of phylogenetic evidence.

The taxonomic and evolutionary importance of pollen morphology may be at specific, generic or higher levels. In many cases the type of pollen of a taxon is characteristic and constant; such a taxon is called *Stenopalynous* or unipalynous

Table 10.3. Embryology of Phytolaccaceae, Aizoaceae, Portulacaceae, Basellaceae, Caryophyllaceae and Chenopodiaceae (Adepted after Maheswari Devi and Pullaiah 1975)

	Phytolaccaceae	Aizoaceae	Portulacaceae	Basellaceae	Caryophyllaceae	Chenopodiaceae
Anther	Tetraporangiate	Tetrasporangiate	Tetrasporangiate	Tetrasporangiate	Tetrasporangiate	Tetrasporangiate
Anther Wall development	—	Basic type	Monoctyledonous	Monoctyledonous	Mono- or dicotyledonous	Monoctyledonous
Endothecium	Fibrous	Fibrous	Fibrous	Fibrous, multiseriate	Fibrous	Fibrous
Tapetum	Glandular, cells bi- or multinucleate	Glandular, cells bi- or four-nucleate	Glandular, cells bi- or multinucleate	Glandular, cells multinucleate	Glandular, cells bi-nucleate	Glandular, cells uni- or binucleatt; occassionally amoeboid
Tetrads	Tetrahedral, iso-bilateral	Tetrahedral, isobi-lateral, decussate	Tetrahedral, isobilateral	Tetrahedral, isobilateral	Tetrahedral, isobilateral	Tetrahedral, isobilateral
Pollen grain	Two- (*Phytolacca*) or three-celled (*Rivina*)	3-celled	3-celled	3-celled	3-celled	3-celled
Ovule	Anacampylotropous, bitegmic, crassinucellate	Anacampylotropous, bitegmic, crassi-nucellate, funicular obturator develops	Anacampylotropous, bitegmic, crassinucellate	Ana- or campylotropous, bitegmic, crassinucellate	Camplotropous, bitegmic, crassinucellate nucellar beak conspicuous	Campylotropous, bi-tegmic, crassinucellate nucellar beak conspicuous
Micropyle	Formed by integument	inner integument	Inner integument	Inner integument	Inner integument	Inner integument
Air-space	Present	Present	Present	Present	Present	Present

	Phytolaccaceae	Aizoaceae	Portulacaceae	Basellaceae	Caryophyllaceae	Chenopodiaceae
between the integuments at their base						
Archesporium	2- or 3-celled	Multicelled	2- or 3-celled	2- or 3-celled	Uni- or Multicelled	2- or 3-celled
Embryo sac	Polygonum type Endymion, Drusa, Adoxa or Penaea	Polygonum,	Polygonum	Polygonum	Polygonum	Polygonum, Allium type in *Sueda maritima*
Antipodal cells	Persistent, secondary multiplication in *Rivina*	Ephemera	Ephemeral	Ephemeral supernumeraty due to divisions	Ephemeral	Ephemeral
Endosperm	Nuclear Type	Nuclear, chalazal haustorium in *Hereroa hesparantha,*	Nuclear	Nuclear	Nuclear in some taxa superificial extensions develop	Nuclear
Embryogeny	Caryophyllad type, massive suspensor	Solanad, occasional suspensor polyexmryony	Onagrad or Caryo-, phyllad, occasional polyembryony	Chenopodiad	Caryophyllad, basal cell of suspensor haustorial massive	Chenopodiad, suspenssor
Seed and endotegmic,	Small, exotestal, perispermous	Exotestal, and endotegmic, perispermous	Small, exotestal, albuminous, perispermous	Small, exotestal,	Small, exotestal, perispermousperispermous	Small, exotestal,

(Nair, 1970) and may be exclusive of that group, e.g. thick walled pollen grains of the Gyrostemonaceae of Australia. In other cases, the types of pollen may vary considerably in size, aperture, stratification of exine etc. Such taxa are termed *eurypalynous* or multipalynous. Stenopalynous taxa are generally considered to be very natural. Ascelpiadaceae, Asteraceae, Poaceae, Brassicaceae and Lamiaceae are some of the stenopalynous families.

Eurypalynous condition, seen in Malvaceae, is primitive while stenopalynous condition, seen in Poaceae, is an advanced character.

The major characters of pollen grains that are used in solving taxonomic and phylogenetic problems are (1) Apertural patterns, (2) Exine stratification and ornamentation, (3) Pollen association and (4) Pollen nuclear number.

Apertural Patterns

By using the NPC system, formulae can be derived to individual taxa and families. Families with more closely similar formulae must be considered more closely related than those with less similar formulae.

Engler and Diels (1936) included Canellaceae in the order Parietales. When NPC formulae of all the families included in the Parietales are observed, they appear to be more or less uniform having $N_3P_4C_5$ (345). But Canellaceae appears to be an exception - with the formula 133. Subsequently Melchoir (1964) divided Partietales into two orders the Violales in which the NPC is essentially 345 throughout and others in Guttiferales.

The genus *Theobroma* provides an example of a genus with two distinct types of pollen. One is small, flattened and is found in the majority of the species including *T. cacao.* The other type is larger, more elongate. This is one of the reasons for the recent removal of the species with larger pollen grains to a distinct genus *Herrania.*

Caltha palustris var. *palustris* and *C. palustris* var. *alba* can readily be distinguished by their aperture.

C. palustris var. *palustris*	*C. palustris* var. *alba*
Tricolpate	Pantoporate
Position of apertures equatorial	Postion of apertures global
Size of paerttures – large	Size of apertures – small

In the genus *Sesamum*, the number of colpi is species specific. Eleven in *S. indicum*, nine in *S. prostratum* and eight in *S. laciniatum*. Similarly *Butomopsis umbellatus* can be differentiated from *B. lanceolata* by the nature of apertures, the former having monocolpate and the later pantoporate grains.

Salix and *Populus* of Salicaceae can readily be distinguished from one another on the basis of aperture. The former has 3-colpate pollen and the latter inaperturate one.

Similarly *Phytolacca* and *Rivinia* of Phytolaccaceae can be distinguished from one another on pollen apertural patterns. *Phytolacca* has 3-zonocolpate pollen while *Rivinia* has 5-colpate pollen.

The genus *Nelumbo* has till recently been included in Nymphaeaceae. *Nelumbo* has 3-colpate pollen while Nymphaeaceae have uniformly monosulcate pollen. Hence *Nelumbo* has been segregated into a separate family Nelumbonaceae.

Bentham and Hooker (1862-83) and other earlier taxonomists have included *Alisma, Butomus* and *Butomopsis* in the family Alismataceae. But Hutchinson (1969, 1973), Cronquist (1988) and Takhtajan (1980) have segregated *Alisma* into Alismataceae and *Butomus* and *Butomposis* into Butomaceae. This is supported on Palynological grounds. *Alisma* have pantoporate pollen while *Butomus* and *Butomopsis* have monocolpate one.

Apertural characters are also useful in understanding the phylogeny. The monosulcate (uniaperturate) pollen found in some Ranales indicate primitiveness, Triaperturate and polyaperturate pollen are found in advanced families like Asteraceae.

Exine Sculpturing

The exposed surface-details of the pollen wall constitute the sculpturing. Some of the more important types are: Psilate (smooth), foveolate (pitted), fossulate (grooved), scabrate (very fine projections), verrucate (warty), baculate (rod- like elements with swollen tips), gemmate (sessile pillar), echinate (spiny), rugate (elongate elements irregularly distributed tangentially over the surface), striate (elongate, more or less parallelly distributed tangentially over the surface), punctate (minute perforations) and reticulate (elements forming open network) (Fig.10.8).

Exine patterns have occasionally been found useful in differentiating species of the genus. In the genus *Bauhinia, B. racemosa* has reticulate, *B. purpurea* has reticulate-tuberculate, *B. malabarica* has spinulate, *B. acuminata* has pilate, *B. krugii* has striate and *B. retusa* has verrucate pollen grains.

The genus *Cicer,* usually placed in the tribe Vicieae of the Fabaceae, has many characters, which suggested that this position is anomalous. Clarke and Kupicha (1976) studied the pollen grains of *Cicer*, of the Viceae and the allied tribes, Trifolieae and Ononideae. They found that *Cicer* has pollen that is more similar to that of the Ononideae than that of Viceae. They suggested the transfer of the genus *Cicer* from the Viceae to the Ononideae.

In general, Angiosperm pollen grains have a massive exine and a thin intine. However certain taxa (e.g., Amaryllidaceae, Cannaceae, Iridaceae, Musaceae, Zingiberaceae, Costaceae, Marantaceae, Strelitziaceae and Heliconiaceae among Monocotyledons) are unique in having a highly reduced exine and elabroate intine. In *Heliconia*, the exine is relegated to a few spinules and a thin layer over intine, which is almost a hundred times as thick.

Pollen Association

In majority of the Angiosperms pollen grains occur singly. However, in Podostemaceae pollen grains occur in pairs. In some Angiospermous families pollen grains are clustered in Tetrads.

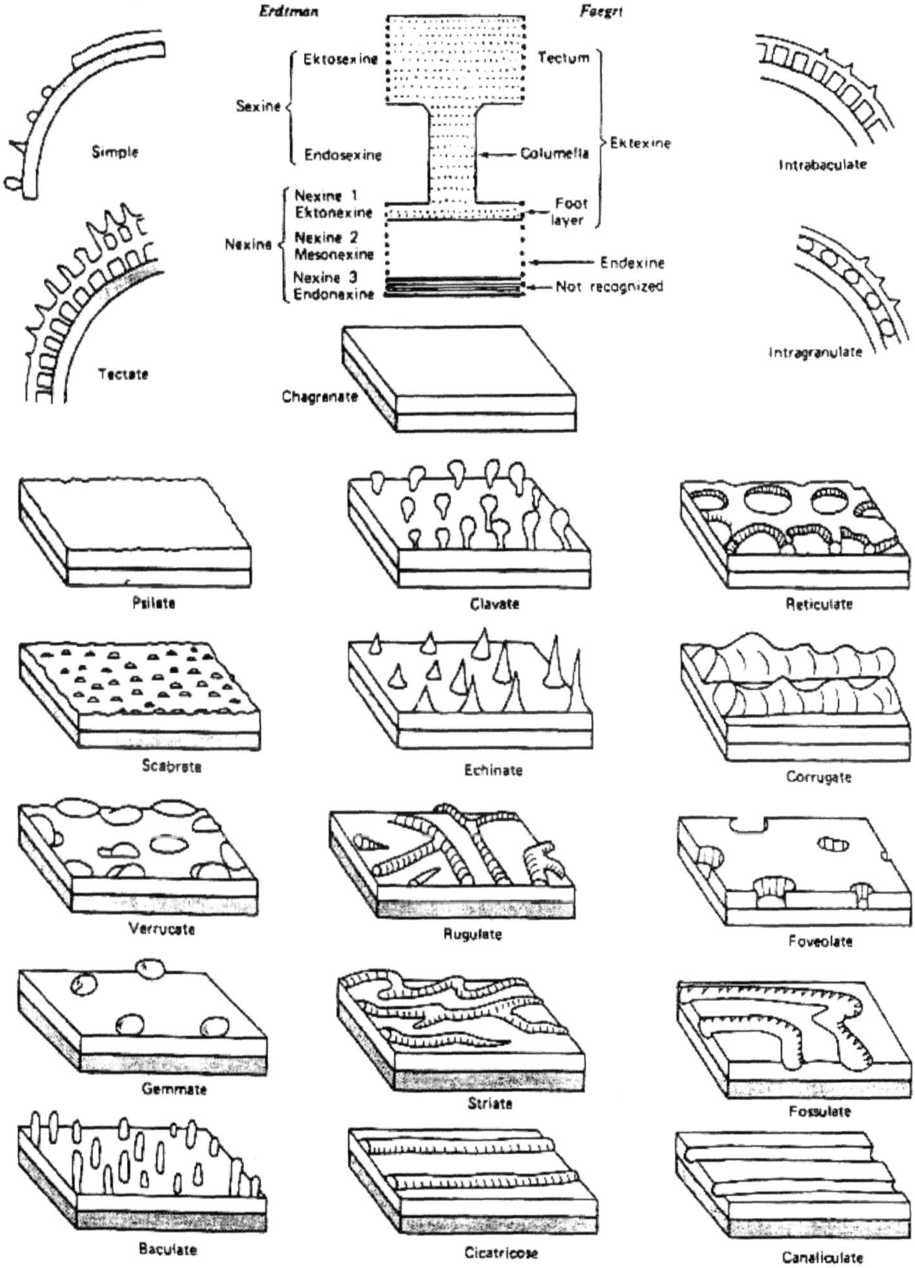

Figure 10.8: Exine Patterns in Angiosperms

In Dicotyledons such associations are seen in such diverse families like Winteraceae, Monimiaceae, Annonaceae, Droseraceae and Mimosaceae. In Monocots, they are restricted to advanced families like Juncaceae. Extreme cohesion of pollen is met with in Asclepiadaceae and Orchidaceae, where they form pollinia.

Vij and Kashyap (1978) pointed out the taxonomic importance of pollen association in the multipalynous family Orchidaceae. Of the 50 species they investigated, single pollen grains were found in the relatively primitive groups (*Paphiopedilium villosum, Cyprepedium cordigerum, Cephalanthera ensifolia*), tetrads in the tribe Neottieae and Epidendrieae and perfect massulae in Orchideae and some Neottieae. Poddubnaja Arnoldi (1976) has also reported that pollen grains are single in *Cypripedium insigne*, tetrads in *Calanthe veitochii* and massulae in *Platanthera bifoliata*. Pollen grains are single in diandrous forms, and some monandrous forms (*Cephalanthera, Vanilla*). They remain attached in tetrads as in *Spiranthes australis, Aa achalesis, Bromhedia finlaysoniana, Calanthe* and *Polystachya flavescens* or form massulae as in *Epigonium roseum, Liparis viridiflora, Platanthera* and *Peristylus spiralis* or pollinia as in Monandrae (see Johri *et al.*, 1992).

Similarly the family Mimosaceae pollen grains are shed singly, in tetrads or pollinia. In *Neptunia, Leucaena, Prosopis* and *Desmanthus* pollen grains are shed singly. In *Mimosa pudica* pollen grains are shed in tetrads but in *Mimosa hamata* pollen grains are shed in single tetrads as well as twin tetrads. They are shed together in masses of eight or sixteen pollen grains in *Acacia dealbata, A. farnesiana* and *A. leucophloea.* All the microspores (in a sporangium) together form a single pollinium in *Albizzia lebbeck, Acacia baileyana* and *Calliandra.*

Pollen Nuclear Number

In Angiosperms pollen grains are shed either at 2-nucleate stage or 3-nucleate stage. This character is valuable in taxonomy and phylogeny of Angiosperms.

Brewbaker (1967) made and extensive survey of about 2000 species belonging to the 265 families. According to Brewbaker two-nucleate pollen grains are more primitive in Angiosperms as a whole and within individual taxa, three-nucleate pollen grains have arisen from the two-nucleate ones.

In the Monocotyledons, the three-nucleate pollen grains occur in highly reduced wind pollinated groups, aquatics and achlorophyllous saprophytes. In the Dicotyledons, however, there is no definite pattern. Certain orders, e.g., Caryophyllales, Plumbaginales, Polygonales and Asterales, have consistently tri-nucleate pollen grains (Dahlgren, 1975). Magnoliales, Laurales and Violales have consistently two-nucleate pollen grains. Most other orders are quite heterogeneous in having both two and three-nucleate ones. Both these types may occur in one and the same genus, as has been observed in *Burmannia, Lobelia* and *Ipomoea* (Brewbaker, 1967).

The important recently discovered connection between bi- and tri-nucleate pollen grains and types of self-incompatibility systems (see Brewbaker, 1957) opens up further possibilities. Particularly in the dicotyledonous taxa the groups with binucleate pollen grains tend to have the gametophytic type of self-incompatibility and those with trinucleate pollen grains the sporophytic type. In Lamiaceae Kooiman (1972) has also demonstrated the correlation between binucleate and tricolpate pollen grains and between trinucleate and hexacolpate pollen grains.

Cytology In Relation To Taxonomy

Cytotaxonomy utilizes cytological characters in solving taxonomical problems. Solbrig (1968) indicated that most Cytotaxonomic studies have largely made use of only two nuclear aspects: (1) number, shape and size of chromosomes and (2), the behavior of chromosomes during mitosis and meiosis. Ehrendorfer (1964) has reviewed the previous work pertaining to the taxonomic role of chromosome studies.

Chromosome Number

There is a great variation in the number of chromosomes in vascular plants. All individuals within a species have the same chromosome number. Somatic cells of *Haplopappus gracilis, Brachyscome dichromosomatica* (Asteraceae), *Zingeria biebersteiniana, Colpodium versicolor* (Poaceae), *Ornithogalum tenuifolium* (Asparagaceae) and *Rhynchospora tenuis* (Cyperaceae) show 4 chromosmes per cell, *Arabidopsis thaliana* has 2n = 10 chromosomes while *Poa littrosa* (Poaceae) show 530 chromosomes, *Kalanchoe* and *Echeveria* (both Crassulaceae) species show 500-520 chromosomes, *Morus nigra* (Moraceae) 308 chromosomes and *Ophioglossum reticulatum* have 1260 chromosomes in their somatic cells.

The great diversity in chromosome number and their constancy in species provide an important character for taxonomic groupings of large number of plants.

In higher plants there are three main different cytological types-diploid, polyploid and aneuploid. Certain groups of vascular plants possess a constant number of chromosomes. Genera such as *Pinus* and *Quercus* have n = 12 chromosomes in all the known species and can be delimited from other related genera on the basis of chromosome number.

When chromosome numbers in various members of a taxon are in the proportion of exact multiples, the series is described as euploidy. Several such examples can be found in the flowering plants and the ferns. In Malvaceae for instance, the somatic numbers in various species range from 10, 15, 20, 25…..40, from 12, 18, 24 to 30; from 14, 28, 42, 56 to 84 and so on. In the genus *Taraxacum* (Asteraceae), there are species with 2n=16, 24, 32, 40 and 48 chromosomes. Various members of such an euploid series may be unified by a basic number of x which is the gametic number of a diploid species and other species can be described triploids (3x), tetraploids (4x), hexaploids (6x)…..Polyploids (nx). Thus one of the series in Malvaceae has x= 5 while the others have x = 6, 7 etc. in *Taraxacum* the basic number is x = 8.

In the living species of *Chlorophytum* (Liliaceae), for example, the chromosome numbers vary from 14 to 28, 42, 56, 84 etc. and also from 16 to 32. This means that *Chlorophytum* has two basic numbers x = 7 and 8. However, from the detailed analysis of the meiotic behaviour of chromosomes in one of the species *C.laxum* Naik (1976) has shown that the two base numbers 7 and 8 can be considered as secondary, most probably derived from primary basic number x = 4.

In certain groups, the closely related species are cytologically distinct, one being a diploid and the other a tetraploid. Such clearly related pairs are termed polyploid pairs. These have been shown experimentally in *Nasturtium officinale* (2n = 32) and *N. macrophyllum* (2n = 64); *Cardamine hirsuta* (2n = 16) and *C. flexuosa* (2n = 32) of the Cruciferae and *Saxifraga hyperborea* (2n = 26) and *S. rivularia* (2n = 52) of the Saxifragaceae.

Hybridization plays a major role in evolution. In Brassicaceae, *Brassica oleracea* (2n=18) and *B. rapa* (2n = 20) have been shown to hybridize in nature to give rise to *B. napus* (2n = 38) after polyploidy. *Raphanobrassica* a hybrid between *Brassica oleracea* (2n = 18) and *Raphanus sativus* (2n = 20), has 2n = 38 chromosomes and was artificially synthesized by Karpechenko (1927, 1928). *Galeopsis tetrahit* (2n = 32) is a hybrid between *Galeopsis speciosa* (2n = 16) and *G. pubescens* (2n = 16).

If the chromosome numbers found within a group bear no simple numerical relationship to each other, then the series is termed as an aneuploid series or simply aneuploidy. Here two distinct phenomena occur. (1) increase or decrease in the number of chromosomes whereby the same genetic material becomes distributed in a different number leading to a change in basic number, (2) the basic number is unaltered but the genetic material present is changed through the addition or loss of chromosomes.

In nature, a large number of plant groups are known to exhibit aneuploidy. Various species of *Carex* (Cyperaceae) show a wide range of chromosome numbers from n = 6 to 112 with multiples of 5, 6, 7 and 8. The genus *Iseilema* also shows aneuploidy. *Iseilema anthephoroides* is having n = 5, *I. venkateswarlui* n= 4 and *I. lubbardii* n = 3 (Satyavathi and Murthy, 1974). In the genus *Vicia* there are chromosome numbers of 2n = 10, 12, 14, 24 and 28 obviously consisting of aneuploids clustered around the diploid and tetraploid levels, and in *Crepis* 2n = 6, 8, 10, 12, 14, 16, 18, 22, 24, 42, 44, 66, 88 (and some other numbers), with less obvious clustering Cytological studies, especially the Karyotype data, have been shown to be useful in taxonomy at all levels up to that of family and even above, but most involve species and groups of species.

Raven and Kyhos (1965) have shown Karyotype affinities for the closely related Magnoliaceae, Himantadraceae and Degeneriaceae and also for Schisandraeae and Illiciaceae while there is marked difference in chromosome numbers of Lauraceae (x = 12) and Hernandiaceae (n = 20, 40). Wiens (1975) has provided cytological evidence, for separation of Viscaceae from Loranthaceae. Various members of Loranthaceae are unified by a single basic chromosome number x = 9 while Viscaceae are characterized by a series of aneuploid numbers ranging from 10 to 14. In most classical systems of classification the genus *Paeonia* is kept in the tribe Paeonieae of the Ranunculaceae. Gregory (1941) noticed that the basic chromosome number in *Paeonia* is five in contrast to seven, eight or nine in the Ranunculaceae, hence advocated its separation into a separate family Paeoniaceae.

The classical example of the use of chromosomal characters in the delimitation of genera can be seen in *Crepis, Cymboseris* and *Youngia* of the Asteraceae (Babcock,

1947). In *Crepis* various species have x = 7, 6, 5, 4 or 3 chromosomes while the basic numbers of other related genera are either x = 8 or x = 9.

Cytological data has been shown to be useful in the taxonomy of *Chlorophytum* (Liliaceae) by Naik (1977). Three species namely *C. bharuchae*, *C. glaucum* and *C. glaucoides* are morphologically rather difficult to distinguish from one another but *C. bharuchae* has 2n = 16 chromosomes while the other two have 2n = 42. Further, species with 2n = 42 differ, in their karyomorphology.

Chromosome Size and Structure

The most commonly utilized aspect of chromosome structure is the position of centromeres, i.e. the arm length ratio of each chromosome in the genome. Chromosomes with centromeres near the middle are known as metacentric, those with centromere near one end are known as acrocentric while the ones with a truly terminal cetromere are known as telocentric.

In many genera and families of flowering plants, conspicuous differences in appearances of Karyotype have been found in species having same chromosome number. In some instances these differences follow definite trends associated with trends of morphological specialization. In the primitive genus *Helloborus* (Ranunculaceae), for example the chromosomes in the Karyotype differ little from each other in size and most of them are V-shaped with median or submedian centromere. Karyotype of this nature are the most common ones. Levizky (1931) considered them as generalized types. Various specialized types have been considered to have been derived from these. Thus, the chromosomes with uniform length may gradually differ in size and shape from V-Shaped (metacentric) to L-shaped (submetacentric) to J-shaped (acrocentric) and ultimately to rod-shaped or I-shaped (telocentric) and become asymmetrical. In the advanced genera *Aconitum* and *Delphinium* (Ranunculaceae), for example, the flowers are zygomorphic with the largest number of J-shaped chromosomes.

Less commonly, the reverse trends of specialization i.e. from telocentric chromosomes to metacentric ones have been demonstrated in certain members of the Commelinaceae (Jones, 1970).

Chromosome structure, together with chromosome size and number have been found extremely useful at all levels of the taxonomic hierarchy (Jackson, 1971). The best known use of the importance of chromosome size is in the Monocotyledonous genera *Yucca*, *Agave* and relatives. *Yucca* and some others have flowers with a superior ovary, while *Agave* and others have an inferior ovary. For this reason the former were previously placed in Liliaceae and the latter in the Amaryllidaceae. Because of the great overall similarity in these genera they are now usually segregated from their respective families and placed in the Agavaceae. This classification is strongly supported by evidence obtained in the 1930s when it was found that the Agavaceae are characterized by a very distinctive bimodal karyotype consisting of five large chromosomes and 25 small ones (McKelvey and Sax, 1933). This finding, indeed, supported Hutchinson's realignment of these families and his separation of both these genera into a distinct family, Agavaceae.

In the genus *Crepis* the use of chromosome number and morphology has solved many taxonomic problems at the generic, sectional and specific levels and show how many species are evolutionary inter-related (Babcock, 1947).

Chromosome Behaviour

The behavior of chromosomes during pairing and subsequent separation is sometimes useful in studying the interrelationships of taxa. Meiosis itself is sometimes provide taxonomic information. For example in Juncaceae and Cyperaceae which have small chromosomes with unlocalized centromeres is considered to be inverted i.e. equatorial division precedes the reductional division instead of vice versa. The phenomenon is not always correlated with unlocalized centomeres and it therefore provide evidence for the close relationship of the above two families.

Many peculiarities of meiotic behavior owe their origin to existence of heterozygosity, so that one is observing pairing of unlike genomes. Some species are known to be consistently heterozygous for certain translocations, which give rise to multivalent formation at meiosis. In the genus *Oenothera* all the species are diploid with 2n = 14. Many of these exhibit normal meiosis, but in the subgenus *Oenothera* the species are heterozygous for translocations involving varying numbers of chromosomes, so that univalents of varying sizes arise at meiosis (Cleland, 1972). In the *O.biennis* group, for example, *O. biennis* itself exhibits a ring of six and ring of eight chromosomes, *O. erythrosephala* exhibits a ring of 12 chromosomes and a bivalent and *O. stringosa* exhibits a single ring of 14 chromosomes.

Banding Patterns

In recent years our ability to distinguish morphologically between chromosomes, atleast in favourable material has been greatly enhanced by new staining techiques using Giesma and fluorochrome dyes which stain chromosomes in a consistent banding pattern instead of with a uniform intensity, as is the case with the usual basic fuchsin (Fuelgen reagent). Schweizer and Ehrendorfer (1976) and Ehrendorfer *et al.* (1977) have applied this technique in the systematics of the genus *Anacyclus* and have found good correspondence with the results obtained by morphological and phytochemical approaches. Their data also correspond well with phylogenetic conclusions derived from other fields of study and suggest a more primitive position for perennial taxa of *Anacyclus* than for the annual ones. *A. officinarum* was shown to be probably a hybrid between *A. pyrethrum* and another annual species *A. radiatus*.

Chemotaxonomy

The subject of chemotaxonomy is concerned with the application of chemical characters to problems of classification and phylogeny. This rapidly expanding discipline of plant taxonomy has been variously called as chemotaxonomy, chemosystematics, biochemical systematics or phytochemistry. Chemotaxonomy has made rapid progress in the last 50 years because of new instruments and newer techniques. Chromatography and elctrophoresis have made the analysis of plants much quicker and simpler. Botanists of late have come to the conclusion that evidence for discussing the relationships and phylogeny must be taken from as many sources as possible. As more comprehensive survey of phytochemistry has become available, taxonomists have shown a growing interest in the application of chemical characters to taxonomic problems.

The earliest attempts to correlate chemistry with the phylogenetic level of development was that of Abbott (1886). According to her the Saponin containing plants occupied the middle level of Hackel's scheme of plant evolution. Greshof (1909) suggested that chemical characters should be included in natural classification. He was also of the opinion that every description of a new genus or species should include a short chemical description of that taxon.

The first successful attempt to combine chemical and morphological evidence in the study of a single genus was the work of Baker and Smith (1920) on the essential oils of *Eucalyptus*. They suggested that the level of relationship should be reflected in chemical similarities (primitive plant chemically primitive). According to the morphological and chemical data collected on 176 *Eucalyptus* species, they divided the genus into three groups differing in both morphological structure and chemical constituents. They concluded that primitive species are those which have feather-veined leaves and high Pinene (I) content in their essential oils (terpenes) while more advanced types have intermediate venation and contain Pinene and Cineole (II). According to their system, the most advanced taxa show butterfly wing venation and contain oils with Phellandrene (III), Piperitone (IV), and Geranyl acetate (V). The work of Baker and Smith was the first comprehensive chemotaxonomic-morphological study of a complex genus, and in their conclusions have been a significant contribution to the taxonomy of *Eucalyptus*.

Considering that the modern development of the topic can be dated quite precisely from the year 1963, the progress since that time has been astonishingly rapid. 1963 saw the publication of the first text book on the subject by Alston and Turner (1963), the publication of the proceedings of the first international meeting on the topic (Swain, 1963) and the launching of Hegnauer's seven volume encyclopedic series on plant chemotaxonomy (Hegnauer, 1962- 1974). Later a four volume guide to flower plant chemotaxonomy by Gibbs (1974) has been published.

Compounds Useful in Plant Taxonomy

Although in theory all the chemical constituents of a plant are potentially valuable to a taxonomist, in practice some sorts of molecules are far more valuable than others. Apart from inorganic compounds, which are of relatively little use, three broad categories of compounds can be recognized: Primary metabolites, secondary metabolites and semantides.

Primary metabolites are parts of vital metabolic pathways and most of them are of universal occurrence, or at least occur in a very wide range of plants. Aconitic acid (first isolated from *Aconitum*) or citric acid (from *Citrus*), for example, participate in the Krebs (tricarboxylic acid) cycle and are present in all aerobic organisms; the presence or absence of such compounds is therefore not of much systematic value.

The same is true of the 22 or so amino acids which are known to be constituents of plant proteins or any of the sugars which figure in the photosynthetic carbon cycle and so on.

In some cases, however, the quantities of such metabolites vary considerably between taxa and this in itself can be taxonomically useful. For example, taxa in which universally occurring substances were first detected (such as the two above) often possess particularly large quantities of the molecules concerned well above the amounts which participate in the essential metabolic pathways often as food-storage materials. Sometimes such compounds are stored in a different form from that in which they are metabolized, for example sedoheptulose, a sugar constituting the carbohydrate food reserve of the genus *Sedum*, which as sedoheptulose diphosphate is part of the photosynthetic carbon cycle Secondary metabolites (or secondary plant products) perform non-vital (or at least non-universally vital) functions and are therefore less widespread in plants. It is of course this restricted occurrence among plants, which renders them valuable as taxonomic information. The most well known groups of compound which have been utilized in this way include alkaloids, phenolics, glucosinolates, amino acids, terpenoids, oils and waxes and carbohydrates.

Secondary plant products are largely waste substances, food stores, pigments, poisons, scents, structural units or water repellents etc. In many cases they obviously do have an essential function but of a general nature so that the precise molecular configuration of the compound is not vital. Thus a yellow pigment with an absorption maximum at 477 nm (which presumably defines its function) might be a betalin or anthocyanin and a poison might be an alkaloid or a glycoside.

Semantides are the information-carrying molecules. DNA is primary semantide, RNA a secondary semantide and proteins are tertiary semantides, following from the sequential transfer of the genetic code from the primary genetic information (DNA). In theory, the sequences nucleotides and amino acids in these substances should provide all the taxonomic information necessary for classification and offer an alternative to the study of secondary metabolites, cytology, morphology and anatomy etc., for the latter are nearly manifestations of the former. However, in practice there are great difficulties in the gathering of the sequence data and the results frequently present puzzling anomalies. Sometimes the semantides, together with the larger polysaccharides are known as macromolecules and the primary and secondary metobolites as micromolecules.

Use of Chemical Characters in Plant Taxonomy

In Table 10.5: some major grouping of plant natural products are listed along with some comment on their botanical distribution.

Table 10.5. A summary of some groupings of natural products, which have been used in chemotaxonomic correlations including some comments on the botanical distribution of these constituents (Radford et al., 1976).

Compound Grouping	Comments on botanical distribution	Citation
Terpenoids	widely distributed	Weissmann (1966, see: Swain)
Monoterpenes	widely distributed, of Taxonomic value primarily below the generic level	Numerous reports, see Mirov (1967) or others
Sesquiterpenoids	rather wide distribution, but particularly important and useful in the taxonomy of the Compositae	(Herz (1968, see: Mabry et al.) Herout and Sorn (1969, see Harborne and Swain)
Asperulosides and Aucubins	Rubiaceae, Scrophulariaceae, related families,	Bate-Smith and Swain (1966, see: Swain)
(Iridoid and glycosides)	Plantaginaceae, Cornaceae and others	Kooiman (1969, 1970)
Ranunculins	found only in the Ranunculaceae Kooiman (1969, 1970)	Bate-Smith and Swain (1966, see: Swain)
Cyanogenic compounds	Ranunculaceae and other families	Ruijgrok (1966, see: Swain)
Polyalcohols	widely distributed but have taxonomic potential	Plouvier (1963, see: Swain)
Sulfur	a chemically diverse grouping of compounds which in their Various forms are widely distributed, the iso-thiocyanate-producing glycosides are characteristic of families in the Rhoeadales	Kjaer (1966, see: Swain)
Amino acids	Liliaceae and related (non-protein) Families, Leguminosae (particularly Papilionatae, and other groups)	Bell (1966, see: Swain)
Alkaloids	a chemically and biosynthetically diverse group of compounds rarely found in lower vascular plants and irregularly distributed among the angiosperms; highly useful in the taxonomy of some groups	Hegnauer (1966, see: Swain) Schafer (1964) Price (1963, see: Swain)
Betacyanins and betaxanthins	Centrospermae ("Betanales")	Mabry (1966, see: Swain)
Alkanes (fatty acids and waxes)	widely distributed, possibly of use in classification below the genus Level, and of use in organic geochemistry	Shorland (1963, see: Swain)
Fatty acid epoxides	found in seven families distributed in the basidiomycetes and atleast 13 angiospermous families, of particular use in the link between the Compositae and the Umbelliflorae	Bu'lock (1966, see: Swain)
Acetylenes		Bu'lock (1966, see: Swain)

Compound Grouping	Comments on botanical distribution	Citation
Betacyanins	Centrospermae ("Betanales")	Mabry (1966, see: Swain) and betaxanthins
Alkanes (fatty acids and waxes)	widely distributed, possibly of use in classification below the genus Level, and of use in organic geochemistry	Shorland (1963, see: Swain)
Fatty acid epoxides	found in seven families distributed in the basidiomycetes and atleast 13 angiospermous families, of particular use in the link between the Compositae and the Umbelliflorae	Bu'lock (1966, see: Swain)
Acetylenes		Bu'lock (1966, see: Swain)
Assorted compounds	Ferns	Berti and Bottari (1968, see: Reinhold and Liwschitz)
Diterpenes	rather wide distribution in the seed plants, taxonomically useful in specific groups such as the conifers	Erdtman (1968, see Mabry et al.), Ponsinet et al 1968, see: Mabry et al.)
Triterpenes	wide distribution in living organisms, Taxonomically useful in several Angiosperm families such as the Cucurbitaceae, and the fungi	Ponsinet et al. (1968, see Mabry et al.)
Carotenoids	the universally distributed Photosynthetic carotenoids useful In algal classification; the non-Photosynthetic carotenoids in fruits of possible limited value as taxonomic markers in the angiosperms	Goodwin (1966, see: Swain)
Flavonoids	very large number of diverse compounds found throughout the vascular plants and in nearly, if not all, angiosperms, of great level and of possible use in the classification of higher categories	Bate-Smith (1963, see: Swain); Harborne (1966, see: Swain); Wagner (1966, see Swain); Williams (1966, see Swain); Alston (1968, see Mabry et al.) (extensive literature)
Lignins and lignans	useful in the classification of higher categories	Erdtman (1968, see: Mabry et al.)
Quinones	widely distributed among living organisms, but compounds of this type with limited distribution have potential value in the classification of a number of angiosperms families	Mathis (1966, see: Swain)
Polysaccharides	universally distributed, probably useful in the classification of higher categories — particularly among the algae, comparative data mostly lacking	Percival (1966, see: Swain)
Plant glycosides	great chemical variation in the non-sugar portion of the molecule, varyingly of great usefulness in taxonomy	Paris (1963, see Swain)
Assorted compounds and Liwschitz)	lichens	Huneck (1968, see Reinhold

Compound Grouping	Comments on botanical distribution	Citation
Vide range of chemical approaches	Umbelliferae	Heywood (1971)
Vide range of chemical approaches	Leguminosae	Harborne et al. (1971)

The special value of the chemotaxonomic approach can be seen when chemical characters correlate well with data obtainable from other sources. To take just one example, the fact that there is chemical discontinuity among the families of the order Rhoeadales as defined by Wettstein (1935) (see Table 10.6) has been used by modern taxonomists as a reason for dividing this assemblage of six or seven families into two orders, the Capparales and the Papaverales (see Cronquist, 1968, 1981). Other reasons for making this split lie in the differences in anatomy (presence or absence of latex system) and in morphology (position of stamens). This new division is also suggested by recent serological and palynological studies of the relevant families. It should be pointed out in this case that there are many morphological similarities between the separated groups; for example common floral features (e.g., perfect hypogynous flowers, compound ovaries with parietal placentation). Thus, the additional chemical evidence may have been crucially decisive in the acceptance of this newer grouping. In this instance, the chemical data are unusually good in the sense that relatively wide chemical surveys have been conducted in the relevant families. Not only have over 300 crucifer species have been examined for glucosinolates with positive results (Kjaer, 1976) but also all Papaveraceae studied have been shown to have complex mixture of alkaloids in their tissues (Manske, 1966).

Table 10.6: Revised Classification of Families of the Order Rhoedales (Sensu Wettstein)

Families	Chemical characters	Biological characters
CAPPARALES		
Brassicaceae	Glucosinolates (mustard	Centrifugal stamens;
Capparaceae	oil glycosides) universal;	no latex system
Resedaceae	isoquinoline alkaloids	
Moringaceae	absent	
PAPAVERALES		
Papaveraceae	Glucosinolates absent;	Centripetal stamens;
Famariaceae	isoquinoline universal	well developed latex system

Examples from Secondary Metabolites

Of all the groups of secondary metabolites used by chemotaxonomy, probably none has provided more taxonomic data than phenolic compounds. Plant phenolics are so numerous that detailed reviews on these compounds have been written by Bate-smith (1962), Harborne (1964, 1967) and Ribereau Gayon (1972).

The phenolic molecules most often studied in Chemotaxonomy fall into a general class known as flavonoids. They are the largest group of naturally occurring phenols. These are further classified into flavones, flavanones, isoflavonoids, flavanols, anthocyanidins etc.

One of the best examples of the taxonomic value of secondary metabolites concerns flower pigments. The flower pigments, which are usually anthocyanins and anthoxanthins, vary greatly and have been shown to be under genetic control. But at the same time all variation in pigment production is not genetic in origin.

Bate-Smith (1948, 1958, 1962) has reviewed the early application of evidence from phenolics to taxonomic problems. According to him, leucoanthocyanins are generally present in woody plants. The herbaceous Primulaceae, however, are exceptionally rich in leucoanthocyanins, and some of the woody members of Oleaceae and Scrophulariaceae lack them completely. Phenolic characters have been used by Bate-Smith (1958) in the rearrangement of the species of *Iris*. Williams (1982) pointed out chemical evidence from the flavonoids in the classification of *Malus* species. On the basis of flavonoid patterns. McClure and Alston (1966) identified highly problematic taxa of Lemnaceae a family in which morphological and cytological characters were observed to be inadequate in distinguish in different taxa of *Spirodela, Lemna, Wolfiella* and *Wolffia*. Williams (1976) investigated leaf flavonoids of Liliaceae, Juncaceae, Cyperaceae and Poaceae and supported the view that all these four families have arisen from Liliaceous ancestors. Williams *et al.* (1981) observed Araceae to be unusual amongst Monocots because of their simple and relatively uniform flavonoid profile. No one subfamily was clearly distinguishable and it was only at the tribal level where some significant taxonomic patterns were observed. Their observations indicated that this family has not been derived from Liliaceous stock since flavone C-glycosides, the characteristic constituents of the Aroids, were rarely encountered in the present day Liliaceous taxa. Moreover, they were observed to be totally absent from the tribe Aspidistreae from which the origin of this family has been suggested by Hutchinson (1959).

Chemistry is useful to taxonomy not only at generic level but also in family assignments. There are still a considerable number of usually small angiosperm families whose affinities are not clear and which are difficult to place with certainly into an ordered system of classification. One such family is the Julianaceae the taxonomic relationships of which have been the subject of controversy for at least a century. Based on wood anatomy and morphological features associated with wind pollination, this family has variously been placed in three different associations, near either the Anacardiaceae, or the Burseraceae or the Juglandaceae.

Chemical analysis of the heartwood flavanoids by Young (1976) provides clear-cut data in support of the first of these associations. Thus the heartwood of *Amphipterygium adstrigens* (Julianaceae) yielded sixteen flavonoids, seven of which were 5-deoxy-flavonoids. These rare substances occur quite characteristically in the Anacardiaceae but are not known in either the Burseraceae or the Juglandaceae. Indeed, all seven of these 5-deoxyflavonoids have been recorded widely in the tribe Rhoeae of the Anacardiaceae and one of them, the aurone rengasin, is only known otherwise from this family. In contrast, no evidence could be found in *Amphipterygium* of Juglone, which is a characteristic quinone of Juglandaceae.

By using the new flavonoid data together with a reappraisal of the anatomical characters, Young (1976) decided to reduce the Julianaceae to subtribal rank and include it in the Anacardiaceae as the subtribe Julianiinae of the tribe Rhoeae. This decision has been accepted by Thorne (1981) in his latest revision of Angiosperm classification and is also in line with serological investigations of these and related families (Petersen and Fairbrothers, 1979).

Analysis of flavonoid, carotenoid and quinonoid pigments in plants of the Gesneriaceae have shown significant correlations with the sub-family division of Burtt, based on geography and anisocotyly, while not fitting in with the earlier division on von Fritsch, based on position of ovary (Harborne, 1967). Similarly phytochemical surveys in the tribe Anthemideae (Asteraceae) for both flavonoids and polyacetylenes have indicated that the morphological splitting of the *Chrysanthemum* complex *sensu lato* into smaller generic groups is justified on chemical grounds (Bohlman *et al.*, 1964; Harborne *et al.*, 1970). In the tribe Genisteae of the Fabaceae a survey of the isoflavones, proanthocyanidins and flavones in the leaf provided results which were closely in accord with Rothmaler's treatment of the genera but were at variance with Hutchinson's groupings of the same taxa (Harborne, 1969).

Considering the taxonomic importance of instances of compounds with unique distribution in plants, the study of the "nitrogenous anthocyanins" is perhaps the best known example. The Betalins, as they are now known, include the red to violet betacyanins and the yellow betaxanthins (Fig. 10.9).

Betalins have been found in nine families of Angiosperms, all members of the order Centrospermae. These are: Aizoaceae, Amaranthaceae, Basellaceae, Cactaceae, Chenopodiaceae, Didieraceae, Nyctaginaceae, Phytolaccaceae and Portulacaceae. Engler and Prantl (1887-1915) kept the Cactaceae in a separate order. But is has now been found that the Cactaceae also shows Betalins and hence they should be included in the Centrospermae. Two families —Molluginaceae and Caryophyllaceae — which are often included in the Centrospermae do not posses Betalins and instead show Anthocyanins.

According to Mabry *et al.* (1976) these two families should be kept in a separate suborder Caryophyllinae in the order Caryophyllales. These should be included in Centrospermae on ultrastructural evidence as they also shoe P III subtype sieve element plastids (see table 10.7).

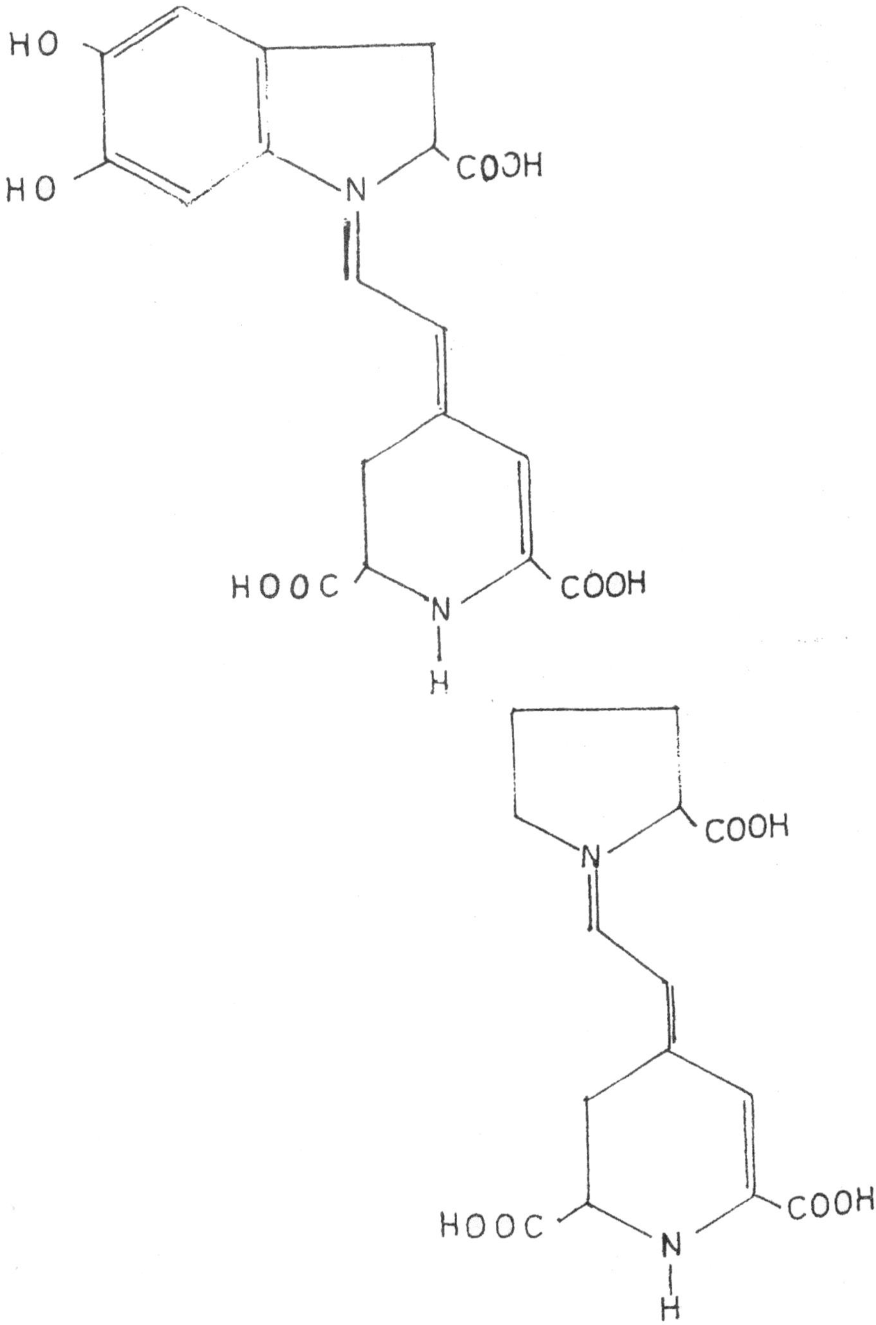

Fig. 10.9: Betacyanin and Betaxanthin.

Table 10.7: Classification of the Order Centrospermae (Caryophyllales) Varying According to the Use of Evidence from Pigments

Structural Classification Engler and Prantl (1897-1915)	Chemical Classification	Compromise Classification
Centrospermae	Chenopodiales	Order: Caryophyllales Suborder: Chenopodinae
Aizoaceae	Aizoaceae	Aizoaceae
Amaranthaceae	Amaranthaceae	Amaranthaceae
Basellaceae	Basellaceae	Basellaceae
Caryophyllaceae	Cactaceae	Cactaceae
Chenopodiaceae	Chenopodiaceae	Chenopodiaceae
Didieraceae	Didieraceae	Didieraceae
Molluginaceae	Nyctaginaceae	Nyctaginaceae
Nyctaginaceae	Phytolaccaceae	Phytolaccaceae
Phytolaccaceae Portulacaceae	Portulacaceae	Portulacaceae
Cactales	Caryophyllales	Suborder : Caryophy
Cactaceae	Caryophyllaceae	Caryophyllaceae
	Molluginaceae	Molluginaceae

Cyanogenic Compounds

Hegnauer (1977) defined the term cyanogenesis as the ability of certain plants to release hydrocyanic acid after injury of cells. Usually cyanophoric plants contain one or several cyanogenic glycosides. Since 1801, when Bohm first detected HCN and amygdalin (the first plant glycoside) in seeds of *Prunus amygdalus* (Rosaceae) many Rosaceous and other plants have been shown to be cynophoric. Till date about 2,056 species of vascular plants are known to be cyanophoric. In Angiosperms they occur erratically. Cyanogenic taxa are relatively frequent in Araceae, Juncaceae, Juncaginaceae, Gramineae and Scheuchzeriaceae among Monocotyledons and several Dicotyledonous families Cyanogenesis, if adequately and carefully used as a character in plant systematics, may prove to be of considerable value in the classification of plants even at the higher systematic levels (Hegnauer, 1977). The distribution of individual cyanogenic compounds in the tracheophytes supports the theory that the Liliopsida (Monocotyledons) evolved from ancestors resembling present day Magnoliidae because cyanogenesis proceeds in exactly the same way in both these groups, with only tyrosine as the precursor (Hegnauer, 1973, van Valen, 1978a).

The subfamilies of Poaceae are characterized by three different tyrosine derived cyanogenic compounds, namely Dhurrin (Andropogonoideae), Triglochinin (Festucoideae and Eragrostioideae) and Taxiphyllin (Bambusoideae) (van Valen, 1978b).

The facts known today suggest that more than one biosynthetic group of cyanophoric compounds occur only in very large genera or families belonging to Dilleniidae, Rosidae and Asteridae. In these taxa, accumulation of Cyanogenic compounds is taxonomically significant at the infra-familiar level. Linamarin for example, is a character of many Loteae, Trifolieae and Phaseoleae in Fabaceae and Calenduleae in Asteraceae.

In India several botanists have contributed to the Chemotaxonomy of plants. Notable among them are S.D. Sabnis from Baroda, L.L. Narayana from Warangal and M. Radhadrishnaiah, from Hyderabad.

Nageswar *et al.* (1984) has studied the distribution of different chemical compounds in *Caesalpinia* and gave a key for identification of the species on chemical grounds.

1. Juglones present.. *C. sappan*
1. Juglones absent:
    2. Aucubin compounds present..*C. coriaria*
    2. Aucubin compounds absent:
        3. Catechol tannins present:
            4. Methylene dioxy compounds present.........*C. pulcherrima* (red)
            4. Methylene dioxy compounds absent.....*C. pulcherrima* (Yellow)
        3. Catechol tannins absent:
            5. Leucoanthocyanins present:
                6. Anthraquinones present............................ *C. gledeschiodes*
                6. Anthraquinones absent:
                    7. Tannins present... *C. cacalace*
                    7. Tannins absent.. *C. sepiaria*
            5. Leucoanthocyanins absent:
                8. Syringaldehyde doubtful................................. *C. bonducella*
                8. Syringaldehyde absent:
                    9. Steroids present.. *C. digyna*
                    9. Steroids absent:
                        10. Tannins present..................................... *C. ferrae*
                        10. Tannins absent.................................. *C. tortuosa*

Radhakrishnaiah *et al.* (1982) on phytochemical grounds supported the separation of *Trapa* from Onagraceae. The phytochemical studies by Radhakrishnaiah *et al.* (1984) lend support to the close kinship among the genera *Sparganium, Typha* and *Pandanus* and their retention under one Englerian taxon Pandanales and negate their segregation. Anuradha *et al.* (1988) studied the chemotaxonomy of Capparaceae. Their study does not lend support the separation of Cleomaceae from Capparidae. The chemical data given by Rajendra Prasad *et al.* (1986) do not warrant the creation of Mimosaceae as an independent family.

Examples from Semantides

Serology and Taxonomy

Boyden (1964) defined Serology as that portion of Biology which is concerned with the nature and interactions of antigenic material and antibodies. The science dealing with immunochemical reactions between, serum antibodies and antigens (Klaus, 1971) has provided valuable taxonomic information because its techniques help to detect homologous proteins. Boyden (1967) and Fairbrothers (1968, 1977) have discussed the importance of this type of study in plant systematics. A protein extract of plant or animal origin (antigen) is injected into the blood stream of an experimental animal, usually a rabbit, to form antibodies. A specific antibody is produced in response to a specific antigen. The serum (termed the antiserum) is then made to react *in vitro* with the antigenic proteins as well as those of other taxa of which the affinities are in question. The amount of precipitation indicates the degree of protein homology and hence is taken as a phylogenetic marker and taxonomic character. Kowarski (1901), Bertarelli (1902) and Magnus (1908) were the first notable serologists. They compared proteins from various grass and legume species showing similarities and differences.

Serological studies using crude plant protein extracts have been widely used in estimating phylogenetic relationships and elucidating the taxonomy of a wide variety of taxa. Thus a serological comparison of major seed proteins from Angiosperms has indicated a close relationship among the Magnoliidae, Hamamelididae and Corniflorae (sensu Dahlgren, 1980) and has refuted the idea of their independent evolution. It has also confirmed the homogeneity of the iridioid — producing Corniflorae and support the inclusion of the Gentinaceae in it (Jensen and Greven, 1984).

Based on serological studies the genus *Liriodendron* had been found to be quite distinct from other members of the family Magnoliaceae, and the genera *Magnolia* and *Michelia* displayed closest affinity within the family (Johnston and Fairbrothers, 1965). Similar studies have also been useful in estimating the relationship of the Nymphaeaceae and Nelumbonaceae (Simon, 1971), generic kinship in the Caprifoliaceae (Hillbrand and Fairbrothers, 1969, 1970) and in the Rubiaceae and related groups (Lee and Fairbrothers, 1978). In the Umbelliferae the findings support the classification of the family into Hydrocotyloideae, Saniculoideae and Apioideae and suggest the Apioideae is more closely related to Saniculoideae than the Hydrocotyloideae (Pickering and Fairbrothers, 1970). Chrispeels and Gartner (1978) found serological evidence for assigning *Phaseolus aureus* and *P.mungo* to the genus *Vigna*. The classification of the Ranunculaceae at tribal and generic levels indicated by serological relationships strongly supported that based on classical evidence, particularly from the chromosomes (Jensen, 1968, 1974). At lower levels Smith (1972) has studied the relationship of annual brome-grasses (*Bromus*) relating immuno- electrophoretic data to those from morphology, cytology and cytogenetics. He was able to throw new light on species problems, diploid-tetraploid relationships and hybrids, and the serological and cytological distinctness of a previously named variety of *Bromus secalinus* led him to recognize it as a new species *B. pseudosecalinus*.

DNA and RNA Hybidization

It has so far not proved possible to sequence DNA or the larger RNA molecules in a routine way and, for this reason nucleic acid chracteristics are inferred by indirect means such as DNA hybridization. In brief, DNA extracted from one organism is treated to convert it to a single stranded polynucleotide chain, and the amount of reassociation (annealing) with similarly treated DNA from another taxon which occurs on mixing the two is taken as a measure of similarity of the nucleotide sequences. Bendich and Botton (1966) discussed the relationships among the Leguminosae as measured by the DNA-Agar technique. In quantitative species comparisons based on DNA hybridization to and for *Secale, Hordeum* and *Triticum* showed, that, of these Cereals, wheat (75%) is more closely related to rye (100%) than is barley (75%).

RNA has been investigated by the equivalent technique of DNA-RNA hybridization, whereby the amount of association occurring between DNA of on organism and a fraction of the RNA (usually ribosomal RNA) of another is taken as a measure of relationship. Mabry (1976) investigated plants of the Centrospermae (Caryophyllales) by this method, and concluded that Caryophyllaceae (which contain anthocyanins instead of betalins) are quite close to the betalin-containing families, but not as close as the latter are to each other.

Amino Acid Sequencing

Amino acid sequencing is a more modern development, aiming to identify pure proteins down to the atomic level. It is now possible to break off the amino acids from the polypeptide chain one by one, identify each in turn chromatographically, and so build up the complete sequence of amino acids step by step. A first step is to break the total polypeptide chain into smaller peptide fragments, each of which is then sequenced separately. In the past these small peptide units were separated by chromatography and/or electrophoresis (finger- printing), and such relatively crude results themselves proved of taxonomic value. Amino-acid sequencing investigates the variation in the precise sequence of amino-acids in a single homologous protein (i.e. one presumable of monophyletic origin) throughout a range of organisms. This relies upon the fact that a particular protein does not have a single invariable structure, but a good relationships strongly supported that based on classical evidence, particularly from the chromosomes (Jensen, 1968, 1974). At lower levels Smith (1972) has studied the relationship of annual brome-grasses (*Bromus*) relating immuno- electrophoretic data to those from morphology, cytology and cytogenetics. He was able to throw new light on species problems, diploid-tetraploid relationships and hybrids, and the serological and cytological distinctness of a previously named variety of *Bromus secalinus* led him to recognize it as a new species *B. pseudosecalinus*.

Amino Acid Sequencing

Aminoacid sequencing is a more modern development, aiming to identify pure proteins down to the atomic level. It is now possible to break off the amino acids from the polypeptide chain one by one, identify each in turn chromatographically, and so build up the complete sequence of amino acids step by step. A first step is to break the total polypeptide chain into smaller peptide fragments, each of which is then sequenced separately. In the past these small peptide units were separated by chromatography and/or electrophoresis (finger- printing), and such relatively crude results themselves proved of taxonomic value. Amino-acid sequencing investigates the variation in the precise sequence of amino-acids in a single homologous protein (i.e. one presumable of monophyletic origin) throughout a range of organisms. This relies upon the fact that a particular protein does not have a single invariable structure, but a good proportion of it may vary without altering its essential functions. In the case of cytochrome C for example (the molecule most used for sequencing purpose), about 79 out of approximately 113 amino-acids vary from species to species, but alteration of even one of the other 34 destroys the functioning of the molecule. Cytochrome C is an ideal molecule as it is relatively small and stable, is coloured and is ubiquitous is aerobic organisms. So far its sequences in at least 25 species of vascular plants have been determined besides others in algae, animals, fungi and bacteria.

In general, the number of differences is closely parallel to the distance apart of the organisms in traditional classifications, suggesting that the method is broadly reliable (Boulter, 1974a, b, Boulter *et al.*, 1972).

Certain amino acids have very restricted distribution and this feature has been fruitfully exploited to characterize certain subfamilies and genera. Canavanine has been observed only in the taxa of the subfamily Lotoideae of Fabaceae (Turner and Harborne, 1967). Lathyrine has been found to be restricted only to the genus *Lathyrus* (Bell, 1971).

Electrophoresis

Electrophoresis is one the most extensively used techniques in protein investigations. The underlying principle is quite simple, proteins, when subjected to an electrical field in a solution of suitable pH, will migrate because of the presence of ionisable molecules on their surface. Under identical conditions the rate of migration of each protein is constant and hence can be reliable character for the detection of homologous proteins. Separation of proteins can be done either on paper (paper electrophoresis) or on a gel medium (gel electrophoresis). The importance of electrophoretic evidence in plant systematics has been discussed at length by Gottlieb (1977). This has been employed in estimating the parental affinities of the interspecific hybrid *Galeopsis tetrahit* (Houts and Hillebrand, 1976). Natarella and Sink (1975) studied the peroxidases and proteins of four species of the genus *Petunia* (viz. *P.axillaris*. *P. inflata*, *P. violacea* and *P. parodii*) and some 11 cultivars of *P.hybrida* by this method. Electrophoretic patterns revealed that all the species are closely related and that *P. inflata*, as well as *P. axillaris* and *P. violacea* is involved in *P. hybrida*. Crawford (1985) has demonstrated that electrophoretic data can be a useful supplement to other data used to infer modes of speciation among plants.

The most valuable electrophoretic results have perhaps been obtained in the cereals related to wheat, in relation to the genomic constitution and ancestry of the tetraploids and hexaploids. For example, Johnson (1972), Johnson and Hall (1965) working on storage proteins, concluded that the hexaploid bread-wheat (*Triticum aestivum*) did indeed contain a sum of the proteins possessed by the diploid species, which had been postulated on morphological and cytological evidence to be ancestral to it.

Numerical Taxonomy

Plants are classified based on their characters. Taxonomists classified plants giving particular reliance to one character, for example one cotyledon (Monocotyledons), two cotyledons (Dicotyledons) etc. A classification with emphasis on one character is called as *a priori* reasoning.

Michel Adonson (1763) for the first time pointed out that equal weightage should be given for all the characters while classifying plants. These principles are called Adansonian principles. The Adansonian principles have received great support since 1960s and have developed new methods in taxonomy included under a general term Numerical taxonomy.

Numerical Taxonomy may be defined as numerical evaluation of the apparent (phenetic) similarity among the groups of organisms and ordering them into higher ranking taxa (Heywood, 1967). Numerical taxonomy is empirical and aims to develop objective methods. They can be tested through repeated experiments for evaluation of taxonomic relationships and in the erection of taxa in hierarchial system of taxonomy. Besides, numerical methods have provided clues in phylogenetic analysis for evaluation of evolutionary rates more accurately.

The branch of botany concerned with numerical approaches towards taxonomy has been variously termed as Numerical Systematics (W.H.Wagner), Mathematical taxonomy (Jardine and Sibson, 1971), Taxometrics (Mayor, 1966), Taximetrics (Roger, 1963) and Numerical Taxonomy (Sneath and Sokal, 1973).

In Numerical Taxonomy first of all individuals are selected and there after their characters are spotted out. On the basis of character analysis the resemblance amongst the individuals are established. These stages are often worked out with the help of computers. The accuracy in the estimation of resemblance between taxonomic groups depends on the appropriateness in character. Choice of character requires a careful approach while working on these lines. The selection of character may be made from either existing literature or original observations from the specimen itself. So far as the character number is concerned there is no limitation but larger the number better is the approach for generalization of the taxa. It is widely regarded that a number of 80-100 characters is desirable in numerical taxonomic studies.

There are two aspects of Numerical taxonomy — the construction of taxonomic groups and their discrimination.

Construction of Taxonomic Groups

For designing natural system of classification, two major bases have been recommended: firstly phenetic and secondly phylogenetic (cladistic).

Phenetic classification is the similarity (relationship) of individual under consideration based on a set of phenotypic (characters related to the appearances) characters. This phenotypic relationship expressed in terms of similarity does not embody any relationship through ancestry.

Cladistic relationship is expressed in terms of correlation amongst individuals with regard to their evolutionary history. The relationship means common ancestry. The word cladistic means study of evolutionary sequence and pathway followed by individuals and the origin of branches from evolutionary tree. The evolutionary tree with anastomosing branches leading to various taxa are termed as cladograms.

Operational Taxonomic Units (OTUs)

The basic unit in Numerical taxonomy is known as operational taxonomic unit (OTU). It can be an individual, species, genus, family, order or class. When the OTU is supra-individual (above the level of an individual), there should be adequate representation of various polymorphic forms i.e. when genera are compared they should be represented by different species, when families are compared, they should be represented by different genera and so on. It should always be borne in mind that comparison of OTUs of equal rank is made in Numerical taxonomy.

Unit Characters

The characters employed in Numerical taxonomy are known as Unit characters. A unit character is that one which cannot be sub-divided logically. There are two types of unit characters.

(a) Those unit characters which exist in two states. They are called binary or two state characters. This is the simplest form of coding where characters are divided into + and - or as 1 and 0. The positive characters are recorded as + or 1 and negative characters as - or 0. In case the organ possessing a character is missing in the organism, the character must be scored NC, which means No comparison. The example for two state character is the presence or absence of spines.

(b) Those unit characters which exist in more than two states are called multistate characters. The multistate characters may be (i) qualitative such as the colour of the flower, the colour of the flower could be in any number of states —white, red, yellow, violet, blue etc. or (ii) quantitative such as the length of the leaf, it could be in any number of states — 1cm, 2cm, 3cm, 4cm etc. The multistate characters can be converted into two state characters like flowers white versus coloured, leaf long versus short. It may be noted that character is an attribute that exists in abstract form, while the character state is an expression of the character that exists in concrete form. The phyllotaxy is a character and alternate, opposite or whorled forms are character states. There is another term characteristic associated with the above two terms. When a character-state is exclusive to a particular taxon/ taxa, it is called characteristic.

The proper selection of characters is a critical point in the application of numerical taxonomy. Certain characters are clearly disqualified for numerical taxonomy and these are listed by Sokal and Sneath (1963) as inadmissible characters. According to these authors it is undesirable to use.

(i) attributes which are not a reflection of the genotypes of the organisms themselves.

(ii) any property which is a logical consequence of another, either partly or wholly; and

(iii) characters which do not vary within the entire sample of organisms.

(iv) To be qualified as an admissible character:

(v) It should be inherent in the organism and should neither be susceptible to environmental changes, nor affected by experimental uncertainties,

(vi) It must be of some diagnostic value,

(vii) It cannot be sub-divided,

(viii) The character that is selected must show variation among the taxa under comparison.

Measurement of Resemblances

There are three methods of estimating phenetic resemblance between the taxonomic groups, namely

(i) coefficients of association,

(ii) coefficients of correlation and

(iii) measurement of taxonomic distance.

Cluster Analysis

Different OTUs are grouped together on the basis of degree of similarity. These groups of OTUs are termed clusters.

There are several techniques to describe structure in matrices of similarity coefficients. One of the common techniques is the differential shading of the similarity matrix. In this method, similarity coefficient are grouped into five or ten evenly spaced classes. Each of these classes are represented by different degrees of shading in the squares of half matrix. The highest value is generally shown darkest and the lowest value the lightest as in Fig. 10.10. Then the half matrix can be seen as a pattern to different shades, limited by diagonal of squares with the darkest shade (Fig. 10.10). On rearrangement of the sequence of OTUs, clusters can be more sharply defined as in Fig 10.10 C.

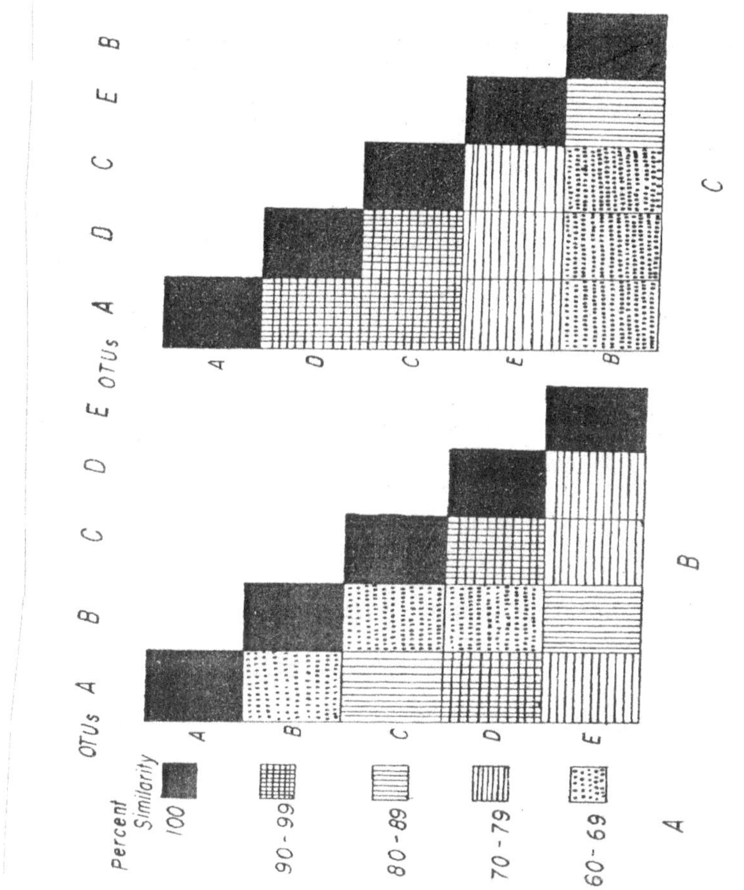

Fig. 10.10: Shaded similarity matrices. A. Percentage smilarity. B. OTUs arranged. C. After rearrangement of OTUs.

The groups of similar organisms organized in this manner are termed phenons. The clusters of phenons are then rearranged in a dendrogram, which summarizes the main features of cluster analysis.

The second way of clustering is by dendrograms. In this case the mutually most similar taxa are paired. The pairs are successively joined by the average similarity. The process is continued till all units have been joined together. It results in a tree like dendrogram with taxa at the tips of branches. The horizontal (phenon line) intersecting the vertical lines of the dendrogram gives number of clusters.

The delimitation of phenons is done by drawing a horizontal line across the dendrogram (Fig. 10.11) at a similarly value. A line at 75% for example creates five 75-phenons 1; 7; 3, 5, 6; 4, 9, 10, and 2, 8; while that at 80% creates six 80-phenons. Such a dendrogram will have a reference to a given taxon and cannot be transferred to any other study. In the above dendrogram, if the OTUs 1 to 10 had been species, an 80-phenon line could indicate 6 subgenera and a 65-phenon line two genera. It should however be remembered that phenons are arbitrary and relative groups.

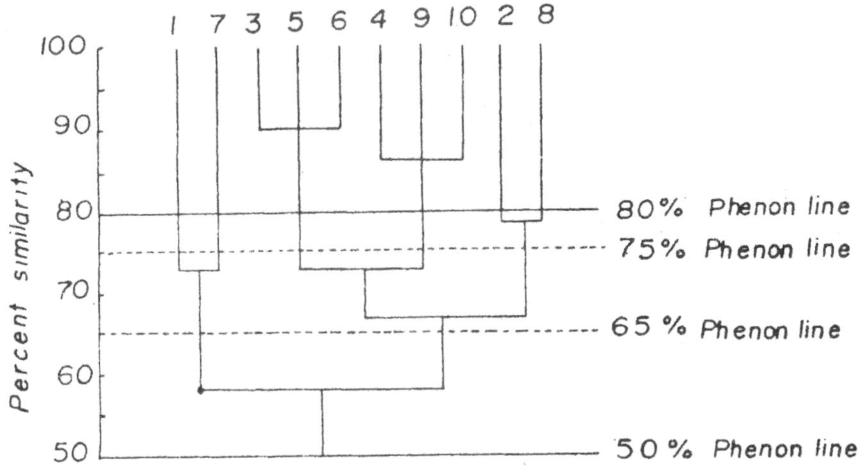

Fig. 10.11: Dendrogram to show formation of phenons.

Discrimination

If taxonomic groups chosen for the study show overlapping of characters, discrimination should be used to select them. Various techniques, such as discriminant analysis, have been devised for such purposes. The best methods for delimiting taxa are based on the utilization of maximum number of characters with similar weightage given to them.

Nomenclature and Numerical Taxonomy

Modern nomenclature does not concern itself with the problems of delimitation of taxa. It serves only as a reference point to the taxonomic names. The limits are debatable subjective and forever changeable. Numerical taxonomy, on the other hand, is very useful in delimitation of taxa by exact estimation of affinities (although phenetic). Thus, there is no scope for personal opinion or decision of taxonomists. The limits may be objective, utilitarian, permanent and fixed by common consent.

References

Batygina, T.B. and M.S. Yakovlev. 1980-1990. *Comparative embryology of Flowering plants*. 5 Vols. St. Petersberg.

Davis, G.L. 1966. *Systematic Embrology of Angiosperms*. John Wiley & Sons. New York.

Erdtman, G. 1952. *Pollen Morpholgy and plant Taxonomy of Angiosperms*. Almqvist & Wicksell, Stockholm.

Herr, J.M. Jr. 1984. Embryology and Taxonomy. In: B.M. Johri (ed.) *Embryology of Angiosperms*. Springer Verlag. Berlin. pp. 647-696.

Johri, B.M. (ed.). 1984. *Embryology of Angiosperms*. Springer Verlag, New York.

Johri, B.M., K.B. Ambegaokar and P.S. Srivastava. 1992.*Comparative Embryology ofAngiosperms*. 2 Vols. Springer-Verlag, Berlin.

Kapil, R.N. and A.K. Bhatnagar. 1991. Embryological evidence in angiosperm classification and phylogemy. *Bot. Jahrb. Syst.* **113**:309-338.

Radhakrishaiah, M. 1996. *Essentials of Plant Taxonomy*. Hyderabad.

Sneath, P.H.A. and R.R Sokal. 1973. *Numerical Taxonomy*. W.H. Freeman & Co., San Francisco.

Sokal, R.R. and P.H.A. Sneath. 1963. *Principles of Numerical Taxonomy*. W.H.Freeman & Co., San Fracisco.

11

Molecular Plant Taxonomy

Taxonomy is the biological discipline producing the classification of the diversity found in the biological domain. Its purpose is to classify the structure of the living world by constructing hierarchies consisting of progressively more general groupings of organisms. Taxonomists can identify most organisms with which its morphological features. The modern science introduced several novel techniques for identifying and classifying the organisms based on minute or molecular characters by the application of newly invented machines and techniques. More recently the evolutionary history, or phylogeny, of a group of organisms, has come to play a major role in identifying the relationships between organisms, and the reconstruction of this phylogenetic history is a main function of taxonomy, the study of biological diversity. The accumulation, over the last few decades, of vast quantities of molecular data from an ever-increasing range of plant, animal, and bacterial species is having a huge impact on phylogenetic analysis. Molecular data come in various forms, but DNA sequence data now account for effectively all new information. Molecular techniques in the field of biology have helped to establish genetic relationship between the members of different taxonomic categories. The classification of plants are on the basis of distribution of chemical substances in them such as DNA, RNA and proteins. Molecular data have comparatively significant differences from traditional morphological data. In molecular taxonomy, more recent technique is DNA barcoding is a novel system designed to provide rapid, accurate, and automated species identifications by using short, standardized gene regions as internal tags. The genomes of living organisms are analogues to bar-codes.

Objectives of Molecular Taxonomy

Through the molecular data to reconstruct the correct genealogical ties among biological entities, estimate the time of divergence between biological origin, chronicle the sequence of events along evolutionary lineages and simplify the classification and identification of biological diversity. The study of evolutionary relationships among biological entities (individuals, populations, species or higher taxa) is being pursued by using a combination of molecular data (such as DNA, RNA and protein sequences) with statistical techniques. Recognize, describe and label the diversity of plants with automated electronic system is an essential task of taxonomists.

Advantages of Molecular Data

Molecular entities are strictly heritable. The description of molecular character is unambiguous. There is some regularity to the evolution of molecular traits. The molecular data are amenable to quantitative treatment. Homology assessment is easier than with morphological traits. The data are abundant and robust to evolutionary distance. The process is less time consuming.

Type of Molecular Markers

The molecular markers targets different regions of the genomes either at coding or non-coding loci. Technically molecular markers are divided into five technical generations such as first generation markers (Polymerase chain reaction, RFLP), second generation markers (Inter Simple Sequence Repeats, RAP-DNA), third generation markers (Amplified Fragment Length Polymorphism, Triple RAPD by using three primers), New generation markers (Recursive Genome Function) and modern Generation markers (Nanopore Genome sequence). Molecular markers can also be characterized based on the functions such as Type I and Type II markers. Type I markers are associated with genes of known function. Type II markers are associated with genes of unknown function. Microsatillites and other neutral markers are Type II markers unless they are associated with genes of some known function. Allozyme markers are Type I markers as the proteins they encode are associated with some functions. Mitochondrial DNA markers are maternally inherited with a haploid genome. The entire genome undergoes transcription as one single subunit. They are not subjected to recombination and thus they are homologous markers.

Molecular Sequences for Plant Taxonomy

The use of micromolecules and DNA sequences to identify organisms has been proposed a more efficient approach than traditional taxonomic practices. It has been recently demonstrated on a large scale through the use of a short DNA sequence of the mitochondrial cytochrome oxidase subunit 1 (Cox1), a chloroplast gene such as matK (maturase K) and a nuclear gene such as ITS (Internal Transcribed Spacer). In flowering plants another approach has been put forward. The use of several plastid loci do discriminate between species level, for example the trnH-psbA intergenic spacer and some more typical phylogenetic markers such as rbcL and trnL-F. The nuclear internal transcribed spacer region and the plastid tranH-psbA intergenic spacer are potentially usable DNA region for applying barcoding identification and phylogenetic studies in flowering plants.

The plastid locus most commonly sequenced by plant systematists for phylogenetic purposes is rbcL, followed by trnL-F intergenic spacer, matK, ndhF and atpB-rbcL. Though, it has generally been used to determine evolutionary relationships at the genetic level and above. Besides rbcL and atpB all of the later plastid loci have been used at the species level with various degrees of success. A total of 17 barcode regions such as matK, rbcL, ITS, ITS2, psbA-trnH, atpF-atpH, ycf5, psbK-1, psbH-trnD, rps16, Cox1, nad1, trnL-f, rpoB, rpoC1, atpF-atpH and rps16 are frequently used in various groups of plants for their identity in different levels.

Basic Steps in Molecular Taxonomy

Molecular taxonomy is a new method for quick identification of any species based on extracting DNA sequence from tiny tissue sample of any organism, is now being applied to taxa across the tree of life. The molecular systematics entails two basic steps: 1. Building the DNA barcode library of known species. 2. Matching the barcode sequence of the unknown sample against the barcode library for identification.

Molecular Methods Available

The fast and complete array of techniques used for analysis of DNA, RNA and protein can be applied to the classification of angiosperms. The type of molecular methods used to measure genetic distances in plants will vary depending upon the magnitude of the genetic differences being assessed. RAPD analysis may be useful for distinguishing different genotypes within a pant cultivar while sequence analysis of the ribosomal genes may allow species or higher taxonomic level. The three most common markers are protein based, hybridization based and PCR based markers. Each marker has its own advantages and disadvantages over the other in different levels of the taxonomic hierarchy as individual, variety, species, genus, family and so on. The following methods are usually used for recognizing taxonomic relationship of plant taxa.

1. Restricted Fragment Length Polymorphism (RFLP).
2. Variable number of Tandem Repeats (VNTRs).
3. PCR based Molecular Markers.
4. Multiple Arbitrary Amplicon Profiling (MAAP).
5. Random Amplified Polymorphic DNA (RAPD).
6. Arbitrary Primed PCR (AP-PCR).
7. DNA Amplified Fingerprinting (DAF).
8. Amplified Fragment Length Polymorphism (AFLP).
9. Sequence Tagged Sites (STS).
10. Sequence Tagged Microsatellites (STMS).
11. Single Primer Amplification Reactions (SPAR).
12. Inter Simple Sequence Repeats (ISSR).

13. Sequence Characterized Amplified Regions (SCAR).
14. Cleaved Amplified Polymorphic Sequences (CAPS).

Gene Sequencing

Sequencing of DNA for getting genes, partial genes or non-coding regions is becoming more and more common and is now widely used in molecular systematics. Sequencing determines the precise order of DNA nucleotides – adenine (A), cytosine (C), guanine (G) and thymine (T) along the stretch of DNA. The common difficulty of sequencing has always been getting enough DNA from the samples to work. It was done initially by cloning genes into bacteria and allowing bacteria to replicate the genes along with their own genomes. This method is quite slow but it is reliable and avoid some of the possible artifacts of more efficient methods. This rapid method has allowed systematists to study the same region in many species of a particular group. One disadvantage of PCR is that the polymerase itself introduces occasional errors, which could in theory affect an estimate of phylogeny, particularly if the sequences being compared are extremely similar.

The first method for determining DNA sequences involved a location-specific primer extension strategy established by chemical degradation method (Gilbert and Maxam, 1973). They reported the sequence of 24 base pairs using a method known as wandering-spot analysis. Advancements in sequencing were aided by the concurrent development of recombinant DNA technology, allowing DNA samples to be isolated from sources other than viruses. Later whole genome sequencing technology developed early 1990s. Several new methods for DNA sequencing were developed in the mid to late 1990s and were implemented in commercial DNA sequencers by the year 2000. Together these were called the "next-generation" or "second-generation" sequencing methods. Recently evolved other methods for sequencing of DNA are de novo sequencing, shotgun sequencing and Bridge PCR. Apart from them rapid methods established for sequencing techniques such as Massively parallel signature sequencing (MPSS), Polony sequencing, 454 pyrosequencing, Illumina (Solexa) sequencing, SOLiD sequencing, Ion Torrent semiconductor sequencing, DNA nanoball sequencing, Heliscope single molecule sequencing, Single molecule real time (SMRT) sequencing, Nanopore DNA sequencing and Tunnelling currents DNA sequencing.

Phylogenetic Analysis

According to modern evolutionary theory, all organisms on earth have descended from a common ancestor, which means that any set of species, extant or extinct, is related. This relationship is called a phylogeny, and is represented by phylogenetic trees, which graphically represent the evolutionary history related to the species of interest (Figure 1). Phylogenetics infers trees from observations about existing organisms using morphological, physiological, and molecular characteristics. The "tree of life" represents a phylogeny of all organisms, living and extinct. Other, more specialized species and molecular phylogenies are used to support comparative studies, test biogeographic hypotheses, evaluate mode and timing of speciation, infer amino acid sequence of extinct proteins which are used to track the evolution.

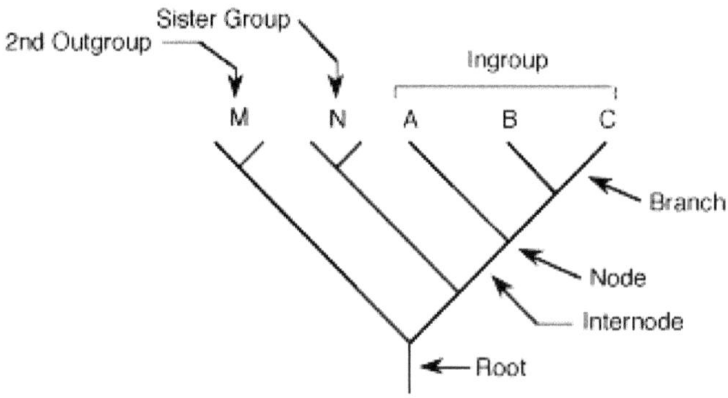

Figure 11.1: Typical phylogenetic tree and its parts

Phylogenetic Trees

Before exploring statistical and bioinformatic methods for estimating phylogenetic trees from molecular data, it's important to have a basic familiarity of the terms and elements common to these types of trees. Phylogenetic trees are composed of branches, also known as edges, that connect and terminate at nodes. Branches and nodes can be internal or external (terminal). The terminal nodes at the tips of trees represent operational taxonomic units (OTUs). OTUs correspond to the molecular sequences or taxa (species) from which the tree was inferred. Internal nodes represent the last common ancestor (LCA) to all nodes that arise from that point. Trees can be made of a single gene from many taxa (a species tree) or multi-gene families (gene trees).

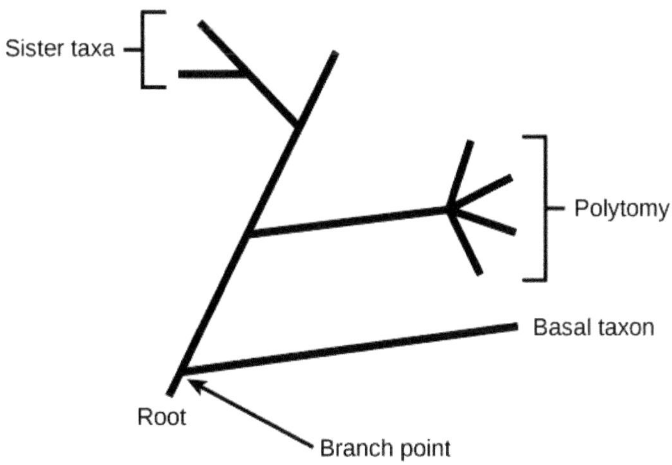

Figure 11.2: Parts of phylogenetic tree

A tree is considered to be "rooted" if there is a particular node or outgroup (an external point of reference) from which all OTUs in the tree arises. The root is the oldest point in the tree and the common ancestor of all taxa in the analysis. In the absence of a known outgroup, the root can be placed in the middle of the tree or a rootless tree may be generated.

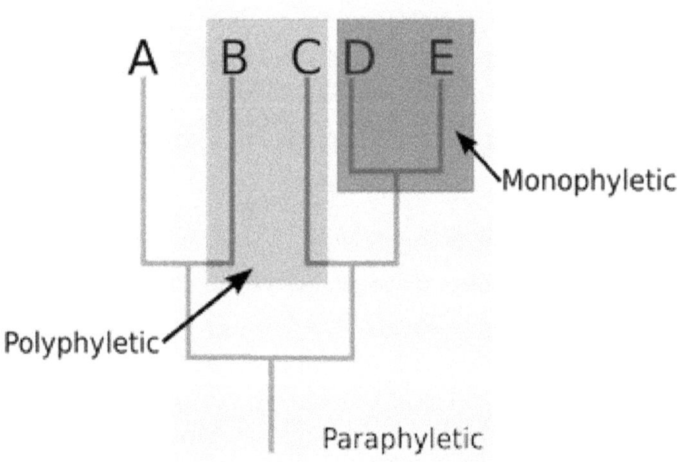

Figure 11.3: Grouping clades of phylogenetic tree

Monophyletic group consists of an internal LCA node and all OTUs arising from it. All members within the group are derived from a common ancestor and have inherited a set of unique common traits. A paraphyletic group excludes some of its descendents and a polyphyletic group can be a collection of distantly related OTUs that are associated by a similar characteristic or phenotype, but are not directly descended from a common ancestor.

Molecular Trees and Homology

Evolution is shaped by homology, which refers to any similarity due to common ancestry. Similarly, phylogenetic trees are defined by homologous relationships. Paralogs are homologous sequences separated by a gene duplication event. Orthologs are homologous sequences separated by a speciation event (when one species diverges into two). Homologs can be either paralogs or orthologs.

Molecular phylogenetic trees are drawn so that branch length corresponds to amount of evolution (the percent difference in molecular sequences) between nodes. Paralogs are created by gene duplication events. Once a gene has been duplicated, all subsequent species in the phylogeny will inherit both copies of the gene, creating orthologs. Interestingly, evolutionary divergence of different species may result in many variations of a protein, all with similar structures and functions, but with very different amino acid sequences. Phylogenetic studies can trace the origin of such proteins to an ancestral protein family or gene.

Estimation of Molecular and Phylogenetic Trees

Molecular phylogenetic trees are generated from character datasets that provides evolutionary content and context. Character data may consist of biomolecular sequence alignments of DNA, RNA, or amino acids, molecular markers, such as single nucleotide polymorphisms (SNPs) or restriction fragment length polymorphisms (RFLPs), morphology data, or information on gene order and content. Evolution is modeled as a process that changes the state of a character, such as the type of nucleotide (AGTC) at a specific location in a DNA sequence; each character is a function that maps a set of taxa to distinct states. Phylogenetic trees and their relationships among the tested taxa are determined by sequential procedure by computer software. Although the nature and scope of phylogenetic studies may vary significantly and require different datasets and computational methods, the basic steps in any phylogenetic analysis remain the same: assemble and align a dataset, build (estimate) phylogenetic trees from sequences using computational methods and stochastic models, and statistically test and assess the estimated trees.

Steps in Phylogenetic Analysis

Assemble and Align dataset: The first step is to identify a protein or DNA sequence of interest and assemble a dataset consisting of other related sequences. There are a number of free, online tools available to simplify and streamline this process. DNA sequences of interest can be retrieved using NCBI BLAST or similar search tools. There are many websites and software programs, such as ClustalW, MSA, MAFFT, and T-Coffee, designed to perform multiple sequence on a given set of molecular data. ClustalW is currently the most mature and most widely used.

Building phylogenetic trees: To build phylogenetic trees, statistical methods are applied to determine the tree topology and calculate the branch lengths that best describe the phylogenetic relationships of the aligned sequences in a dataset. Many different methods for building trees exist and no single method performs well for all types of trees and datasets. The most common computational methods applied include distance-matrix methods, and discrete data methods, such as maximum parsimony and maximum likelihood. There are several software packages, such as Paup, PAML, PHYLIP, that apply most popular methods. PAML (Phylogenetic Analysis by Maximum Likelihood) is open-access set of programs for phylogenetic analysis and evolutionary model comparison. PHYLIP is another large suite of open-access program s for phylogenetic inference that estimates trees using numerous methods, including pairwise distance, maximum parsimony, and maximum likelihood. The maximum likelihood programs can handle a few simple stochastic models and have good tree searching capabilities. PHYLIP is generally considered good educational software for novice phylogeneticists.

Distance-Matrix method: Distance matrix methods compute a matrix of pair wise "distances" between sequences that approximate evolutionary distance. Distance-based methods tend to be in polynomial time and are quite fast in practice. These methods use clustering techniques to compute evolutionary distances, such as the number of nucleotide or amino acid substitutions between sequences, for all pairs of taxa.

UPGMA method: UPGMA produces rooted trees for which the edge lengths can be viewed as times measured by a molecular clock with a constant rate. This method uses a sequential clustering algorithm to identify two OTUs that are most similar (meaning they have the shortest evolutionary distance and are most similar in sequence) and treat them as a single new composite OTU. This process is repeated iteratively until only two OTUs remain.

Discrete data methods: Discrete data methods examine each column of a multiple sequence alignment dataset separately and search for the tree that best represents all this information. Although distance-based methods tend to be much faster than discrete data methods, they typically yield little information beyond the basic tree structure. Discrete data analyses, on the other hand, are information rich. These methods produce a separate tree for each column in the alignment, so it is possible to trace the evolution for specific elements within a given sequence, such as catalytic sites or regulatory regions.

Maximum Parismony: Among the most widely used tree-estimation techniques, maximum parsimony applies a set of algorithms to search for the tree that requires the minimum number of evolutionary changes observed among the OTUs in the study.

Maximum likelihood: The maximum likelihood method requires a probabalistic model of evolution for estimating nucleotide substitution. This method evaluates competing hypotheses (trees and parameters) by selecting those with the highest likelihood, meaning those that render the observed data most plausible. The likelihood of a hypothesis is defined as the probability of the data given that hypothesis. In phylogeny reconstruction, the hypotheses are the evolutionary tree (its topology and branch lengths) and any other parameters of the evolutionary model.

Stochastic Models of Evolution: Evolutionary changes in molecular sequences result from mutations, some of which occur by chance, others by natural selection. Rates of change can also differ among OTUs, depending on several factors ranging from GC content to genome size. To accurately estimate phylogenetic trees, assumptions must be made about the substitution process and those assumptions must be stated in the form of a stochastic evolutionary model. A few popular stochastic models of evolution include the single parameter Jukes-Cantor (JC) method, Kimura 2-parameter (K2P), Hasegawa-Kishino-Yano (HKY), and Equal-Input. Some software programs, such as Paup, will automatically use a default mode l for the tree estimation method chosen. The JC method is the easiest one to comprehend, because it assumes that if a site changes its state, it changes with equal probability to the other states.

Bootstrap analysis: A bootstrap is a statistical method for assessing trees that takes its name from the fact that it can "pull itself up by its bootstraps" and generate meaningful statistical distributions from almost nothing. Using bootstrap analysis, distributions that would otherwise be difficult to calculate exactly are estimated by repeated creation and analysis of artificial datasets. In a Non-parametric bootstrap is artificial datasets generated by resembling from original data. In a parametric bootstrap, data is simulated according to hypothesis tested. The objective of any bootstrap analysis is to test whether the whole dataset supports the tree.

Case Studies for Molecular Taxonomy

The generic names of *Caralluma* (*sensu lato*) are still confusion among botanists. The phylogenetic study reported with help of molecular systematics, genus *Caralluma* groupings among 13 species and 4 varieties (total of 17 taxa) and to provide taxonomic revisions for the Carallumas occuring in India. Nuclear ITS, chloroplast *trnL* intron and *trnL-F* intergenic spacers were used and resolved that there are two separate genetic lineages of *Caralluma s.l.* within India based on ITS sequences, "*Caralluma*" clade and *Boucerosia* clade. This investigation supported Plowes view (Plowes, 1995) who classified the Carallumas on two groups, Group A - stems slender, tapering apically, inflorescences racemose, leaves usually lanceolate. Group B - stems robust, not tapering, inflorescence terminally umbellate or laterally displaced by further stem-growth, leaves broad, ovate-acute. The "*Caralluma*" clade falls in Group A and "*Boucerosia*" clade falls in Group B according to Plowes' revision. According to this molecular data Indian *Caralluma* segregated in to two subgenera *Caralluma* and *Boucerosia* (Karuppusamy *et al.*, 2013).

The Euphorbiaceae (*sensu lato*) are distributed mainly in the tropics, in various types of vegetation and habitats, being one of the largest, most complex and diverse families of angiosperms. It has recently been divided into four families, according to modern classification systems based on molecular phylogeny: Euphorbiaceae (*sensu stricto*), Phyllanthaceae, Putranjivaceae, and Picrodendraceae (APG IV). There is a proposition to establish Peraceae still under discussion. There were also changes in the taxonomic position of genera widely distributed such as *Amanoa, Drypetes, Pera, Phyllanthus, Podocalyx, Pogonophora*, and *Richeria*, among others (Recardo de S. Secco *et al.*, 2012). In addition, many new species have been proposed and the limits of taxa distribution are expanding worldwide. Thus, phylogenetists provide an overview of recent studies and advances in the taxonomy of Euphorbiaceae *s.l.* concentrating on review works and regional floras, as well as the changes that resulted in setting a new taxonomic family.

Applications of Plant Molecular Taxonomy and Barcoding

The plant DNA barcoding offers taxonomists the opportunity to greatly expand, and eventually complete, a global inventory of life's diversity. DNA barcoding and plant identification are of great utility to users of taxonomy. It provides more rapid progress than the traditional taxonomic practices. The plant DNA barcoding allows taxonomists to rapidly sort specimens by highlighting divergent taxa that may represent new species. The advocates of DNA barcoding says that it have revitalize biological collections and speed up species identification and inventories, however the opponents argue that it will destroy traditional systematic and turn it into a service industry. By harnessing technological advances in electronic and genetic, DNA barcoding will help many people quickly and cheaply recognize known species and retrieve information about them, and will speed discovery of thousands of species yet to be named.

Recent Advances

Global DNA barcoding efforts have resulted in the formation of the consortium for the Barcode of Life (BOL). In January 2013, the Barcode of Life Database (BOLD) contained more than 2.7 million specimen records, with 2 million having barcodes belonging to over 1,70,000 species. Several small databases containing groups like Fungal Databases, Algal Databases, Genome Databases for Rosaceae, etc are also existing.

Limitations

DNA-based species identification depends on distinguishing intra-specific from inter-specific genetic variation. The ranges of these types of variation are unknown and may differ between taxa. It seems difficult to resolve recently diverged species or new species that have arisen through hybridization. There is no universal gene for DNA barcoding, no single gene that is conserved in all domains of life and exhibits enough sequence divergence for species discrimination. The validity of DNA barcoding therefore depends on establishing reference sequences from taxonomically confirmed specimens. This is likely to be a complex process that will involve cooperation among a diverse group of scientists and institutions. Barcode sequences are, in general, short (approx. 500–1000 bp) and this fundamentally limits their utility in resolving deep branches in phylogenies. Some controversy exists over the value of DNA barcoding, largely because of the perception that this new identification method would diminish rather than enhance traditional morphology-based taxonomy, and species determinations based solely on the genetic divergence could result in incorrect species recognition. However, we must keep open the possibility that the barcode sequences per se and their ever increasing taxonomic coverage could become an unprecedented resource for taxonomy, systematic biology and diagnosis, and may be equally useful. As the advantages and limitations of barcoding become apparent, it is clear that taxonomic approaches integrating DNA sequencing, morphology and ecological studies will achieve maximum efficiency at species identification. DNA barcoding may help speed the work of taxonomists and others interested in species identification. The evidence from a number of studies largely confirms the feasibility of such a system. Despite some drawbacks of using DNA barcoding, the reported success of using the barcoding region in distinguishing species from a range of taxa and to reveal cryptic species is remarkable. Efforts should therefore be made to develop nuclear barcodes to complement the barcoding regions that are currently in use.

Future Perspectives of Molecular Taxonomy

The future perspective of DNA barcoding and molecular approach will be to provide a bio-literacy tool for the general public and it will also help in opening to the treasury of biological knowledge, which is currently underused partly because of the weak taxonomic expertise for species identification. DNA barcoding will also relieve the enormous burden of species identification by taxonomists, so they can focus on more pertinent to discovering and describing the new species. DNA barcoding can be linked to aerial photography, in that it provides an efficient method for mapping the extant species though in sample space rather than physical space. Barcoding links biological identification to advancing frontiers in DNA sequencing, miniaturization in electronics, and computerized information storage. Integrating these links will lead to portable desktop devices and ultimately to hand-held barcoder. A hand-held barcoder such as the one envisioned here, will have many uses. Promoting technology development of portable devices for field uses will be a major goal of this initiative.

References

APG-IV. 2016. Angiosperm Phylogenic Group-IV 2016 An update of the Angiosperm Phylogeny Group classification for the orders and families of flowering plants. *Botanical Journal of the Linnean Society*, 181: 1–20.

Gilbert, W, and Maxam, A.M. 1973 The Nucleotide Sequence of the lac Operator. *Proceeding National Academy of Science*, U.S.A. 70 (12): 3581–4.

Karuppusamy, S., Ugraiah, A., Pullaiah, T. 2013. Caralluma (sensu lato) Antiobesity Plants. Astral Internationl Pvt. Ltd. New Delhi.

Plowes, D.C.H. 1995. A classification of Caralluma R.Brown (Stapeliae: Asclepiadaceae). Haseltonia 3: 49-70.

Recardo de S. Secco, Cordeiro, I. and de Senna-vale, L. 2012. An overview of recent taxonomic studies on Euphorbiaceae s.l. in Brazil. Rodriguésia vol.63.

12

Computer Applications in Plant Taxonomy

To most contemporary biologists, the concept of taxonomy conjures up visions of dusty museum cabinets filled with pressed plant parts, insects mounted on pins and bottles of formalin-fixed soft-bodied animals; gathered by naturalists in bygone times and serving the sole purpose of introducing natural science students to the wonders of science. Yet, taxonomy, is a vibrant field that continues not only to serve as the underpinnings of all modern biology, but also serves to reshape our thinking about the possible relationships of all living and extinct species. Now its possible to reach everyone because of modern biology depends on computational biology and computing all the data stored, displayed and manipulated in appropriate usage. Computers can be used for identification of plant specimens and construction of identification keys, parallels between specimen identification and diagnosis, plant disease diagnosis, information problems in paleontology, plant distribution mapping, biometrics and statistical analysis, probabilistic approaches to specimen identification, numerical taxonomy, the modeling of genetic and evolutionary systems, possible computerized nomenclatural aids, biodiversity informatics, biogeographical analysis, and aids to teaching systematic biology.

The Need for Algorithmic Approaches to Taxonomy

Systematic biology is still relying up on using of tools that would be familiar to Linnaeus. Computers are used mainly to create databases of characters relevant to relatively few taxa or, more recently, to try and encompass biodiversity by building and distributing large lists of taxon names. The compilation of these lists has made it clear that, because algorithmic approaches to classification, and even identification, have been frequently rejected by practitioners of taxonomy, the number of taxa to be described far exceeds the existing capacity for describing new taxa. Estimates of described eukaryotic taxa are in the range from 1.7 million species to 3.5 million. However, these estimates must be viewed with caution as there is no universal concept of either a biological species or higher taxon. Taxon description is rendered more complicated by the need for qualified practitioners to do routine identifications of the already described species and by the need to identify the same taxa under different names (synonymies). Computer applications could be reduced this gap considerably. For example, described taxa could be identified using universally applicable methods based either on morphology or on molecular sequences (barcode methods) by using pattern recognition software. Such algorithms could readily detect and provide a preliminary filter for taxa that conform to previously described taxa, leaving experts free to concentrate on the real issues of circumscribing and describing new taxa. At present, more than 25% of the validly published names applied to eukaryotes are synonyms and the incidence of synonymy in higher plant names is expected to be considerably higher, but much more difficult to ascertain as the relevant codes of nomenclature do not provide for a centralized control of names. The recent computer database technologies would seem well-adapted to dealing with these issues, providing that a good model for such occurrences either exists or can be constructed and that transitive closure and semantic resolution of such events can be readily accomplished in a consistent manner.

Algorithms designed to automate classification and identification requires data and these tend to be distributed piecemeal across the scientific community. Furthermore, most of the current databases are generally not interoperable. Tools that integrate data handling and data analysis have become crucial to the successful practice of taxonomy. In this sense, many of the problems facing systematic biology are informatics problems, such as developing flexible and comprehensive data models, providing methods for dealing with constantly changing nomenclatures and ensuring database interoperability. Automated systems for identifying and classifying the higher plants or animals appear to be following two trajectories: one is based on digital image analysis and the second on molecular sequence comparison. Sequence-based efforts, like Barcodes of Life (http://barcoding.si.edu/index_detail.htm), are still in incomplete state, so it might be instructive to examine the experience of systematic biologists in developing integrated identification and classification systems.

Electronic Resources for Taxonomy

Electronic resources for taxonomy may be divided into two classes. The first one provides information about taxa, including taxonomic information that the user already knows the identity features of the taxon. The information found here could be used to identify an unknown (it may be images or textual descriptions), but no algorithms are supplied. In the second class, we place those resources that enable an investigator to identify an unknown (which would allow it to be placed into a reference classification) or to create a classification that includes the unknown and extant taxa.

Database Resources for Taxonomy

In order to explore biological diversity and arrive at a definitive picture of 'what is out there', researchers have recognized authoritative and non-redundant list of previously described species. The authentically identified comprehensive list of taxa has led to the establishment of several databases and data portals. The simple establishment of what are often described as lists of species has proven, perhaps, to be a much more daunting task than was initially foreseen. These sites do not generally provide real classification and identification services. Rather, they serve up lists of names and provide some rudimentary information about the covered taxa. Neither portals nor databases, in our definition, supply tools to aid in the classification of organisms, that is, they are static. A few of the more prominent services are listed in Table 12.1. They provide access to information about organisms, including a taxonomy, geographic distribution, where the voucher specimens exist, databases containing information relevant to the organism, etc. [e.g. Global Biodiversity Information Facility (GBIF)]. Some (e.g. Digital Taxonomy) also link to information about data standards, software and other basic information. Databases are similar to portals, but are more focused on specific taxa (e.g. AlgaeBase) or on a certain aspect of systematics (e.g. The International Plant Names Index).

Table 12.1: A list of online resources for plant taxonomy

Database and Portals	Online id
Angiosperm Phylogeny Website	http://www.mobot.org/MOBOT/Research/APweb/welcome.html
The Plant List	www.theplantlist.org/
SPECIES 2000	http://www.usa.sp2000.org/
Global Biodiversity Information Facility (GBIF)	http://www.gbif.org/
ITIS—Integrated Taxonomic Information System	http://www.itis.usda.gov/
The International Plant Names Index (IPNI)	http://www.ipni.org/
Index Fungorum	http://www.indexfungorum.org/
Kew Scientific Research and Data	https://www.kew.org/science/data-and-resources/data
AlgaeBase	http://www.algaebase.org/
The Universal Biological Indexer and Organizer (uBio)	http://www.ubio.org/
The NCBI Taxonomy database	http://www.ncbi.nlm.nih.gov/Taxonomy/taxonomyhome.html/
Biodiversity and Biological Collections Web Server	http://www.biocollections.org/
Digital Taxonomy	http://digitaltaxonomy.infobio.net/?Home
The Tree of Life	http://www.tolweb.org
Indian Biodiversity Portal (IBP)	indiabiodiversity.org/

Web Tools

The computerization of taxonomical data is leading to the widespread application of taxon names that are superfluous, ambiguous or incorrect, resulting in mismatched records and inflated species numbers. The ultimate consequences of misspelled names and bad taxonomy are erroneous scientific conclusions and faulty policy decisions. The lack of tools for correcting this 'names problem' has become a fundamental obstacle to integrating disparate data sources and advancing the progress of biodiversity science. However, biologists try to develop appropriate web tools for establishing taxonomic usage. Some example for the web tools of taxonomy, Taxonomicon - http://sn2000.taxonomy.nl/Taxonomicon/Default.aspx and Phylomatic- http://www.phylodiversity.net/phylomatic/

Identification and Classification

We divide the software applications that follow according to whether they provide identifications or classifications. Computers began to be used in taxonomy soon after they became more generally available in the 1950s and 60s and were quickly embraced for both purposes. Many of the uses envisioned for computers at that time have only recently become feasible. It seems obvious that, in principle, any software that can identify an organism could be extended to classify organisms as well, but few applications go beyond the identification, based on the previously established criteria.

Identification: Phenotype-based

DELTA—DEscription Language for TAxonomy (http://delta-intkey.com/). These tools for creating classifications and identification keys have been very popular, but development of DELTA was discontinued in 2000. Designed as a single-user tool, it was felt that DELTA would not be able to meet the demands of the collaborative, multi-user efforts that are seen as the future of systematics. An open source version, Free DELTA is actively under development and a proposed replacement, BioLink (http://www.biolink.csiro.au/), is available. Linnaeus II (http://www.eti.uva.nl/products/linnaeus.php) was developed by ETI Bioinformatics to support the development of taxonomic databases and identification keys by researchers. There are automated identification applications available for certain taxa. For example, Digital Automated Identification SYstem (DAISY) (http://chasseur.usc.edu/pups/projects/daisy.html) was initially developed for insect identification, and the Automated Bee Identification System (ABIS). DAISY is now being developed as a universal identification system, but ABIS is not available. Both use machine learning approaches to pattern recognition, training the software to recognize taxa based on digital images. There are drawbacks to the pattern recognition approach. Although the approach approximates the morphometric methods used by humans to identify taxa, computers are less flexible in the way images are interpreted and consequently are more likely to fail to arrive at the same identification. The way the software is trained is critical because a poorly chosen training set can lead to incorrect identifications.

Molecular Sequence-based

Recently, the Consortium for the Barcodes of Life has proposed an alternative to classical identification system that is DNA barcoding. It is relying on the sequencing of one or a few universally distributed genes to identify the organism. DNA barcoding has been proposed as a rapid, inexpensive and generally applicable way of identifying a broad range of organisms. A concerted effort is now underway to establish the necessary databases, tools and information management systems needed to make this approach successful (http://barcoding.si.edu/index_detail.htm). The Ribosomal Database Project (RDP; http://rdp.cme.msu.edu/index.jsp) has been collecting ribosomal RNA sequence data for about 12 years and supplies tools for sequence comparison, phylogenetic analysis and probe matching, among others. Molecular systematic of higher plants are highly conserved with chloroplast gene *rbcL*, mitochondrial gene *matK* and ribosomal gene ITS sequences, now its expanding to *ndhF*, *rpl16* and *atpB* for verifying the phylogeny of recent identification. The database is currently available at http://bfw.ac.at/200/1859.html and at http://bfw.ac.at/rz/bfwcms.web?dok=4977.

DNA barcoding, which relies on the use of a standardized DNA region for accurate and rapid species identification, has been used more and more by ecologists. Since the last decade the international initiative CBOL (Consortium for the Barcode of Life, http://barcoding.si.edu) aims in global standards for DNA barcoding. But, in plants the situation is controversial and many strategies have been proposed. The mitochondrial cytochrome c oxidase subunit I gene (COI), which serves as the standard barcode for animals, is not suitable for species identification in plants, due to low levels of variability. Previous studies on DNA-based plant identification were primarily focusing on the plastid genome but there is a lack of consensus regarding the most universal, informative and technically practical DNA region(s). The suitability of a molecular marker strongly relies on the questions to be answered.

Classification: Phenotype-based

Recently molecular sequence-based analysis has come to dominate most discussions of classification though a strong case can be made for the continuing importance of some morphological features in classification. There are a few widely used software packages that can be employed to analyse phenotypic data using a cladistic/phylogenetic approach. These include PAUP, PHYLIP, MacClade and Mesquite. The phenetic approach has fallen from favour, there are at least two software packages that allow researchers to take this approach to morphological data, even though general statistics packages like S-Plus (Insightful, Inc., Seattle, WA) offer similar functionality. These packages are NTSYSpc (Numerical Taxonomy System 2.2) which takes a phenetic approach, and can be used to discover pattern and structure in multivariate data, and SYN-TAX, another suite of software for multivariate data analysis.

Molecular Sequence-based

Now-a-days, most classifications are produced using cladistic/phylogenetic analyses of molecular sequence data. Most of the software packages that are explicitly intended to integrate phylogenetic analysis and classification. The data used in these analyses are readily available from the major sequence databases in GenBank, EMBL and DDBJ. All three of these databases use the National Institute for Biotechnology Informatics (NCBI) Taxonomy for annotating their sequence data. This taxonomy is regularly updated by NCBI, with up to 100 new species added each day. Corrections based on internal research and outside information are also made, but NCBI makes no claims for the authority or accuracy of the taxonomy and classification. Nevertheless, this classification and taxonomy provides a convenient way to start exploring various taxonomic groups and to assemble data sets for phylogenetic analysis. All three databases provide tools for browsing the classification and taxonomy (http://www.ncbi.nlm.nih.gov/Taxonomy/taxonomyhome.html/index.cgi), TAXY (http://www.ebi.ac.uk/taxy/index.html) and TXSearch (http://sakura.ddbj.nig.ac.jp/uniTax.html). The NCBI taxonomy is being widely adopted by databases such as PIR (http://pir.georgetown.edu/home.shtml) and BIND (http://www.bind.ca/Action) that use the taxon ID as a permanent identifier. This could, however, be problematic as NCBI does not regard these identifiers as permanent and indicates that taxon IDs can disappear as a result of various taxonomic acts. Another one is PhyloCode (http://www.ohiou.edu/phylocode/) a set of rules or procedures for producing taxon names that are consistent with the phylogeny. An emerging trend in classification that is worth-mentioning is the use of whole genome phylogenies.

Computers in Numerical Taxonomy

The application of the concepts of numerical taxonomy was made possible only through the use of computers, because of the heavy load of routine calculations. However, the principles can easily be illustrated in hand-worked examples. In addition, two problems had to be solved: the first was to decide how to weight different variables or characters; the second was to analyze similarities so as to reveal the taxonomic structure of groups, species, or clusters. For handling the large quantities of taxonomic data by a computer so as to yield the outline of a classification based on equally weighted characters. This enables similarity to be expressed numerically and would allow taxonomic rank to be measured in terms of it.

Geographic Information System (GIS) in Plant Taxonomy

The latest technique used as a tool in Biosystematics is Remote sensing and the use of computer. Remote sensing is carried out by satellites to get an idea of the particular phytogeographical area. It was first used in 19th century by using camera. But now the technology is well developed and pictures are provided by the satellites. This system gives the information about an area. Computer and software is required for interpreting results. A Geographic Information System (GIS) captures, stores, analyzes, manages, and presents data, which is linked to a location. It includes mapping software and its application with remote sensing, land surveying, aerial photography mathematics, photogrammetry, geography, and tools that can be implemented with GIS Software. The technology is used for scientific investigation, resource management archaeology, environment impact assessment, cartography, criminology and topography etc. The technology is also used in taxonomy for mapping the taxa and their distribution in the geographical ranges. It provides the concentration of vegetation of the area, identity of vegetation which is observed by aerial survey include identification and distribution of individual species and their population structure. GIS helps in investigation of an area and allow the taxonomists to improve management of the resources.

Digital Taxonomy

Designing a convenient and automatic recognition system of plants is necessary and useful since it can facilitate fast classifying plants, and understanding and managing them. Beginning of the 21st century has started to develop the digital technologies in all counterparts of sciences including plant taxonomy. Digitization is useful to storage of large data, retrieve and reproduce, transform the data wherever necessary, management, manipulation and magnification. At present most of the popular herbaria manage their deposited specimens in digital forms in computers. It makes easy to view the stake holders and make use of research and learning purposes. The digital forms of herbaria is now called as virtual herbarium, it concerns a collection of high resolution digital images of preserved plants or plant parts. Additional information about each specimen, such as the location, the collector, and the botanical name, synonyms available, even colour photographs, attributed literature are attached to every specimen. Virtual herbaria are established in part to increase the longevity of specimens. Major herbaria participate in international loan programs, where a researcher can request specimens to be shipped in for study. This shipping and transport contributes to the wear and tear of specimens. If, however, digital images are available, images of the specimens can be sent electronically or through internet. These images may be a sufficient substitute for the specimens themselves, or alternatively, the researcher can use the images to "preview" the specimens, to which ones should be sent out for further study. This process cuts down on the shipping, and thus the wear and tear of the specimens, as well as the wait times associated with transport. Major online virtual herbaria in the world are given in Table 12. 2.

Table 12.2: Popular virtual herbaria of the world.

Name of the Herbarium	Web link
Kew Herbarium Catelogue, Kew	http://apps.kew.org/herbcat/gotoSearchPage.do
Tropicos (Missouri Botanical Garden)	http://www.tropicos.org/
Australias Virtual Herbarium	http://avh.ala.org.au
Botanic Garden Meise Virtual Herbarium	http://www.br.fgov.be/RESEARCH/COLLECTIONS/HERBARIUM/
University of Connecticut Virtual Herbarium	http://bgbaseserver.eeb.uconn.edu/
Virtual Herbarium of the Botanic Garden and Botanical Museum Berlin-Dahlem	http://ww2.bgbm.org/herbarium/

Name of the Herbarium	Web link
The Virtual Herbarium at the New York Botanical Garden	http://sweetgum.nybg.org/science/vh/
Reflora Virtual Herbarium, Brazil	http://floradobrasil.jbrj.br/reflora
Moscow Digital Herbarium	https://plant.depo.msu.ru/

Digital Floras or *e*Floras

One of the primary challenges for the creation of digital libraries is to enhance the value of paper-based publications by providing digital access to the materials. The time-tested internal structure of paper oriented documents and multivolume collections have evolved to address specific information needs. These structures can be more fully exploited in the process of digital conversion of the material while some structural aspects of paper-based publishing are helpful, if difficult, to bring to the electronic medium. Recently the Global Taxonomic Initiative encourage the projects to create electronic floras for better understanding and managing the world's heritage. Now the digital floras are called eFloras and it is motivated by several national and international agencies. Many countries have already been completed the eFlora projects which are readily available and accessible online to the public. For example, the Flora of North America, eventually included 31 volumes and over 18,000 species. Six hundred authors may contribute this project and many other flora projects are of comparable size. The Flora of China is projected to include 25 volumes and 30,000 species and the Flora of Australia 59 volumes and 26,000 species. Information science needs to provide computer-supported collaborative environments to support these efforts. The eFloras facilitates the easy accessible for identification, taxonomy and ecological studies. With the growth of the internet it is natural that many of the digital floras available over the World Wide Web as well as on paper to provide improved access.

Table 12.3: eFloras available online.

Name of the Flora	Web link
Flora of North America	http://www.efloras.org
Flora of China	www. flora.huh.harvard.edu/china/
Flora of Pakistan	www.efloras.org/flora_page.aspx?flora
Flora of India	http://efloraindia.nic.in/efloraindia/eFloraHomePage.action
Flora of Australia	www.anbg.gov.au/abrs/online-resources/flora/
Flora of Tropical East Africa	https://plants.jstor.org/collection/FTEA
Flora of Europe	www.morphle.com/indexflora.html
A catalogue of plants of Madagascar	www.efloras.org/flora_page.aspx?flora
Flora of Karnataka	www.florakarnataka.ces.iisc.ac.in
All old Floras available in electronic form	https://www.biodiversitylibrary.org/

Biodiversity Informatics

The application of informatics techniques to biodiversity information for improved management, presentation, discovery, exploration and analysis is called Biodiversity informatics. It is mainly built up on a foundation of taxonomic, biogeographic, or ecological information stored in digital form with the application of modern computer techniques. This technique can yield a new ways to view and analyse existing information, as well as predictive models for information that does not yet exist (diversity prediction modeling). Biodiversity informatics is a relatively modern discipline but has hundreds of practitioners worldwide, including the numerous individuals involved with the design and construction of taxonomic and biogeographic databases. The term "Biodiversity Informatics" is generally used in the broad sense to apply to computerized handling of any biodiversity information; the somewhat broader term "bioinformatics" is often used synonymously with the computerized handling of data in the specialized area of molecular biology. One major issue for biodiversity informatics at a global scale is the current absence of a complete master list of currently recognised species of the world, although this is an aim of the Catalogue of Life project which has ca. 1.65 million species of an estimated 1.9 million described species in its 2016 Annual Checklist. A similar effort for fossil taxa, the Paleobiology Database documents some 100,000 plus names for fossil species, out of an unknown total number.

Table 12.4: Popular Biodiversity informatics projects of the world.

Biodiversity informatics projects	Web link
Global Biodiversity Information Facility (GBIF)	https://en.wikipedia.org/wiki/Global_Biodiversity_Information_Facility
Global names	http://globalnames.org/
Encyclopedia of Life	http://www.eol.org
Map of Life	https://www.mol.org/
Universal Biological Indexer and Organizer	http://www.ubio.org/
Index Nominum Genericorum	http://botany.si.edu/ing/
International Plant Name Index	http://www.ipni.org

In modern days many individuals and organizations developed multi-access computer keys for identification of plants through digital images of plants and herbarium sheets in different aspects like trees, medicinal plants and food plants etc. FRLH- Herbarium and Raw Drug Repository is a nationally recognised herbarium developed the online digital library for medicinal plants about 1709 high resolution scanned herbarium sheet images (www.envis.frlht.org/digital_herbarium.php). These are all the achievements in taxonomy without the help of computer, in future still the field is waiting for global level challenges.

13

Biosystematics

The term 'Biosystematics', coined for the first time by Camp and Gilly (1951), as Biosystematy' which, with passage of time, has changed to Biosytematic. The central theme of this term is systematics with the help of cytology and genetics, help in checking the limits of taxonomic and systematic categories.

The main purpose of this branch of study is to delimit evolutionary units and devise newer taxa and check the limits of those that are already existing. The main objective is to work out the operative processes of speciation rather than the speciation itself.

At a particular time interval, phenotype is the final results of action, reaction and co-action between and among the factors of the environmental complex on one hand and genotypic milieau on the other. Biosystematics emphasizes on this relationship and evaluate the phenotypic expression, the result of the variation in the factors of the environmental complex and genotypic compound.

Biosystematics is a phase of botanical research that endeavors, by study of living populations, to delimit the natural biotic units, and to classify them objectively as taxa of different orders of magnitude. This necessitates use of data from the fields of ecology, genetics, cytology, morphology, phytogeography and physiology, particularly as observed from plants grown under artificial and natural conditions of environment. The systematist strives to determine objectively whether a taxon (or a population) belongs in the category of genus, species, subspecies etc. In this approach emphasis is placed on cytogenetics and cytotaxonomy supplemented by the classical approaches of morphology, ecology, and phytogeography.

A first step in biosystematic investigations is a thorough sampling of the taxon (it may or may not be a species) and its populations and the cytological study of the chromosomes of many populations within geographic races, species, genera and so on. Differences in chromosome number, their morphology and behaviour at meiosis usually indicate genetic differences of taxonomic significance.

A second step includes the determination of the ability of the different population to hybridize, and a study of the vigour and fertility of the hybrids. This disclosed the presence or absence of breeding barriers between groups and is of taxonomic importance as indicating the natural limits of the taxa of various levels or order.

A third step studies the homologies of the chromosomes in the hybrids, as determined at meiosis. These are important indicators of the degree of genetic relationship in the material.

Information obtained from these 3 steps is compared with the data obtained from comparative morphology and geographical distribution. The resultant classification of the taxa to which it is applied (within the category of genus and taxa of lower level) has an increased objectivity over one obtained through a consideration of morphology and distribution alone but may not in all instances be an acceptable substitute for the classification resulting from the synthesis of data obtained from all sources.

Biosystematic Categories

The major objective of biosystematic studies is to arrive at a better understanding of the natural relationships of plants, particularly those of the rank of genus and below. These categories are not intended as substitutes for the units used in classical or practical taxonomy, and they are necessarily the equivalent of these, although they may be counterparts of them. Each provides a single-word term for biosystematic situation, and in no case should the term be applied to a plant or a population unless the situation for which the term stands has been proved experimentally to exist for the particular taxon. The four most widely accepted categories of the biosystematics are, in order of ascending phyletic value, ecotype, ecospecies, coenospecies and comparium.

Ecotype

It is the basic unit in biosystematics. The term ecotype proposed first by Turesson (1922) was well defined by Gregor et al. (1936) as "a population distinguished by morphological and physiological characters, most frequently of a quantitative nature; interfertile with other ecotypes of the escospecies, but prevented from freely exchanging genes by ecological barriers".

Ecospecies

The ecospecies was defined first by Turesson "as a group of plants comprised of one or more ecotype, within the coenospecies, whose members are able to interchange their genes without detriment to the offspring". Related ecospecies are usually separated by incomplete genetic barriers, which, in addition to ecological barriers, are adequate to preclude free interchange of genes with any other ecospecies. When ecospecies of one coenospecies are crossed, the resultant hybrids are either partially sterile, or if fertile, they produce many weaklings in the F_2 generation (as slow growing dwarfs, individuals highly susceptible to disease against which parents enjoyed immunity, and teratological misfits). Such weaklings are unable to compete, and fail to reproduce. A few such hybrid segregates may possess sufficient vigor to survive. These may be reabsorbed by interbreeding into one or the other parental ecospecies. Related ecospecies generally inhabit different but often contiguous ecological or geographical areas, thus retaining a relative genetic purity. In general, the ecospecies approximates the conventional and conservative taxonomic species.

Coenospecies

The coenospecies is a group of plants representing one or more ecospecies "of common evolutionary origin, so far as morphological, cytological and experimental facts indicate". Coenospecies of the same comparium are separated by genetic barriers so nearly absolute that all the hybrids between them are sterile unless amphiploidy (amphidiploidy) occurs. For this reason, distinct coenospecies may exist in a single environment without genetic intermixing. Rather often the coenospecies parallel the taxonomic sections or subsections of the genus.

Comparium

The comparium is the biosystematic unit that often is comparable to the genus. It is composed of one or more coenospecies that are able to intercross. Distinct comparia are unable to intercross and complete genetic incompatibility prevails between them. There are, however numerous taxa, accepted by orthodox systematist as genera, that may contain two or more comparia (e.g., in the Fabaceae). In some families the accepted and conventional genera are not equivalents of comparia or even of coenospecies (e.g., some Crassulaceae, Orchidaceae, Brassicaceae).

Two separate schemes have been devised by Turesson (1922) and Danser (1929). In Turesson's system there are three ranks (Coenospecies, ecospecies, ecotype). In Danser's system the definitions of the terms are such that none of the six is synonymous with any other (Comparium, Commiscuum,, Convivium). Danser's lowest term convivium covers the definition of both ecospecies and ecotype while his higher terms Comparium and Commiscuum define the higher and lower aspects of coenospecies. In addition, Turesson employed a lower term still, the ecophene, which denotes the ecological variant, purely the product of environmental modification of the phenotype. F.E. Clement used the term ecad for such variation.

Deme Terminology

Gilmour and Gregor (1939) proposed a new system of terminology designed to provide an infinitely flexible series of categories which could be used to define any group of individuals on the basis of any set of criteria. This system, the deme terminology is at least in its original concept, non-hierarchial and it falls outside the scope of formal taxonomic categories (genus, species etc.). For that reason it avoids the use of root-words such as 'species' and 'type' which are associated with the latter. Central to the idea is the use of a natural root, deme, which implies nothing in itself except a group of related individuals of a particular taxon. The precise meaning of the terms are provided by various prefixes of which only three were originally proposed.

Topodeme : a deme occurring within a specified geographical area.

Ecodeme : a deme occurring within a specified kind of habitat.

Gamodeme : a deme composed of individuals which interbreed in nature.

Later Gilmour and Heslop-Harrison (1954) expanded these suggestions and provided seven further basic terms.

Phenodeme : a deme differing from others phenotypically.

Plastodeme : a deme differing from others phenotypically but not genotypically

Genodeme : a deme differing from others genotypically.

Autodeme : a deme composed of predominantly self fertilizing (autogamous) individuals.

Endodeme : a deme composed of predominantly closely inbreeding (endogamous) but dioecious individuals.

Agamodeme : a deme composed of predominantly apomictic (non-sexually reproducing) individuals.

Clinodeme : a deme which together with other such demes forms gradual variational trend over a given area.

Cytodeme : a deme composed of individuals all with the same karyotype.

Gilmour and Heslop-Harrison (1954) also proposed some second order terms plastoecodeme, gamoecodeme, hologamedeme, coenogamodeme and synamodeme which presently are regarded as equivalent to ecophene or ecad, ecotype, ecospeices, coenospecies and comparium respectively.

Methods in Experimental Taxonomy

The techniques of experimental taxonomy comprise methods of testing to determine to which of the above-described biosystematic units a population belongs. The most reliable evidence and conclusions are obtained from application of all methods: those of orthodox taxonomy as well as those of cytology and genetics, combined with cultivation in uniform and in varied environments. Ordinarily the methods of testing in uniform and in contrast environment are applied simultaneously to a number of populations of the same and of different taxonomic rank within a genus.

1. Growth in uniform environment provides a means of studying variability in heredity. Population samples of the same taxonomic category (e.g., species, subspecies) are procured from different environments, and are grown in a common experimental plot or under controlled environmental conditions. This permits comparison of behaviour of plants of unlike heredity in the uniform environment, thereby distinguishing between hereditary variation and environmental modification.

2. Growth in varied environments, as afforded by a series of field stations or artificially controlled environments, is of equal importance as a means of observing the interplay between heredity and environment.
3. Cytogenetic analysis may involve the 3 following types of techniques.
   (a) Cytological studies are made of as many populations as practical to determine the possible correlation's between visible chromosome differences and differences in external morphology and geographic distribution within the species complex.
   (b) selected forms are crossed and the hybrid progeny grown to determine the fertility of F_1 generation, the fertility and vigor of F_2 generation, and to analyze the genotypes of related ecotypes and ecospecies.
   (c) The chromosomal homologies of these natural units are analyzed through studies of the chromosome pairing in their hybrids.
4. Crossing programmes to test the presence or absence of sterility barriers are a part of biosystematic study. The crossing test is applied to any taxa suspected of possessing these barriers, and in instances where the barrier is found not to be absolute the F_1 and F_2 generations must be studied statistically to determine their vigour and fertility. Such tests are conducted also with plants of populations presumed to be of close affinity to determine the genetic mechanism that may be allowing them to retain distinctness. The crossing tests are applied also to geographical or ecological extremes to determine if such extremes are separated by genetic barriers.

14

Botanical Survey of India

Introduction

The rich plant resources of the vast Indian region has attracted the attention of many early explorers, traders and writers. Even before the Swedish Naturalist Carolus Linnaeus provided some kind of organized approach to classification and naming of plants in 1753, Garcia d' Orta in 1565 and Van Rheede during 1678-1703 has written on Indian plant wealth. During the 18th and 19th centuries there was great interest about Indian economic plant products in the west and middle-east. This, augmented with the enthusiasm of many naturalists in botanical exploration of unexplored regions of Asia, resulted in the establishment first of a large botanic garden near Calcutta and subsequently of Botanical Survey of India.

The Royal (now Indian) Botanic Garden at Sibpur was established in 1787, largely due to the interest of Col. Robert Kyd who was appointed as its first superintendent. He was succeeded by eminent botanists and naturalists like W. Roxburgh, F. Buchanan-Hamilton, N. Wallich, W. Griffith, T. Thomson, T. Anderson, C.B. Clarke, G. King, D. Prain, K. Biswas, D. Chatterjee and others. The garden was originally established for commercial and horticultural purposes, but is soon became a centre of taxonomic and phytogeographic research, and an agency for exchange of plant material of economic importance, between India and the farthest corners of the world.

The herbarium at Calcutta was started in late 18th century. In 1883, the collections were transferred to a new building.

Early History of the Botanical Survey of India

The Botanical Survey of India was established on 13th February 1890 in the Royal Botanical Garden, Howrah (Calcutta) and Sir G. King, the then superintendent of the Garden, was appointed its first ex-officio director. Incidentally, till then, five of the seven volumes of the *'Flora of British India'* by J.D. Hooker (1872- 1896) had been published and it had come to be realized that very little was known about many regions and plants of the floristically rich sub-continent.

Regional offices were established at Saharanpur with J.F. Duthie as Director for northern India, at Pune with T. Cooke as Director for western India and at Madras with M.A. Lawson as Director for southern India. In 1911, the office of the Reporter of economic products and the Industrial section of the Indian Museum at Calcutta were also brought under the Botanical Survey of India.

The Survey organized collections, established herbaria and published several volumes of the Records of the Botanical Survey of India, Annals of the Royal Botanic Gardens, Calcutta, provincial Floras for Bombay, Bengal, Upper Gangetic plains and Madras. However, in course of time due to lack of funds and manpower, the activities of the regional centres gradually shrunk and almost came to a stop by early 20th century. After retirement of C.C. Calder in 1939, even the post of Director of Calcutta remained in abeyance and the organization survived almost in name. The scientific personnel of the survey then comprised only the Curator, Industrial section and a systematic assistant in the herbarium at Sibpur.

Reorganization of the Botanical Survey of India

After independence, the Government decided to revive the Botanical Survey of India. In 1952, E.K. Janaki Ammal was appointed as Officer-on-special duty to reorganize it. The plan proposed by her was accepted by the Government of India and orders for the reorganization of the survey were passed on March 29, 1954.

The Central Botanical Laboratory was the first unit to be set up, with Janaki Ammal as the Director. During 1955-56 four regional centres were started broadly on phytogeographical basis at Dehra Dun, Shillong , Pune and Coimbatore. In 1957, the herbarium at Calcutta was re-designated as the Central National Herbarium. The old herbaria of Botanical Survey at Coimbatore and Pune and the forest herbarium of Assam at Shillong were transferred to the Botanical survey of India.

Consequent on shifting of the Central Botanical Laboratory from Allahabad to Calcutta in 1962, a new regional station for Central India was set up Allahabad. In 1963, the Indian Botanic Garden was transferred to the Botanical Survey of India. The Andaman and Nicobar circle at Port Blair and the Arid Zone circle at Jodhpur were established in 1972.

A field station at Itanagar in Arunachal Pradesh started working from 1977. Another circle, the Sikkim Himalayan circle was set up at Gangtok in December, 1979.

The main aims and objectives of the Botanical Survey of India are as follows:

1. Exploration of unexplored and under explored regions for new plant resources,
2. Preparation of National and Regional Floras,
3. Floristic and taxonomic studies,

4. Preservation of types and other authentic specimens by establishing herbaria and musea and exchange of plant specimens in India and aboard,
5. Maintaining germplasm of economic, rare and interesting plants in botanic gardens for study, conservation and education,
6. Study of phytogeography and ecological changes in flora and vegetation,
7. Study of threats to floras and habitats and conservation of ecosystem and
8. Advising Government on all matters relating to plant resources of the country.

Head quarters and Kolkata/Howrah Based Laboratories

The head quarters office of the survey at Kolkata plans, guides and co-ordinates the scientific and technical activities of all units in accordance with the policies of the Government and consultation with the senior scientists of the survey. It maintains liaison with other research institutions in the country like CSIR, ICAR, ICMR and Universities. One senior scientist of the survey works at Royal Botanic Gardens, Kew, U.K. to help Indian workers on Plant Taxonomy and Systematic Botany.

Many scientists are working on families and genera for new Flora of India of which Volumes 1-5, 12, 13 and 23 and 27 fascicles have already been published.

The publication section projects the work of the survey through its publications such as the Bulletin of the Botanical Survey of India (now renamed as NELUMBO in 58th volume), Records of Botanical Survey of India, Annual reports and News letters. It has also published District Floras, State Floras, books on ethnobotanical and Endangered plants.

The Indian Botanic Garden is also maintained by Botanical Survey of India. Spread on an area of 110 hectares it has nearly 12,000 living specimens. A detailed account about the garden is given elsewhere in this book.

Initiated in 1793, the Central National Herbarium today houses more than two million specimens; it includes about 15,000 type specimens and over 12,000 Wallichian specimens. Nearly 12,000 photographs of Indian and other authentic specimens form Kew Herbarium, British Museum and Linnean Herbarium, about 4000 drawings/illustrations and microfiches from 26 European herbaria have been procured for reference works. Collections in this herbarium formed the basis of, or contributed to, many monographic studies in the world.

The Central Botanical Laboratory is engaged in some biosystematic studies involving cytotaxonomy and palynology, anatomy, Plant physiology and ethnobotany. In the Ecology unit, studies have been undertaken in North and South Kanara, Coorg and Coastal regions. Some work has been done on leaf sclereids and ecoanatomy. Emphasis is now being laid on the study of Biosphere reserves and endemic and threatened taxa.

In Cryptogamic unit, studies have been undertaken of fern flora of South India, mosses of Himalayas, lichens of West Bengal and taxonomic revisions of some genera. The pharmacognosy unit has studied bark drugs and conducted a detailed survey of *Coptis teeta* in Eastern Himalayas. The later study helped in regulating the export trade in this important medicinal herb. Limited facilities of chemical screening of plants have been organized. The high solasodin content in *Solanum viarum* was found in these laboratories.

The Industrial section at the Indian Museum concerns itself mainly with the collection, preservation, proper maintenance and display of all plant products to illustrate the vegetable wealth and resources of India in relation to agriculture, forestry, medicine, commerce and industry. About 15,000 exhibits of commercial vegetable products attract visitors, students and researchers. The exhibits include such plant products as crude drugs, foods, fodder, dyes, tans, gums, resins, timbers, fibres, narcotics, oils etc. depicting the vegetable resources of the country and their modes of processing. They also form reference material for thousands of technical enquiries from drug dealers, exporters, researchers and excise and customs department, etc.

Regional Offices

Eastern Circle, Shillong

The Eastern circle at Shillong is entrusted with exploration in the states of Assam, Meghalaya, Mizoram, Tripura, Manipur and Nagaland. The Herbarium has about 1,15,000 specimens. The National Orchidarium at Shillong and Experimental Garden at Barapani have many rare and interesting orchids and other plants. The current major programme include study of orchids of Assam, Meghalaya and Nagaland, ecosystem in Namdapha area and Cryptogams of eastern India. About 5000 orchid specimens belonging to 350 species are under cultivation in the Orchidarium.

Northern Circle, Dehra Dun

The Northern circle includes the States of Jammu & Kashmir, Punjab, Haryana, Himachal Pradesh, Northern hilly parts of Uttaranchal and Delhi. The Herbarium contains about 60,000 specimens. It has an experimental garden at Pauri. The major programmes include Flora of Himachal Pradesh and ecological impact in some development projects, Valley of Flowers, Corbett National Park and Nanda Devi and Dachigan sanctuaries.

Southern Circle, Coimbatore

The Southern circle covers Tamilnadu and Kerala. The Herbarium has about 2,00,000 specimens. The major programmes include study of the Flora of Tamil Nadu, Idukki area, Kerala, Silent valley and Nallamalai hills. They also worked on Ethnobotany of Eastern Ghats and brought a book on that.

It has a National Orchidarium and Experimental Garden at Yercaud founded in the year 1963. This orchidarium is situated in the mixed subtropical and evergreen forests of Shevaroy hills and provide scope for studies on Orchid systematics, hybridization and culture through introduction and acclimatization. systematics, hybridization and culture through introduction and acclimatization.

Deccan Circle, Hyderabad

In the year 2006 Botanical Survey of India established Deccan Circle at Hyderabad covering Andhra Pradesh and Orissa.

Western Circle, Pune

The Western circle covers the states of Maharashtra, Karnataka and Goa and Laccadive and Minicoy Islands. The herbarium has about 1,25,000 specimens. The circle had developed a collaborative programme on chemistry and cultivation of Iphigenia, a source of Colchicine. The major programmes include study of grasses of Western India, revision *Ceropegia* and flora of certain regions. The circle has already published the Flora of Goa and Flora of Maharashtra. An experimental garden at Mudhiva near Pune is being developed.

Central Circle, Allahabad

The Central circle covers Madhya Pradesh and South Eastern parts of Uttar Pradesh. The herbarium has about, 50,000 specimens and experimental garden is attached with the office. The major programmes include study of some district floras and grasses and sedges of central India. The circle is currently engaged in Flora of Madhya Pradesh.

Andaman and Nicobar Circle, Port Blair

The Andaman and Nicobar circle is engaged in preparing a flora of these islands and in developing an experimental garden near Dhanikheri dam.

Arid Zone Circle, Jodhpur

This circle comprises the state of Rajasthan, certain adjoining arid districts of the states of Punjab, Haryana, Uttar Pradesh, Madhya Pradesh and Gujarat. The circle has published the Flora of Rajasthan.

Other circles like Sikkim Himalayan circle and Arunachal Pradesh circle are working on the flora of respective areas.

The reorganization of the Botanical Survey of India has stimulated the floristic, taxonomic and other botanical activities in India. During the last 45 years, the survey has organized collections of more than 3,00,000 field numbers and herbaria of Botanical Survey hold over 2 million plant specimens.

More than 150 new taxa have been described and several new records reported from India by the scientists of the survey.

Current Activities of Botanical Survey of India

1. Botanical Survey of India is currently engaged in Flora of India project. 27 fascicles have been published. Flora of India is being published in 34 volumes, of which 1, 2, 3, 4, 5, 12, 13 and 23 have already been published while the rest are under preparation.
2. The Survey is preparing state floras. Floras of the states Tamil Nadu, Kerala, Madhya Pradesh, Maharashtra, Goa, Rajasthan, West Bengal,

Bihar, Manipur, Mizoram, Tripura, Arunachal Pradesh, Himachal Pradesh, Jammu & Kashmir and Andaman and Nicobar Islands have been published while Floras of other states are underway.

3. Botanical Survey of India has financed about 70 District Flora projects to the Universities. Funds were given to university teachers for studying district floras. Many of these Floras have been published.
4. Scientists of Botanical Survey of India themselves are working on district floras. Some these have already been published. These include Flora of Tonk district (Rajasthan), Flora of Banswara district (Rajasthan), Flora of Pachmarh and Bori reserves etc., while some are in progress. They are also working on Flora of National parks. Many of these have been published.
5. Botanical Survey of India is currently engaged in the Ethnobotany of different tribes in different tribes in different regions of the country.
6. The Survey has been working on Endangered, Endemic and Rare plants. They have brought out several lists of threatened plants of India.

15

Selected Families of Angiosperms

Annonaceae (Custard apple family)

Distribution: This family comprises of 200 genera and 2,500 species distributed exclusively in the tropics. It is the most diverse family of primitive angiosperms. About 25 genera and 200 species have been reported from India. The dominant genera like *Annona* have 120 species and *Xylopia* by about 150 species world over.

Habit: These are usually aromatic shrubs or trees and some of them are scandent shrubs, straggling with the help of leaf-opposed hooks as in *Artabotrys*.

Vegetative Characters

Root system is of the normal tap root.

Shoot system is seen in two different types of growth as monopodial in *Polyalthia* and *Cananga*, while in *Artabotrys* as sympodial development due to the conversion of the terminal bud into an inflorescence. A rhizome covered by scale leaves is produced in *Annona rhizantha*.

The **leaves** are simple, alternate, distichous, exstipulate, petiolate with usually wavy margin, and reticulate venation. Oil passages occur in the parenchyma of leaf and sometimes in the tissues of young stem and flowers.

Floral Characters

The **inflorescence** is axillary or terminal. Sometimes terminal inflorescence pushed to leaf-opposed position by the development of the axillary shoots. It may bear a solitary flower or a few flowers in fascicles or umbels. Cauliflory is also seen in *Polyalthia fragrans*.

The **flowers** are bisexual, but are polygamous in *Popowia* and dioecious in *Stelechocarpus*. They are hypogynous, actinomorphic (zygomorphic in *Monodora*) and hemicyclic. The torus is usually convex (*Artabotrys*), conical (*Annona*), hemispherical (*Alphonsea*), flat (*Xylopia*), or concave (*Unona*). Stamens and carpels are arranged in a spiral manner on the torus, whereas the perianth lobes are in regular whorls.

The **perianth** is usually in 2 or 3 whorls with 3 members in each whorl but dimerous in *Disepalum*. The perianth is usually sepaloid but can be distinguished into sepals and petals on the basis of size and position.

The **calyx** consists of 3 sepals which are distinct or connate at the base. Aestivation is usually valvate.

The **corolla** consists of 6 petals in 2 whorls of 3 each, but the inner whorl is suppressed in many species of *Annona*. The petals are free, sepaloid, thick, sometimes with a concavity at the base. Aestivation is valvate but slightly imbricate in *Uvaria*.

The **androecium** consists of numerous (only six in *Orophea zeylanica*) stamens which are spirally arranged on an elongated torus. The filaments are short and thick with dithecous, dorsifixed, extrorse anthers that dehisce by longitudinal slits. The connective of the anther is prolonged beyond the lobes to form a truncate (*Annona*), conical (*Artabotrys*), or leafy (*Uvaria*) hood that is characteristic of the family.

The **gynoecium** consists of numerous carpels which are free throughout (apocarpous). The carpels are arranged in a spiral manner on the torus. Ovary is superior with terminal style and stigma. Ovules are anatropous, one to numerous on ventral or basal placenta. In *Monodora*, gynoecium is syncarpous, and ovary is unilocular with parietal placentation. The gynoecium is monocarpellary in *Kingstonia*.

Pollination is by the agency if insects. The flowers are large and fragrant.

The **fruit** is an aggregate of berries, the berries may coalesce during ripening with one another by their pericarps and with the enlarging floral axis to form a pulpy body, the numerous seeds being embedded in the pulp, as in *Annona*. The individual berries are free and shortly stalked and they contain one (in *Polyalthia*) to many seeds. In *Unona* there are constrictions in between seeds as in lomentum.

The **seeds** are dicotyledonous and albuminous. The tegmen of the seed encroaches into the endosperm and forms ruminate endosperm (as in *Polyalthia*).

Key Characters of the Family

1. Arborescent habit.
2. Exstipulate, distichous leaves.
3. Hemicyclic trimerous flowers.
4. Elongated torus in flowers.
5. Valvate aestivation of perianth lobes.
6. Hooded anthers.

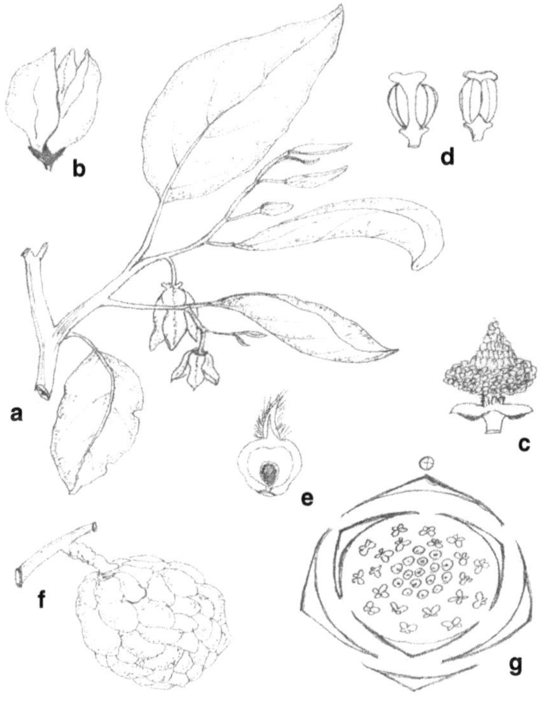

Fig. 15.1: *Annona squamosa* L. a. Flowering twig; b. flower; c. flower devoid of petals; d. stamens; e. carpel; f. fruit; g. floral diagram.

7. Apocarpous pistil.
8. Aggregate fruit.
9. Ruminate endosperm.

Important Plants

Annona is an important genus, many species yield delicious edible fruits such as *A. cherimola* (cherimoyar), *A. reticulata* (bullock's heart), *A. muricata* (sour-sop) and *A. squamosa* (sweet-sop). *Artabotrys hexapetalus* is a common straggler cultivated for its fragrant flowers. *Cananga odorata, Polyalthia longifolia* and *P. pendula* are common avenue trees in gardens. *Cananga odorata* flowers are source of the perfume 'ylang ylang'. Seeds of *Monodora myristica* and fruits of *Xylopia aromatica* are used sometime for spices. Wood fibre is obtained from *Unona pannosa* and *Goniothalamus* sp.

Systematics and Phylogeny

Annonaceae and Magnoliaceae are closely related families possessing large solitary flowers, thick fleshy perianth in hemicyclic whorls, spirally arranged stamens and apocarpous pistils on elongated torus, aggregate fruits and endospermous seeds.

However, Annonaceae differs from Magnoliaceae with exstipulate leaves, extrorose stamens, ruminate endosperm and sarcotesta in the seeds. The above mentioned similarities and differences between these two families indicate that both the families are closely related and have been derived from common ancestry. Hutchinson (1973) has placed these two families under separate orders Magnoliales and Annonales which is not advisable. Cronquist (1981) treated these two families under the order Magnoliales, and Thorne (1983) erected sub-order Annonineae of order Annonales, because some characters of Annonaceae are more advanced than Magnoliaceae. Engler and Prantl (1887-1899) have categorised the family Annonaceae into five tribes, viz, Uvarieae, Miliuseae, Hexalobiae, Xylopeae and Monodoreae.

The modern molecular systematic studies on pantropic flowering plant family Annonaceae well resolved with four major subclades such as Anaxagoreoideae, Annonoideae, Ambavioideae and Melameoideae. Monophyletic relationships are well-supported for Annonaceae with their inter-related families like Eupomatiaceae, Magnoliaceae, Myristicaceae, Degeneriaceae and Himantandraceae. The Magnolid group is deep rooted from the orders Canellales, Piperales and Laurales (APG IV). In Annonoideae, seven tribes have recognized. But still some genera are overlapping with even higher rank level, in other relationships among and within the genera remained unresolved.

Brassicaceae (Cruciferae) (Cabbage family)

Distribution: This family consists of 372 genera and about 4,060 species which are worldwide in distribution, but their chief centre of distribution is in the northern hemisphere, especially in the Mediterranean region. In India, it is represented by 50 genera and 140 species. Some of the larger genera are *Draba* (440 spp.), *Erysimum* (261 spp.), *Lepidium* (234 spp.), *Cardamine* (233 spp.), *Alyssum* (207 spp.), *Sisymbrium* (90 spp.) and *Rorippa* (70 spp.).

Habit and **Habitat:** The family members are usually mesophytic, annual or perennial herbs or rarely shrubs. *Subularia aquatica* is a hydrophyte.

Vegetative Characters

In biennial forms, the tap **root** stores food materials and becomes tuberous. The tuber is napiform in *Brassica rapa* (turnip) and fusiform in *Rhapanus sativus* (radish). In *Brassica oleracea* var. *gongylodes* (knolkohl) the basal part of the stem is tuberous.

The **leaves** are radical or cauline, simple or pinnately or palmately lobed, usually lyrate, exstipulate. Venation is reticulate. Rarely glands or scales (in *Cochlearia armoracia*) are present in the place of stipules. Epiphyllous buds are seen in *Cardamine pratens* and bulbils are present in the axils of leaves of *Dentaria bulbifera*.

The plant body is usually covered by unicellular or multicellular or stellate hairs which are of taxonomic importance. A special type of secretary cells called myrosin, present in the plant body which is an important anatomical feature. The guard cells of stomata are surrounded by three cells, one of which is smaller.

Floral Characters

The **inflorescence** is terminal or axillary, usually of the racemose type, corymb (as in *Iberis*); rarely of the cymose type.

The **flowers** are bisexual, actinomorphic but zygomorphic in *Iberis* by an enlargement of the anterior petals of the outer flower, hypogynous (perigynous in *Subularia*), dichalmydeous, heterochlamydeous, ebracteate and ebracteolate.

The **calyx** consists of 4 sepals which are free in dimerous whorls, outer whorl occupy the median position. Aestivation is valvate or imbricate. The sepals are often gibbous at the base to store the nectar.

The **corolla** consists of 4 petals which are free, arranged in diagonal position in a whorl, alternating with the sepals. They are usually clawed, upper portions spreading out in the form of a cross (cruciform). Aestivation is valvate. Flowers are apetalous in *Lepidium ruderale*.

The **androecium** consists of six stamens; there are two short stamens in the outer whorl in the transverse plane, and four longer ones are in the inner whorl in pairs in the median plane, it is called tetradynamous. In *Lepidium* only 2 median stamens are produced, while in *Megacarpaea* the stamens are as many as sixteen. Anthers dithecous, introrse, dehiscing by longitudinal slits.

The **gynoecium** is bicarpellary (tricarpellary in *Lepidium sativum* and 4-carpellary in *Tetrapoma*), syncarpous. The carples are transversely placed. Ovary is superior, with style (short, terminal), and stigma capitate, usually commissural. Ovary is originally unilocular but later becomes bilocular due to the formation of false septum (replum). The numerous amphitropous or campylotropous ovules are arranged on parietal placentas.

Pollination is usually entamophily by variety of insects. Flowers are showy and outer flowers spread zygomorphic petals more attractive for insects. Nectar stored in gibbous bases of the sepals. Cleistogamous flowers occur in *Subularia aquatica*.

The **fruit** is a dry dehiscent one breaking open from the base upwards and leaving the seeds attached to the replum, it is called siliqua. The fruit is constricted between the seeds in *Raphanus raphanistrum* (articulate siliqua). In *Isatis*, the fruit is samara and explosive one in *Cardamine impatiens*. Geocarpy is seen in *Cardamine pratense*.

The **seed** is dicotylendonous and exalbuminous.

Key Characters of the Family

1. Exstipulate, lyrate leaves.
2. Corymb inflorescence.
3. Ebracteate flowers.
4. Cruciform corolla.
5. Tetradynamous stamens.
6. Parietal placentation.
7. Siliqua fruit.

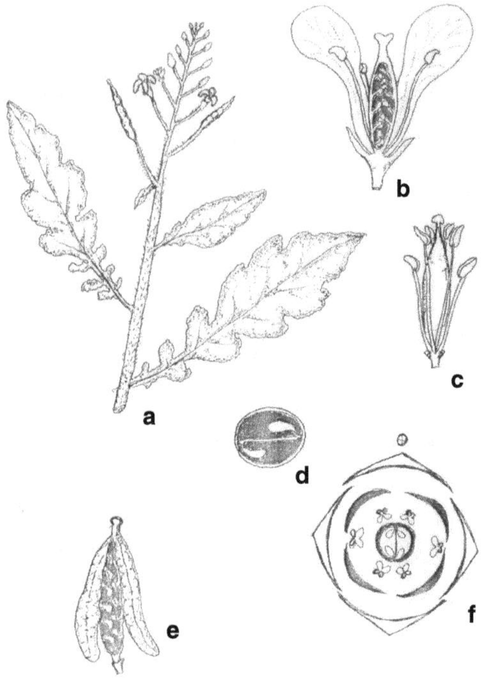

Fig. 15.2: B*rassica juncea* (L.) Czern. a. Flowering twig; b. l.s. of flower; c. Androecium and gynoecium; d. c.s. of ovary; e. fruit; f. floral diagram.

Important plants

Brassica is an important genus with many species and varieties of very high economic value. Many of them are cultivated vegetables; *Brassica oleracea* is the wild cabbage, var. *capitata* (cabbage), var. *botrytis* (cauliflower), var. *gongylodes* (knolkhol), var. *gemmifera* (Brussels sprouts). The root tubers of *Brassica rapa* (turnip) and *Raphanus sativus* (radish) are vegetables. *Lepidium sativum* (garden cress) and *Nasturtium officinale* (water cress) are used as leafy vegetables. The edible oil is extracted from the seeds of *Brassica nigra* and *B. napus*. The seeds of *B. juncea* is used as spice and root stock of *Cochlearia aromatica* is used as condiment. The leaves of *Isatis tinctoria* is the source of Indigo-like dye. Many are beautiful cultivated garden ornamentals, and these include *Alyssum sextile* (goldentuft), *Lobularia maritima* (sweet alyssum), *Cheiranthus cheiri* (wall-flower), *Ibris amara* (candytuft), *Lunaria annua* (honesty). Some members like *Capsella, Thlaspi, Camelia* are weeds and are avoided by cattle on account of their acrid juice.

Systematics and Phylogeny

Brassicaceae (Cruciferae) has been placed in order Parietales by Bentham and Hooker while Engler and Prantl placed under order Rhoedales. Hutchinson (1959) has erected separate order Cruciales while Cronquist (1981) and Thorne (1983) treated under Capparales. Many of the taxonomists believed that Brassicaceae has been derived from Papaveraceae whereas others believe it to be originated from Capparidaceae. The families Capparidaceae and Brassicaceae are closely related families. *Cleome* and its allied genera are usually more related to Capparidaceae and being traditionally included by some taxonomists but doing so results in a paraphyletic. Therefore, in APG II system erected a new family Cleomaceae for *Cleome* and its related genera at the same time Capparidaceae is merged with Brassicaceae. In APG III it has been followed as Brassicaceae *sensu stricto*, the system treating the Cleomaceae and Capparidaceae separately. At present APG IV, Brassicaceae is placed under the clade Malvids, order Brassicales. It includes 15 families like Akaniaceae, Bataceae, Brassicaceae, Capparidaceae, Caricaceae, Cleomaceae, Gyrstemonaceae, Koeberliniaceae, Limnathaceae, Moringaceae, Pentadiplandraceae, Resedaceae, Salvadoraceae, Setchellanthaceae, Tovariaceae and Tropaeolaceae.

Caryophyllaceae (Pink or Carnation Family)

Distribution: The family comprises of 81 genera and 2,625 species. They are distributed widely from Arctic through temperate and tropical to the Antarctic regions. The members of silenoideae occur frequently in the north-temperate zones, where the Mediterranean region may be regarded as their centre of distribution. In India, the family is represented by about 20 genera and 100 species. Some of the larger genera of the family are *Silene* (500 spp.), *Dianthus* (300 spp.), *Arenaria* (250 spp.), *Gypsophila* (125 spp.), *Stellaria* (100 spp.) and *Cerastium* (100 spp.). *Colobanthus* quitensis, the world's southernmost dicot belongs to this family, which is one of only two flowering plants found in Antarctica.

Habit: The plants of the family are annual or perennial herbs like *Spergula fallax*, *Saponaria vaccaria*, *Silene conoidea* and *Stellaria media*. *Acanthophyllum* is an undershrub. *Pycnophyllum* is a caespitose herb with scale leaves.

Vegetative Characters

The **root** is normal tap root.

The **stem** is usually herbaceous, erect or prostrate, often swollen at nodes. The branching of stem is sympodial type but dichotomous branching in *Lyallia*.

The **leaves** are simple, opposite and decussate, rarely alternate as in *Dysphania*. The leaves are usually exstipulate, but sometimes small scarious stipules occur at the base of leaves as in *Spergula* and *Corrgiola*. In *Polycarpaea corymbosa* and *Pycnophyllum*, the leaves are reduced to scaly membranous structure. In *Spergula*, the leaves are needle-like, radial and arranged in false whorls at nodes because of condensation of axillary branches.

Floral Characters

The flower is solitary in *Arenaria, Drymaria* and *Githago segetum*. Most of the members have dichasial cymes, ending with monochasial form. It is characteristic to this family known as Caryophllaceous type.

The **flowers** are bracteate, bracteolate, bisexual, rarely unisexual (*Lychnis dioica*), complete, pentamerous, rarely tetramerous (as in *Sagina*), hypogynous, rarely perigynous. Some members have gynodioecism, small pistillate flowers bloom earlier to the large bisexual flower at the beginning of the flowering season.

The **calyx** with four or five sepals, polysepalous, short and spreading, but in the members of Alsinoideae have long tubular calyx. Sometime gamosepalous condition is seen in Silenoideae. Sepals are fused below and lobes are free in *Gypsophila*. The aestivation is imbricate and sepals persistent.

The **corolla** is with four or five petals, polypetalous, short and spreading in *Stellaria, Arenaria*, etc., often differentiated into claw and limb in the plants of Selenoideae (*Silene, Saponaria*), usually petals with lobes, deeply divided in *Stellaria media*, such condition gives appearance of ten petals. The floral axis slightly elongated in between calyx and corolla especially in Selenoideae. In some genera petals are reduced or absent in *Polycarpaea, Lyallia*, and *Sagina*. The aestivation of petals is imbricate or twisted.

The **androecium** consists of ten stamens in two whorls of five each. Outer whorl is antipetalous and inner whorl is antisepalous. This condition is known as obdiplostemonous. In *Polycarpaea* only five stamens of inner whorl present, while in *Paronychia* only five stamens of outer whorl is present. In some cases there is different degrees of reduction in the number of stamens; for example *Stellaria media* with 5-8 stamens and *Lyallia* has only three stamens. Anthers are dithecous, basifixed, introrse and dehisce longitudinally.

The **gynoecium** is pentacarpellary and syncarpous. The carpels are two in *Dianthus*, three in *Stellaria* and four in *Sagina*. The carpels may be antisepalous in *Viscaria* and *Spergula* or antipetalous in *Githago*. The ovary is superior with axile or free central placentation. Single basal ovule is seen in *Carrigiola littoralis* and *Paronychia*. The styles are united below, but free above. Number of styles are same as number of carpels. *Lyallia* has a single style with bilobed stigma.

The **fruit** is a capsule dehiscing apically by valve or a berry like as in *Cucubalus baccifer*. Indehiscent achene is seen in *Carrigiola*.

The **seeds** are endospermic with a small curved embryo. Straight embryo is found in *Dianthus*. *Spergula* has winged seeds.

Key Characters of the Family

1. Herbaceous plants.
2. Caryopyllaceous inflorescence
3. Deeply divived corolla lobes.
4. Obdiplostemonous stamens.
5. Valvular dehiscing capsule.

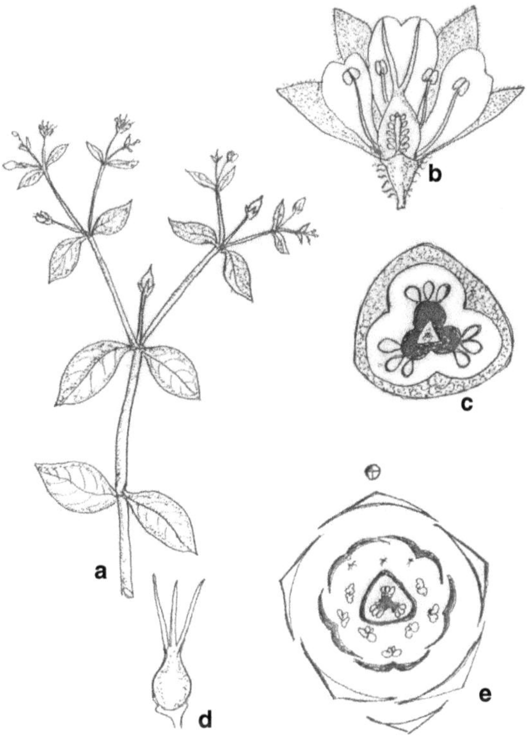

Fig. 15.3: *Stellaria media* (L.) Vill. a. Flowering twig; b. l.s. of flower; c. c.s.of ovary; d. gynoecium; e. floral diagram

Important Plants

Many of the family members are cultivated for garden ornamentals like *Dianthus caryophyllus* (carnation), *D. barbatus* (sweet William), *D. plumarius* (pink), *D. sinensis* (rainbow pink), *Gypsophila paniculata* (baby's brath), *Saponaria vaccaria* (cowcockle), *Silene pendula* (catchfly), *Lychnis alba* (white campion), *Arenaria grandiflora*, *Cerastium arvense*, etc. Roots of *Saponaria vaccaria* and *Gypsophila* yield saponin which is used in food supplements and cattle feed industries.

Systematics and Phylogeny

The family is placed under Caryophyllales (Bentham and Hooker and is usually classified into two subfamilies such as Alsinoideae (free petals) and Silenoideae (fused petals). Tutin *et al.* (1964) recognized three subfamilies i.e. Alsinoideae, Paronychioideae and Silenoideae. Cronquist (1981) categorised this family under

Caryophyllales with 11 other families whereas Thorne (1983) placed Caryophyllaceae under order Chenopodiales of Centrospermae. Different opinions have been given on the origin of this family from different root stocks. Eichler considered it to have originated form Phytolaccaceae but Dickson (1936) believed that Caryophyllaceae originated from Geraniaceae. Caryophyllaceae were considered the sister family to all remaining members of the suborder Caryophylleneae because of the presence of Anthocyanin in most of the members. Modern cladistic analyses showed that Caryophyllaceae evolved from betalain containing ancestors, therefore it shows synapomorphy. Traditional subfamilies of Alsinoideae and Paronychioideae are polyphyletic and Silenoideae indicates monophyletic relationships. In APG IV system, order Caryophyllales consists of 34 families including Aizoaceae and Amaranthaceae along with Caryophyllaceae.

Malvaceae (Mallows Family)

The current system of plant classification APG IV placed the families Tiliaceae, Sterculiaceae and Bombacaceae under Malvaceae. In this book Malvaceae (*sensu stricto*), Tiliaceae and Sterculiaceae are described separately with comparative account of the all those three families. Bombacaceae is included in the family Malvaceae.

Distribution: About 244 genera with 4,425 species are accounted under the family Malvaceae. The plants are abundantly seen in the tropics and a few are natives of the temperate region of the old world. The larger genera of the family are *Hibiscus* (200 spp.), *Sida* (200 spp.) and *Pavonia* (200 spp.).

Habit: They are mostly shrubs (*Hibiscus rosa-sinensis*) or herbs (*Sida*) and a few are trees (*Thespesia*).

Vegetative Characters

Root system is normal tap root.

Stem is herbaceous (*Sida*) or woody (*Hibiscus, Thespesia*, etc.), branching monopodial or sympodial (*Gossypium*).

The **leaves** are simple, palmately lobed, rarely compound, alternate, stipulate. Venation is multicostately reticulate.

The plant body is covered by stellate hairs. Mucilage is present in isolated cells or epidermis or in the lysigenous cavities scattered in the parenchyma. Phloem is found in wedge-shaped patches that can be differentiated into hard and soft parts. Pericycle fibres are found outside the phloem.

Floral Characters

The **inflorescence** is terminal or axillary; flowers are solitary or in fascicle or cymes.

The **flowers** are bisexual (polygamous in *Kydia* and Dioecious in *Napaea*), actinomorphic, hypogynous, pentamerous, bracteate and bracteolate. The bracteoles vary from 3 to many in number and are arranged in a ring froming an epicalyx. But this character is absent in *Abutilon*, *Sida* and *Thespesia*. The bracteoles may be scaly or broad and leaf-like as in *Malope* and *Gossypium*.

The **calyx** consists of 5 sepals, gamosepalous froming a cup. Aestivation is valvate.

The **corolla** consists of 5 petals, free or slightly connate at the base, and adnate to the staminal tube. Aestivation is twisted. Petals are absent in *Cullenia*.

The **androecium** consists of numerous stamens which are united by the filaments to form a tube around the style, condition is called monadelphous (polyadelphous in *Bombax*). The anthers are reniform, monothecous, extrorse. Pollen grains are big and spinous, but smooth in *Bombax*.

The **gynoecium** is 3- to many-carpellary, syncarpous. Ovary is superior; style terminal, single or divided apically to bear as many or twice as many stigmas as the carpels. The chambers in ovary are as many as the carpels.. There are one to many anatropous ovules in each chamber in one or two rows on axile placenta.

Pollination is through the agency of insects. Self pollination noted in *Malva rotundifolia* and Cleistogamous flowers seen in *Pavonia*.

The **fruit** is a loculicidal capsule or a schizocarp (*Sida* and *Abutilon*), rarely berry (*Malva* and *Durio*).

The **seeds** are dicotyledonous with scanty or no endosperm. Hairy seed coat in *Gossypium* and arillate seeds in *Bombax*.

Key Characters of the Family

1. Hairs on the plant body and mucilage tissue in the leaves.
2. Alternate phyllotaxy and stipulate leaves.
3. Multicostate reticulate venation.
4. Presence of epicalyx.
5. Monadelphous stamens with monothecous, reniform anthers.
6. Spinous pollen.
7. Axile placentation.

Important Plants

Many of the family members are cultivated as avenue trees like *Adansonia digitata*, *Thespesia populnea* and *Pseudobombax ellipticum*. Source of cotton fibres is from the genus *Gossypium* which include *G. arboreum* (tree cotton), *G. hirsutum* (upland cotton) and *G. barbadense* (sea-island cotton). Silk cotton fibres from the seeds of *Bombax malabaricum* and *Ceiba pentandra* are used for stuffing pillows. Bast fibres are obtained from several members such as *Hibiscus cannabinus* (deccan hemp), *H. esculentus* (okra hemp), *H. sabdariffa* (roselle hemp), *H. tiliaceus* (majagna hemp), *Sida rhombifolia* (cuba-jute), *Urena lobata* (avamina jute) and *Abutilon avicennae* (china jute).

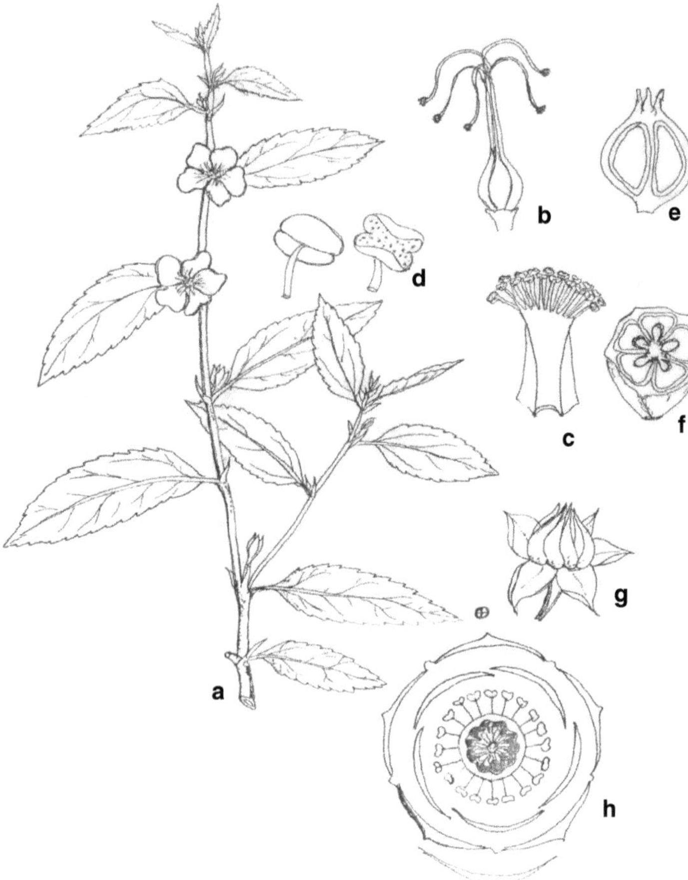

Fig. 15.4: ***Sida spinosa*** **L. a. twig; b. Gynoecium; c. Anther column; d. Stamens; e. Ovary l.s.; f. Ovary c.s.; g. fruit; h. Floral diagram**

Several members are cultivated for their ornamental values for example *Hibiscus rosa-sinensis, H. schizopetalus, H. mutabilis, Althaea rosea, Lavetera trimestris* and *Malope grandiflora*. Leaves of *Hibiscus cannabinus* are used as leafy vegetable.

Sterculiaceae

Distribution: This family comprises 60 genera and 750 species which are distributed exclusively in the tropical countries. The larger genera of the family are *Sterculia* (250 spp.) and *Dombeya* (250 spp.)

Habit: They are mostly trees and shrubs, some are lianes and a few are herbs (*Melhania*). *Heritiera* is found in marshy localities.

Vegetative Characters

Root system is normal type. Breathing roots found in *Heritiera*.

The **shoot system** is usually with monopodial branching.

The **leaves** are simple or palmately lobed (*Sterculia foetida*) or compound; stipulate. Venation is multicostately reticulate.

The plant body is usually covered with stellate hairs. Mucilaginous ducts are absent.

Floral Characters

The **inflorescence** is terminal or axillary. In *Theobroma*, flowers are produced on the old stem which condition is known as cauliflory. Usually inflorescence is cymose type.

The **flowers** are bisexual (unisexual and polygamous in *Sterculia*), actinomorphic (zygomorphic in *Helicteres*), hypogynous, pentamerous. In some *Sterculia* species androgynophore is present.

The **calyx** consists of 5 sepals, free and aestivation is valvate. Epicalyx is totally absent.

The **corolla** consists of 5 petals, free and aestivation is twisted. Apetalous flowers are seen in *Sterculia* and *Cola*. Petals are unequal in *Helicteres*.

The **androecium** consists of 10 stamens (5 in *Melochia*) in 2 whorls of 5 each in obdiplostemonous arrangement. The filaments are usually connate at the base resulting in monadelphous condition but in *Guazuma* filaments form broad staminal tube and anthers are divided. Sometimes stamens are reduced in various degrees and appeared as staminodes. Anthers are dithecous, extrorse, longitudinally slit open.

The **gynoecium** consists of usually 5 carpels, syncarpous, pentalocular (unilocular in *Waltheria*), axile placentation with one to many ovules in a chamber. The ovary is superior with terminal style and free stigmas.

Pollination is entamophily.

The **fruit** is generally dry, dehiscent type. In *Sterculia*, the individual carpels separated out to form each a follicle. In *Helicteres*, it is spirally twisted. Berry is seen in *Theobroma* and Schizocarp observed in *Pentapetes*.

The **seeds** are dicotyledonous and endospermic. Seeds are winged in *Pterospermum*.

Key Characters of the Family

1. Shrubs and trees.
2. Pentamerous flowers.
3. Dithecous, extrorse anthers.
4. Presence of staminodes.
5. Endospermic seeds.

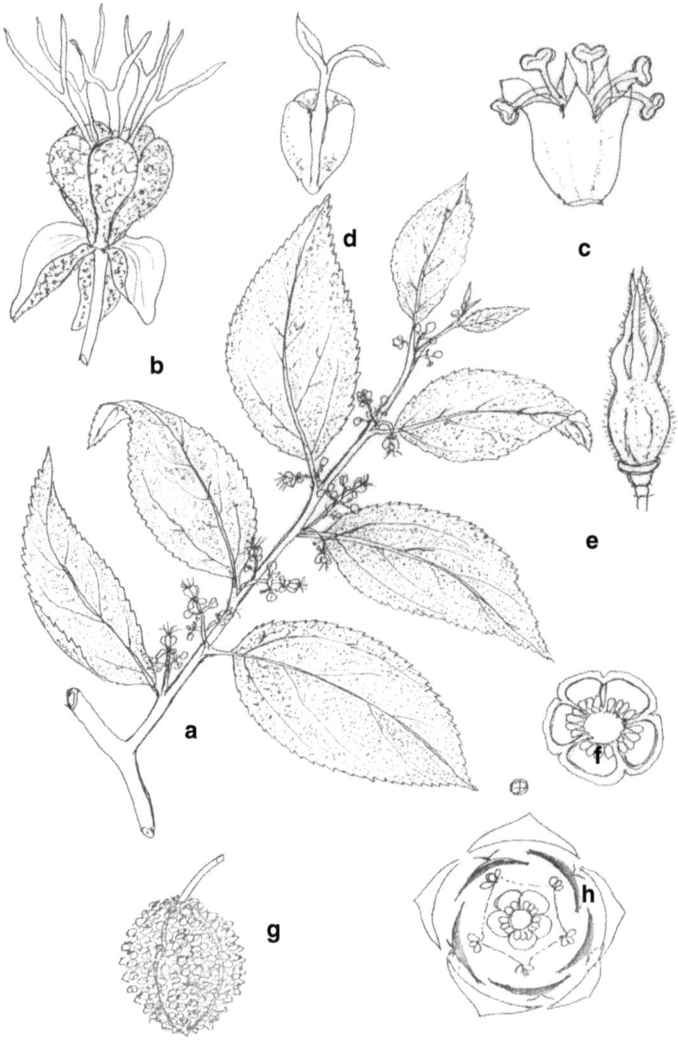

Fig. 15.5: *Guazuma ulmifolia* Lam. a. Twig; b. Flower entire; c. Staminal column; d. Petal; e. Gynoecium; f. Ovary c.s.; g. Fruit; h. Floral diagram.

Important Plants

Guazuma ulmifolia, Kleinhovia hospita and *Pterospermum heyneanum* are cultivated as avenue trees. *Theobroma cacao* is cultivated for its coco fruits. Gum karaya is obtained from *Sterculia urens*. *Helicteres isora* is a medicinal shrub and several *Sterculia* species are found growing in dry forests of southern India. Common weeds are *Melochia corchorifolia, Melhania incana* and *Waltheria americana*. *Dombeya* sp. is a garden ornamental shrub.

Tiliaceae

Distribution: This family is accounted for about 400 species under 45 genera. They are mostly distributed in the tropics and a few are found in temperate region.

Habit: They are trees or shrubs and a few are herbs (*Corchorus*).

Vegetative Characters

Root system is the normal tap root type.

The **branching** of shoot system is monopodial or sympodial.

The **leaves** are simple, stipulate, alternate, usually in two rows, sometimes asymmetrical. Venation is palmately reticulate.

Floral Characters

The **inflorescence** is axillary or terminal, usually of the cymose type.

The **flowers** are bisexual, actinomorphic, hypogynous, pentamerous (tetramerous in *Sparmannia*). Winged bracteole is found in *Tilia*.

The **calyx** consists of 5 sepals, usually free. Aestivation is valvate.

The **corolla** consists of 5 petals, free. Aestivation is valvate.

The **androecium** consists of 10 to numerous stamens which are free or rarely united into bundles as in *Apeiba*, opposite to the petals. There are 10 to 15 stamens in *Triumfetta*. Anthers are dithecous, introrse, dehiscing by longitudinal slits.

The **gynoecium** consists of 2 – 5 carpels, syncarpous, 2-5-locular with axile placenta. Ovary is superior. Style is terminal with capitate or lobed stigma. In *Grewia*, ovary is situated on the elongated axis.

Pollination is through the agency of insects. The flowers are usually protandrous.

The **fruit** is a capsule or berry. Fruits of *Triumfetta* bear hook like outgrowths. They are carried by animals.

The **seed** is dicotyledonous and endospermic.

Key Characters of the Family

1. Asymmetrical stipulate leaves.
2. Stamens numerous and free.
3. Anthers dithecous and introrse.

Important Plants

Cochorus capsularis and *C. olitorius* are cultivated for jute extraction. About 40 species of *Grewia* are growing in southern India and many of their fruits are wild edible, for example *G. abutilifolia, G. tiliifolia, G. umbellata*, etc. *Triumfetta* is a common herb.

Fig. 15.6: *Corchorus aestuans* Lam. a. Twig; b. Flower l.s.; c. Stamen; d. Ovary l.s.; e. Ovary c.s.; f. Floral diagram.

Comparison of the families Malvaceae, Sterculiaceae and Tiliaceae

Malvaceae	Sterculiaceae	Tiliaceae
Herbs, shrubs, trees.	Trees or shrubs, a few or herbs.	Herbs and shrubs. A few are trees.
Leaves simple, alternate, stipulate, often lobed.	Simple or palmately compound.	Simple, alternate, stipulate.
Flowers bisexual, actino morphic, hypogynous, pentamerous.	Bisexual, rarely unisexual, hypogynous, pentamerous.	Bisexual, actinomorphic, hypogynous, pentamerous.
Sepals 5, united, valvate.	Sepals 5, united at the base, valvate.	Sepals 5, free, valvate.
Epicalyx present.	Epicalyx absent.	Epicalyx absent.
Petals 5, twisted.	Petals 5, free, twisted.	Petals 5, free, never twisted.
Stamens numerous, monadelphous, anthers monothecous, pollen spinous.	10 in two whorls, often branched, anthers dithecous, pollen smooth.	10 to many, free, anthers dithecous, pollen smooth.

Malvaceae	Sterculiaceae	Tiliaceae
Carpels 5, united, 1-many ovules on axile placenta.	5, united, ovules 1- many, placention axile.	2-many, united, ovlules 1-many on axile placenta.
Fruit capsule or schizocarp.	Capsule.	Capsule or berry.
Seeds comose with scanty endosperm	Seeds with fleshy endosperm.	Seeds with fleshy endosperm

Systematics and Phylogeny

Malvaceae is placed under order Malvales (Bentham and Hooker) and all other taxonomists retained the same position including Cronquist (1981) and Thorne (1983). Schumann (1964) has classified the family into four tribes i.e. Malopeae, Malveae, Ureneae and Hibisceae. Sterculiaceae are treated under Tiliales by Hutchinson. This family differs from Malvaceae by absence of epicalyx and presence of definite number of stamens. Whereas Tiliaceae is also more close to Malvaceae but differs with dithecous anthers and absence of epicalyx. Tiliaceae and Sterculiaceae are more affine with habit and hairy nature. All the classical literature showed the Bombacaceae included in Malvaceae. Cronquist (1981) and Throne (1983) have separated the family Bombacaceae from Malvaceae because of the family members being exclusively arborescent, prickly trunk, dithecous anthers and smooth pollen grains. Takhtajan (1969) and Cronquist (1981) noted the close relationships of Bombacaceae with Malvaceae, Sterculiaceae and Tiliaceae.

The traditional Malvaceae *sensu stricto* comprises a very homogeneous and cladisitically monophyletic. But recent Malvaceae *sensu lato* based on molecular techniques recognized that families of Bombacaceae, Sterculiaceae and Tiliaceae, which have been considered closely allied to Malvaceae, are polyphyletic. At present APG IV system circumscribed 9 subfamilies under Malvaceae, i.e. Byttnerioideae, Grewioideae, Sterculioideeae, Tilioideae, Dombeyoideae, Bownlowioideae, Helicteroideae, Malvaoideae and Bobacoideae. Some of the genera are treated earlier in the 'new' Malvaceae *sensu lato*, but with *Gyrostemon* moved to family Gyrostemonaceae, *Muntingia* to family Muntingiaceae and *Sloanea* to family Elaeocarpaceae.

Rutaceae (Citrus Family)

Distribution: This family comprises about 150 genera and 1,500 species which are distributed in the temperate and the tropical region of the northern and southern hemisphere. About 25 genera and over 80 species have been so far been reported in India. Some of its larger genera represented in world over are *Fagara* (250 spp.), *Zanthoxylum* (200 spp.), *Ruta* (60 spp.), *Glycosmis* (60 spp.), *Eriostemon* (35 spp.) and *Atalantia* (20 spp.).

Habit: They are shrubs or trees.

Vegetative Characters

Root system is of the normal tap root type.

Stem is usually woody, characteristic bark developed in *Toddalia*. Most of the cases axillary buds are modified into thorns or spines.

The **leaves** are pinnately or palmately compound, (bipinnate in *Ruta*), rarely simple as in *Diosma*, alternate, rarely opposite as in *Evodia*, exstipulate, venation pinnately reticulate. The leaves are gland dotted. In *Citrus*, leaves are regarded as unifoliolate due to the presence of winged petiole.

Floral Characters

The **inflorescence** is axillary or terminal, cymes or panicles, rarely recemes and fascicles.

The **flowers** are bisexual, unisexual as in *Toddalia*, polygamous in *Vebris*, hypogynous, actinomorphic, rarely zygomorphic in *Dictamnus*, 4-5-merous, bracteate and bracteolate. In *Ruta*, terminal flowers are pentamerous and lateral flowers are tetramerous.

The **calyx** consists of 4-5 sepals, usually united. Aestivation is valvate or imbricate.

The **corolla** consists of 4-5 petals, absent in *Zanthoxylum*, free, united in *Correa*. Aestivation is imbricate.

The **androecium** consists of 8-10 stamens, arranged in 2 whorls, outer whorl obdiplostemonous, outer whorls of stamens suppressed in *Toddalia*, numerous in *Aegle marmelos*. Stamens are free and slightly flattened but are united to form one bundle in *Atalantia* and several bundles in *Citrus*. Anthers are dithecous, introrse, connective gland dotted, dehiscing by longitudinal slits.

The **gynoecium** is 4-5-carpellary (8-10 in *Aegle*), syncarpous, ovary superior, often raised on the disc look like gynophore. Style is stout and stigma capitate. Anatropous ovules are on axile placentation, 1-many in each chamber. Ovary is unilocular with parietal placentation in *Limonia*.

Pollination is entamophilous. The flowers are strongly protandrous.

The **fruit** is a hesperidium. In *Dictamnus* the individual carpels of the fruit separate from each other, and dehisce like follicles. Samara is seen in *Ptelea* and drupes in *Murraya* and *Toddalia*.

The **seed** is dicotyledonous, endospermic (non-endospermic in *Diosma*), with mucilaginous outer covers. Perisperm is also present. Polyembryony noted in *Citrus*.

Key Characters of the Family

1. Gland-dotted compound leaves.
2. Obdiplostemonous stamens.
3. Well developed disc between androecium and gynoecium.

Characteristic fruit as hesperidium.

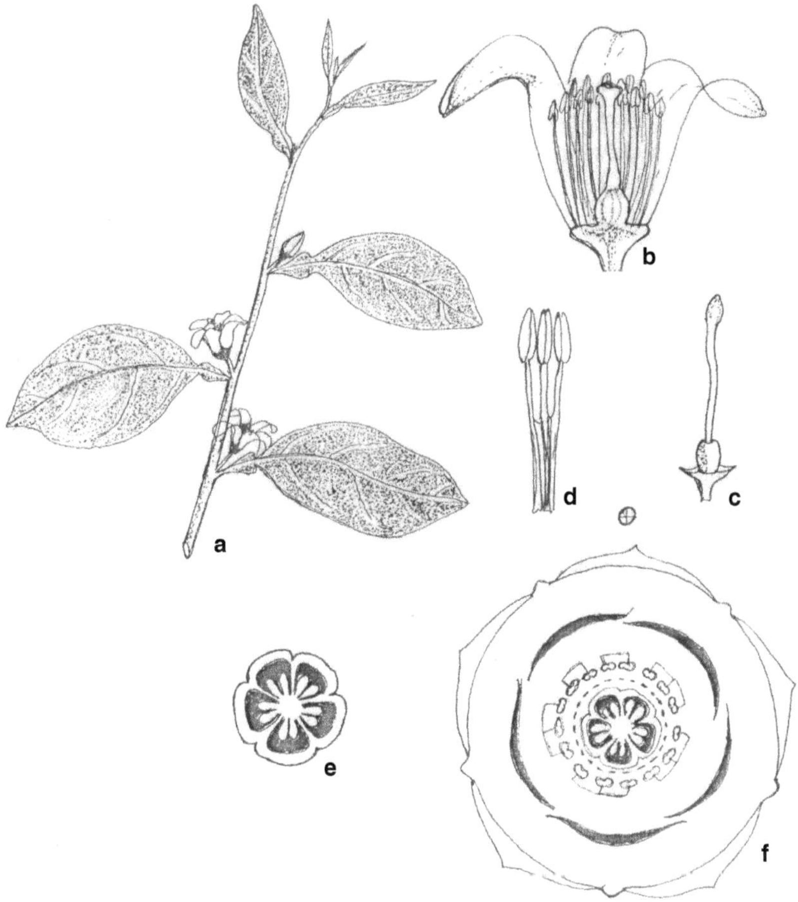

Fig. 15.7: *Citrus medica* L. a. Flowering twig; b. Flower l.s.; c. Gynoecium; d. Staminal bundle; e. Ovary c.s.; f. Floral diagram

Important Plants

Many species of *Citrus* yield edible fruits; *C. aurantium* (orange), *C. decumana* (pumelo), *C. limettoides* (sweet lime), *C. nobilis* (king orange), *C. paradisi* (grape fruit), *C. reticulata* (loose jacket), *C. sinensis* (Sweet orange). The fruits of *C. aurantifolia* (lime), *C. medica* (citron) are of medicinal value. Essential oil of *Ruta graveolense* as narcotic. *Aegle marmelos* is good remedy for dysentery. Wood of *Zanthoxylum armatum* is used as good remedy for tooth ache. Fruits of *Limonia acidissima* used for culinary purpose. *Murraya koenigii* leaves are used for flavouring dishes.

Systematics and Phylogeny

Rutaceae is categorized under the order Geraniales and classified into four tirbes namely Aurantieae, Ruteae, Toddalieae and Zanthoxyleae (Bentham and Hooker). Takhtajan (1969) and Thorne (1983) have placed it under order Rutales whereas it is placed under Sapindales by Cronquist (1981). Hicky and King (1988) have reclassified the family into five subfamilies, i.e. Rutoideae, Dictyolomatoideae, Spathelioideae, Toddalioideae and Aurantioideae and also they stated that Rutaceae is closely related to Zygophyllaceae, Cneoraceae, Meliaceae, Burseraceae and Simaroubaceae, all included under the order Sapindales.

The family Rutaceae is closely related to the Sapindaceae, Simaroubaceae, and Meliaceae, although some systems separate that order into Rutales and Sapindales. The families Flindersiaceae and Ptaeroxylaceae are sometimes kept separate, but in APG system these families are placed in Rutaceae, as are the former Cneoraceae. The subfamilial organization has not been fully resolved, but the subfamily Aurantioideae is well supported by molecular tools; though the placement of several genera remains unclear. Recent APG IV system circumscribed the family Rutaceae under the order Sapindales with 9 families, which include Biebersteniaceae, Nitrariaceae, Kirkiaceae, Burseraceae, Anacardiaceae, Sapindaceae, Simaroubaceae and Meliaceae.

Rhamnaceae (Buckthorn Family)

Distribution: The family comprises 55 genera and 950 species which are distributed in Indo-Malayan and also found in Mediterranean region, Africa, Australia and Tropical America. In India, the family is represented by 15 genera and 85 species. The larger genera of the family are *Rhamnus* (150 spp.), *Phylica* (150 spp.), *Ziziphus* (100 spp.), *Ceanothus* (55 spp.), *Frangula* (50 spp.) and *Colletia* (17 spp.).

Habit: The plants are usually shrubs and trees. Genera like *Ventilago* and *Gouania* are climbers which climb either by hook or thorns. *Berchemia* is a twiner and *Helinus lanceolatus* is a twining shrub without spines.

Vegetative Characters

Root system is of normal tap root.

The **stem** is woody, branched, sometimes climbing.

The **leaves** are simple, usually alternate, rarely opposite or sub-opposite as in *Sageretia,* petiolate and stipulate. The stipules are deciduous and often modified into spines. In *Ziziphus* many nerves arise from the base.

Floral Characters

The **inflorescence** is usually cyme, but flowers may arrange in panicles as in *Ventilago* and *Sageretia*. Flowers are either in axillary clusters or in simple or compound racemes. In *Microrhamnus,* the flowers are solitary and axillary.

The **flowers** are bracteate, bisexual, rarely unisexual as in *Rhamnus catharticus*, pentamerous, rarely tetramerous, actinomorphic and either peri or epigynous, hypogynous in *Berchemia scandens* and have well developed intrastaminal disc. The floral receptacle cup-like, usually it is united with ovary or not.

The **calyx** consists of 5 sepals, rarely 4, free or basal part united with receptacle to form hypanthium, as such only upper part is free. Aestivation is valvate.

The **corolla** consists of 4-5 petals, polypetalous, usually small in size, concave and often clawed. Petals are absent in *Colletia* and *Pomaderis*. Aestivation in usually valvate.

The **androecium** consists of 4-5 stamens, usually obdiplostemonous. They arise from the intrastaminal disc. Anthers are dithecous, introrse, dehisce by longitudinal slits.

The **gynoecium** is bi- to tetracarpellary, syncarpous. The ovary is free or fused with hollow receptacle, superior or half inferior and bi- to tetralocular, usually only one ovule in each chamber on basal placenta. The style is simple, terminal into bifid or bilobed stigma.

Pollination is usually entamophilous.

The **fruit** is a drupe in *Rhamnus*, capsule in *Pomaderis*, samara in *Ventilago* and nut in *Paliurus*.

The **seeds** are large, dicotyledonous, embryo straight, greenish and endospermous.

Key Characters of the Family

1. Thorny shrubs, trees and climbers.
2. Flowers with hollow, cup-shaped intrastaminal disc.
3. Stamens in obdiplostemonous condition.
4. Ovary bi- to tetralocular.
5. Varied fruits of drupe, capsule, samara and nut.

Systematics and Phylogeny

Rhamnaceae is placed under order Celastrales, series Disciflorae by Bentham and Hooker. Majority of the taxonomists erected the separate order Rhamnales (Takhtajan, 1969; Cronquist, 1981; Thorne, 1983) and it shows close affinity with Vitaceae in respect of floral morphology. Many others believe Rhamnaceae is closely related with Celastraceae but main difference is obdiplostemonous condition in the family.

Important Plants

The genus *Ziziphus* yields edible fruits; *Z. jujuba* (jujube), *Z. rotundifolia* (Jharberi), etc. The fruit of *Rhamnus persica* is used in medicine. Some plants are cultivated as garden ornamentals such as *Pomaderis apetala* and *Colletia paradoxa*. *Ventialgo maderaspatana* and *Gouania* are common plants in deciduous forests of southern India.

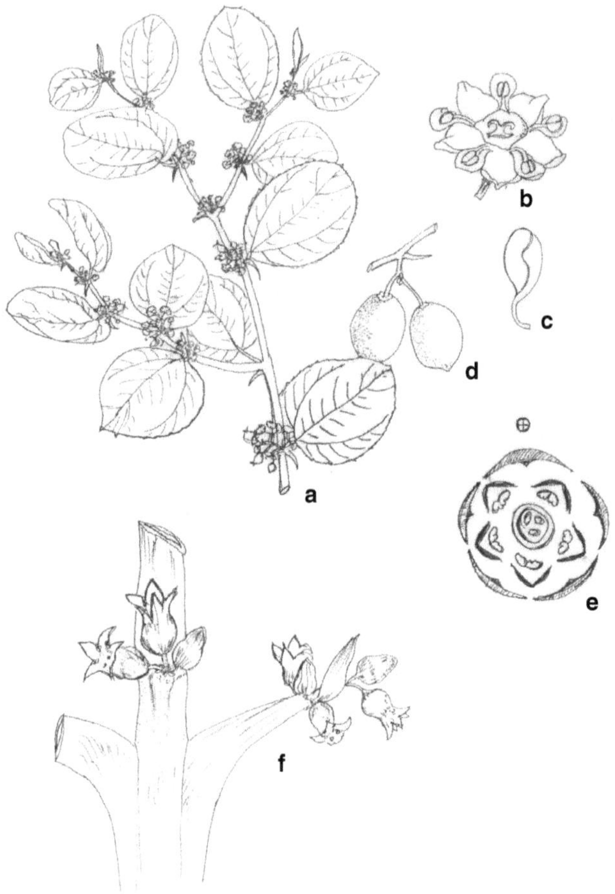

Fig. 15.8: *Ziziphus jujuba* Mill. a. Twig; b. Flower entire; c. petal; d. fruit; e. Floral diagram; f. *Colletia*-flowering twig.

In APG IV classification, Rhamnaceae is placed under the clade Rosids, order Rosales, it includes 9 families - Rosaceae, Barbeyaceae, Dirachmaceae, Elaegnaceae, Ulmaceae, Cannabaceae, Moraceae and Urticaceae. Rhamnaceae is further divided into three subfamilies, i.e. Ampeloziziphoids, Rhamnoids and Ziziphoids. Ziziphoids are devided into two tribes namely Colletieae and Pomaderreae.

Anacardiaceae (Cashew Family)

Distribution: This family comprises 86 genera and 850 species, distributed in tropics of both hemispheres, extending to the north temperate region. About 23 genera and 115 species have been reported in India. Some of the larger genera have been estimated world over are *Rhus* (250 spp.), *Semecarpus* (50 spp.), *Mangifera* (40 spp.), *Schinus* (30 spp.), *Anacardium* (15 spp.) and *Pistacia* (10 spp.).

Habit: They are mostly trees and shrubs.

Vegetative Characters

Root system is of the normal tap root type.

The **stem** is woody and branched.

The **leaves** are simple (as in *Mangifera*), or compound (pinnately in *Spondias*), trifoliolate (*Rhus*), alternate, exstipulate and not gland-dotted. Resin canals are characteristically present in the parenchyma.

Floral Characters

The **inflorescence** is terminal or axillary, usually panicle or sometimes cymose.

The **flowers** are bisexual (*Gluta* and *Buchanania*), polygamous in *Anacardium* and *Mangifera*, dioecious in *Pistacia* and actinomorphic. Ovary is superior, half inferior in *Holigarna*. A conspicuous disc is produced between androecium and gynoecium. It may be flat, annular, concave or deep cup-like and it may be cut into glandular structures.

The **calyx** consists of 3-5 sepals, free. Aestivation is imbricate.

The **corolla** consists of 3-5 petals, polypetalous. Petals are absent in *Pistacia*. Aestivation is imbricate.

The **androecium** consists of 10 stamens in two whorls of 5 each. In *Buchanania*, all the stamens are fertile, in *Anacardium*, there are 7-10 stamens in varying degrees of which only one is large and fertile. *Rhus* has only 5 stamens. In *Mangifera indica*, there are only five antisepalous stamens, but only one is sometimes fertile. Stamens are monadelphous in *Anacardium*. The anthers are dithecous, introrse, dehiscing by longitudinal slits.

The **gynoecium** is usually 3-carpellary (pentacarpellary in *Buchanania*), syncarpous, superior ovary. Out of 3 or 5 carpels, only one is functional and contains single ovule. Style may be free or united, terminal or lateral, stigma capitate or lobed.

Pollination is entamophilous. Insects are attracted by the nectar produced by the disc. Flowers of *Lannea* are anemophilous.

The **fruit** is drupe in *Mangifera* or nut in *Anacardium*, rarely samara in *Swītonia*. In *Anacardium*, thalamus and flower stalk develops into a fleshy pear-shaped juicy structure.

The **seed** is large, dicotylendonous and non-endospermic, rarely endospermous.

The Key Characters of the Family

1. Non-gland-dotted leaves.
2. Resin passage in the parenchyma.
3. Well developed intrastaminal disc.
4. Lateral style.
5. Drupaceous fruit.

Important Plants

Some are fruit yielding plants such as *Mangifera indica* (mango), *Anacardium occidentale* (cashewnut), *Pistacia vera* (pistachio). *Metopium* yields resin and some species of *Rhus* yield tannins, namely *R. glabra, R. typhina, R. caopalliana* and *R. coriaria*. *Semecarpus anacardium* (marking nut) fruits are a source of tannin in southern India. Edible seeds obtained from *Buchanania axillaris*. *Lannea coramandelica* and *Spondias pinnata* are common trees in deciduous forests of southern India.

Systematics and Phylogeny

Most of the taxonomists classified the family Anacardiaceae under order Sapindales (Bentham and Hooker; Engler and Prantl; Cronquist) while Takhtajan (1969) and Thorne (1983) have placed it under order Rutales. The Famlily Anacardiaceae shows close affinity towards Sapindaceae, however other families are also closely related such as Aceraceae, Hippocastanaceae, Meliaceae, Rutaceae and Burseraceae. Hickey and King (1988) divided the family into 5 tribes, i.e. Anacardieae, Spondiadeae, Rhoeae, Semecarpeae and Dobineae. In APG IV system, it is included under Sapindales with 8 other families - Sapindaceae, Rutaceae and Meliaceae.

Fabaceae (Leguminosae) (Legume Family)

This is the third largest family of the seed plants comprising about 750 genera and 19,000 species. They occur wherever climate favours growth of seed plants that often form characteristic vegetation of the area and are cosmopolitan in distribution. Mimosoideae and Caesalpinioideae are abundantly seen in tropical forests, while the members of Faboideae are seen in sub-tropical regions. Many of the herbs of the Arctic and Alpine regions belong to Faboideae.

Sub Family Mimosoideae

Distribution: This sub-family comprises 30 genera and 1,350 species which are distributed in tropical and sub-tropical regions.

Habit and Habitat: There is a great variation in the habit and habitat of the members. They are trees, shrubs or herbs. Some are stragglers and a few are twining forms. *Neptunia* is a hydrophyte.

Vegetative Characters

Root system is of the normal type. Root-nodules are characteristically present in several genera of the subfamily.

The **stem** is herbaceous or woody, usually branching. In *Neptunia*, the prostrate stem is floating on the water with spongy internodes.

The **leaves** are usually bipinnately compound (sometimes reduced), stipulate, stipules modified into spines as in *Prosopis* and *Acacia nilotica*. Leaf base is usually pulvinate. Petiole is leaf-like in *Acacia melanoxylon*. The petiole is modified into flat leaf-like in *Acaia holosericea* and *A. auriculata*. Leaves show characteristic sleep movements. In *Mimosa pudica*, the leaf movement is spectacular.

Floral Characters

The **inflorescence** is axillary, a racemose one, usually a globose head, sometimes a much condensed spike as in *Prosopis*.

The **flowers** are bisexual, sometimes polygamous, actinomorphic, hypo- or perigynous, bracteate, 4-5-merous, rarely 3-6-merous.

The **calyx** consists of 4-5 sepals, united to form a tubular structure. Aestivation is valvate but in *Parkia*, it is quincuncial.

The **corolla** consists of 4-5 petals, united, rarely free. Aestivation is valvate. Petals are not attractive and sometimes caducous.

The **androecium** is haplo (in *Mimosa*) or diplostemonous (in *Leucaena*); stamens numerous, free, monadelphous in *Pithecellobium*, exerted and sometimes filaments are brightly coloured. There are two-coloured filaments in the same flower in *Dichrostachys*. Anthers are dithecous, versatile and sometimes gland tipped.

The **gynoecium** is monocarpellary, rarely bicarpellary as in *Inga*, apocarpus in *Affonsea*; style terminal and stigma capitate. Ovules are numerous on marginal placenta.

Pollination is usually entamophilous and their flowers are brightly coloured with groups of heads conspicuous.

The **fruit** is a legume or lomentum.

The **seed** is dicotyledonous, usually non-endospermous, sometimes arillate as in *Pithecellobium*.

Important plants

Acacia is a genus of shrubs or trees, bearing bipinnate leaves and globose heads or cylindrical spikes with several common species like *A. nilotica, A. horrida, A. planifrons, A. chundra, A. catechu*, etc. *Acacia auriculiformis, A. mangium* and *A. holosericea* are cultivated for their woods. *Albizia lebbeck, Samanea saman, Parkia biglandulosa* are common avenue trees. The arils of *Pithecellobium dulce* seeds are edible. *Prosopis juliflora* is a common invasive tree, now menace to all over India. *Dichrostachys cinerea* is a handsome shrub with ornamental value and some species of *Inga* are cultivated for their showy flowers.

Mimosa pudica and *Neptunia oleracea* are common weeds, the former one is terrestrial and later one is aquatic respectively.

Sub Family Caesalpinioideae

Distribution: This subfamily consists of 80 genera and 750 species which are mostly distributed in tropical region.

Habit: They are mostly shrubs and trees, a few are being herbs. Some are lianes as *Bauhinia vahlii*.

Vegetative Characters

Root system is of normal tap root type.

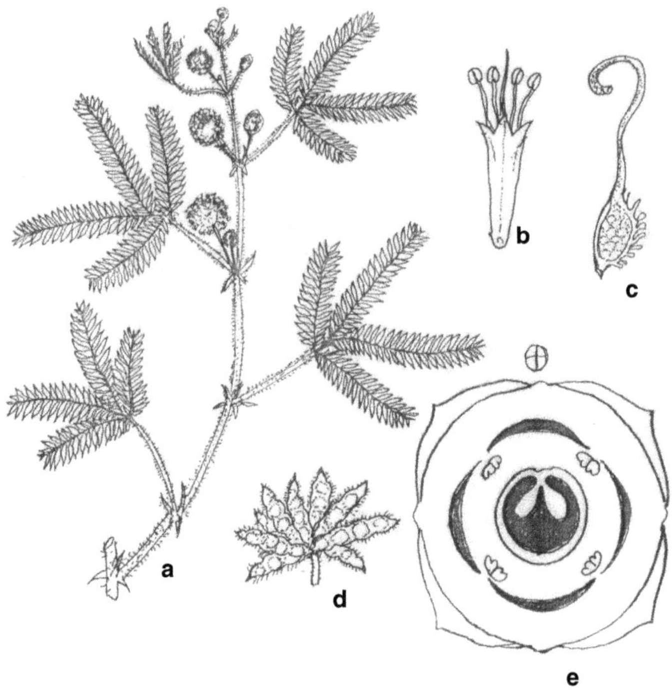

Fig. 15.9: ***Mimosa pudica*** **L. a. Twig; b. Flower – entire; c. Carpel; d. Fruit; e. Floral diagram**

Shoots are woody or herbaceous with branching, sometimes climbing habit with anomalous secondary thickening as in *Bauhinia*.

The **leaves** are usually pinnate or bipinnately compound, alternate, stipulate, stipules often spinous. Simple with characteristically bilobed leaves found in the genus *Bauhinia*. Leaf base is pulvinate. Rachis of the leaf becomes flat and photosynthetic in *Parkinsonia*.

Floral Characters

The **inflorescence** is usually axillary or terminal. It is racemose or corymb, rarely cymose type.

The **flowers** are usually bisexual, slightly zygomorphic, perigynous, bracteate, often bracteolate, 4-5-merous.

The **calyx** consists of 4-5 sepals, usually free. Aestivation is imbricate, but valvate and united in *Cercis*.

The **corolla** consists of 4-5 petals, free. Aestivation is ascendingly imbricate. The anterior petals are suppressed in *Amherstia* and *Tamarindus*. Petals are absent in *Saraca*.

The **androecium** consists of 10 stamens, diplostemonous (haplostemonous in *Ceratonia*), free, monadelphous in *Tamarindus*. The great degree of reduction in number of stamens in various genera, out of 10 stamens, only 3 are fertile in *Tamarindus*, 7 fertile and 3 staminodes in *Cassia*. Anthers are dithecous, basifixed or versatile, dishiscing by longitudinal slits (but pores in *Cassia*).

The **gynoecium** is monocarpellary (rarely bicarpellary, apocarpous in *Saraca*). Ovary is cylindrical with terminal style and capitate stigma.

Pollination is entomophilous. Flowers of *Delonix* is visited by birds.

The **fruit** is a legume, but rarely samara in *Pterolobium*.

The **seeds** are dicotyledonous and non-endospermous, rarely endospermous.

Important Plants

Cassia is a big genus with varied habit and widely distributed; common species are *Cassia auriculata, C. absus, C. italica, C. occidentalis, C. sophora*, etc. The species of *Bauhinia* are trees, shrubs and climbers, some are cultivated for its ornamental purpose *B. variegata, B.acuminata, B. tomentosa* etc. *B. vahlii* leaves are used as packaging material for eatables. *Tamarindus indica* is cultivated for its fruits. *Delonix regia, D. alata, Cassia siamea, C. roxburghii, C. tymoriensis, Saraca indica, Peltophorum pterocarpum* are planted as avenue trees. Some are garden ornamentals like *Cassia fistula, Amherstia nobilis, Caesalpinia pulcherrima*, etc. Some common shrubs are *Parkinsonia aculeata, Pterolobium hexapetalum, Caesalpinia bonduc* in the wild.

Sub-Family: Faboideae

Distribution: This subfamily comprises 450 genera and 12,725 species. The members are cosmopolitan in distribution. Several of the members are distributed in Arctic and Alpine regions especially its herbaceous forms.

Habit and Habitat: They are mostly herbs, shrubs or trees, growing sometimes in arid localities. Some are twiners (*Dolichos*), climbers with tendrils produced from leaflets as in *Pisum* or entire leaves in *Lathyrus*. Species of *Mucuna* are lianes.

Vegetative Characters

Roots are normal type and associated with *Rhizobium* produced characteristic root nodules.

Stem herbaceous or woody, sometimes axillary branches modified into thorns.

The **leaves** are simple or compound, pinnately or palmately, alternate, stipulate; stipules are sometimes spinous, leaf-like as in *Lathyrus* and *Pisum*. Leaf base is pulvinate. The leaves show peculiar movements. The leaves are spinous in *Ulex*.

Floral Characters

The **inflorescence** is usually axillary, raceme or spike or panicle, rarely a condensed head.

The **flowers** are bisexual, rarely polygamous, zygomorphic, perigynous, usually bracteate and bracteolate.

The **calyx** consists of 5 sepals, gamosepalous, irregular. Aestivation is valvate.

The **corolla** consists of 5 petals, free, irregular, papilionaceous. Aestivation in descendingly imbricate. The posterior petal is standard petal, overlapping along the edges of 2 lateral wing petals, which are usually horizontally arranged, and these in turn overlap the two lower ones which are vertical and connate at the lower margin to form the keel. In *Amorpha*, standard petal alone developed.

Pollination is through the agency of insects. A few members of *Erythrina* are visited by birds.

The **fruit** is a legume. In *Pterocarpus*, fruit is samara.

The **seed** is dicotyledonous and non-endospermic.

Important Plants

Arachis hypogaea (ground nut), *Cajanus cajan* (red gram), *Cicer arietinum* (chick pea), *Dolichos lablab* (common bean), *Glycine max* (bean) *Phaseolus trilobatus* (Mung bean), *Pisum sativum* (pea), *Vigna radiata* (pea) are cultivated for their edible seeds and fruits. Seeds and leaves *Trigonella foenum-graecum* are used as pulse and leafy vegetable. Valuable woods are obtained from *Dalbergia latifolia, D. sissoo, Pterocarpus santalinus* and *P. marsupium*. Indigo dye is obtained from the leaves and roots of *Indigofera tinctoria*. *Pongamia pinnata* tree yields fuel wood. *Crotalaria juncea* is cultivated for its green manure. *Erythrina* and *Sesbania* species are cultivated as shade trees in plantations. *Sesbania grandiflora* leaves are used as vegetable green. *Abrus precatorius, Clitoria ternatea, Desmodium gyrans, Psoralea corylifolia* are medicinal plants. Some plants are cultivated for garden ornamentals as *Clianthus dampieri, Lupinus hirsutus, Robinia pseudoacacia, Sesbania grandiflora, Sophora japonica* and *Wisteria sinensis*.

Systematics and Phylogeny

Traditionally the legumes are treated heterogenous group of large family Leguminosae under the order Rosales. Bentham and Hooker divided the legumes into three subfamilies, i.e. Papilionoideae, Caesalpinioideae and Mimosoideae). However, they were treated as separate families as Mimosaceae, Caesalpiniaceae and Fabaceae under order Fabales by Takhtajan (1980) and Cronquist (1981) or Leguminales by Hutchinson (1969). The Fabales are an order of flowering plants included in the Rosid group of the Eudicots in the Angiosperm Phylogeny Group II classification system. In the APG II circumscription, this order includes the families Fabaceae or legumes (including the subfamilies Caesalpinioideae, Mimosoideae, and Faboideae), Quillajaceae, Polygalaceae or milkworts (including the families Diclidantheraceae, Moutabeaceae, and Xanthophyllaceae), and Surianaceae. Under the Cronquist system and some other plant classification systems, the order Fabales contains only the family Fabaceae. In the classification system of Dahlgren the Fabales were in the superorder Fabiflorae (also called Fabanae) with three families corresponding to the subfamilies of Fabaceae in APG II. The other families treated in the Fabales by the APG II classification were placed in separate orders by Cronquist, the Polygalaceae within its own order, the Polygalales, and the Quillajaceae and Surianaceae within the Rosales. APG IV system treated the Fabales under clade Rosids with four families such as Quillajaceae, Fabaceae (Leguminosae), Surianaceae and Polygalaceae.

Myrtaceae (Myrtle Family)

Distribution: The family consists of 143 genera and 3,400 species that are distributed mostly in the tropical regions. Several of them are native to Australia. In India, the family is represented by about 15 genera and 170 species. Some of the larger genera of Myrtaceae are *Eugenia* (1000 spp.), *Eucalyptus* (500 spp.), *Syzygium* (500 spp.), *Myrica* (500 spp.), *Psidium* (140 spp.), *Myrtus* (100 spp.) and *Melaleuca* (100 spp.).

Habit: These are mostly trees or shrubs; a few are creepers. Trees of *Eucalyptus regnans* reach to a height of 100-130 meters.

Vegetative Characters

Root system is of normal tap root type.

The **stem** is woody, branching with characteristic bark which peels off in the form of scaly patches. Bicollateral vascular bundles are characteristically present.

The **leaves** are simple, opposite (alternate in older parts of *Eucalyptus*), exstipulate, entire, leathery, rarely scaly with gland dotted and reticulate venation. A prominent intramarginal vein is usually present.

Floral Characters

The **flowers** are bisexual, actinomorphic, epigynous (perigynous in *Tristania*), 4-5-merous, bracteate and bracteolate. The thalamus is fused with the ovary in varying degrees and for hypanthium. In *Darwinia*, the thalamus tube is beyond the ovary.

The **calyx** consists of 4 or 5 sepals, usually free, often closed in the bud. Aestivation is usually imbricate. Sepals are pappus like in *Darwinia*. They are reduced or even suppressed in *Eucalyptus*. In some species sepals are persistent as in *Myrtus*, *Eugenia*, etc.

The **corolla** consists of 4 or 5 petals, usually free (*Callistemon*) or united (*Eucalyptus*), caducous or shed in the form of lid. In *Eucalyptus citriodora*, the corolla and calyx fuse to form a cup-shaped structure called operculum. Aestivation is imbricate.

The **androecium** consists of numerous stamens arranged in one or two whorls along the margin of the disc and infolded in the bud. They are free or in bundles placed opposite to the petals as in *Melaleuca*. The anthers are dithecous, versatile, introrse, dehiscing longitudinally by slits and gland-tipped.

The **gynoecium** is 4-5-carpellary, syncarpous; ovary is inferior (rarely half-inferior in *Callistemon*). Style is terminal and stigma captiate. Ovary is with many chambers and uniovuled, with axile placentation. Ovary is unilocular and placentation is parietal in *Rhodamnia*.

Pollination is entamophilous. Flowers are highly attractive due to coloured stamens as in *Callistemon* and some species of *Eucalyptus*.

The **fruit** is a simple one, commonly berry (*Psidium, Eugenia*) or a loculicidal capsule (*Callistemon, Eucalyptus*), rarely a drupe (*Pimenta*) or a nut in *Darwinia*.

The **seed** is dicotyledonous and non-endospermous; seeds are winged in some species of *Eucalyptus*. Polyembryony is seen in *Syzygium*.

Important Characters

1. Peeling bark and bicollateral vascular bundles in stem.
2. Gland-dotted simple leaves with intramarginal vein.
3. Epigynous flower with caducous perianth.
4. Numerous stamens bent inwards in the bud.
5. Inferior ovary with axile placentation.

Important Plants

Valuable wood is obtained from *Eucalyptus globulus* (blue gum), *E. leucoxylon* (iron bark), *E. robusta, Melaleuca leucodendron* and *Syzygium cumini*. Volatile oil is extracted from leaves of *Eucalyptus* species and flower buds of *Eugenia caryopyllata* (clove oil). Flower buds of *Eugenia caryophyllata* and fruits of *Pimenta officinalis* are used for flavouring dishes. The fruits of *Psidium guajava, Rhodomyrtus tomentosa, Syzygium cumini* are eaten.

Callistemon linearis (bottle brush), *Leptospermum laevigatum* (Australian tea tree), *Melaleuca leucadendron* (cajuput tree), *Tristania conferta* (Brisbane tree) and *Myrtus communis* (myrtle) are beautiful garden trees.

Systematics and Phylogeny

All taxonomists invariably placed the family under Myrtales. The family is closely related with Rhizophoraceae, Lythraceae, Combretaceae and Onagraceae, and to some extent to Melastomataceae and Sonneratiaceae. Myrtaceae is traditionally divided into two subfamilies Myrtoideae and Leptospermoideae. The molecular phylogenetic studies were first focused on the largest tribe of Myrtaceae, the fleshy fruited and primarily Neotropical tribe Myrteae. Initial results focusing on subtribal relationships were published in 2005 and 2007 and have demonstrated the paraphyly of traditional subtribes Myrtiinae and Eugeniinae, the monophyly of the Myrciinae *sensu stricto* excluding *Myrceugenia* and the monophyly of a group of previously hard-to-place genera, *Myrceugenia, Luma* and one species of *Blepharocalyx*.

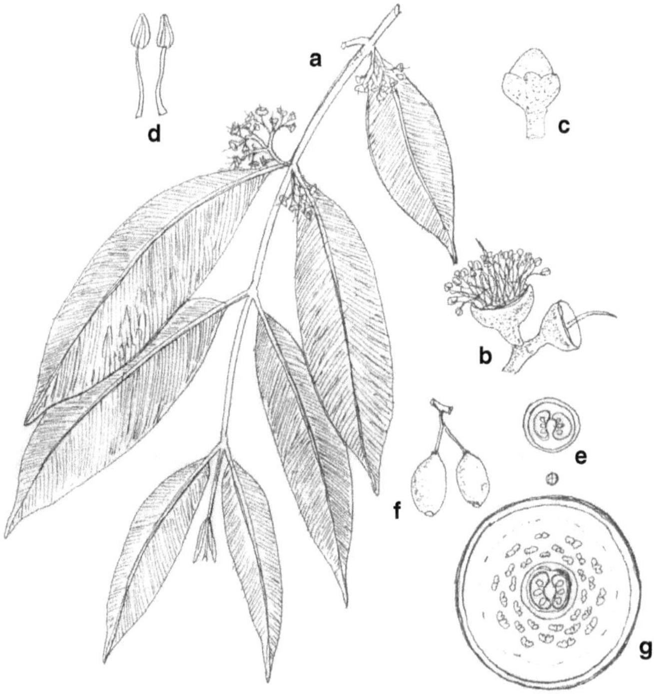

Fig. 15.10: *Syzygium cumini* (L.) Skeels. a. Flowering Twig; b. Flower- entire; c. Floral bud; d. Stamens; e. Ovary c.s.; f. Fruits; g. Floral diagram.

In APG IV system, order Myrtales is included under Rosids and includeds 8 families - Combretaceae, Lythraceae, Onagraceae, Vochysiaceae, Myrtaceae, Melastomataceae, Crypteroniaceae, Alzateaceae and Penaeaceae. Recently many studies revealed the infrafamilial classification of Myrtaceae but until now have not resolved the delimitation of several generical relationships.

Cucurbitaceae (Gourd Family)

Distribution: The family comprises of about 95 genera and 960 species, which are mostly distributed in tropical and subtropical regions. About 37 genera and 100 species have been reported in India. Some of the larger genera distributed widely are *Momordica* (45 spp.), *Cucumis* (25 spp.), *Cucurbita* (15 spp.), *Lagenaria* (7 spp.) and *Luffa* (6 spp.).

Habit: They are mostly annual, rarely perennial, climbing or prostrate herbs. *Dendrosicyos* is a tree and *Acanthosicyos* is a shrub.

Vegetative Characters

Root system is of the normal tap root type.

The **stem** is commonly climbing; five ridged, fistular with bicollateral vascular bundles. Usually plant body is covered by multicellular hairs.

The **leaves** are simple, pinnately or palmately lobed or divided, alternate, exstipulate (stipulate in *Kedrostis* and *Acanthosicyos*), venation palmately reticulate, petiolate and usually cordate at base. In the axil of leaf or slightly extra axillary tendril is constantly present in all genera except in *Ecballium*. Morphology of the tendril is unique in each genus.

Floral Characters

The **inflorescence** is axillary and usually with solitary female flower and several male flowers in various cymose or racemose types.

The **flower** is unisexual (rarely bisexual in *Schizopepon*), actinomorphic, epigynous, rarely perigynous in *Actinostemma*. Thalamus formed a cup above the ovary.

The **calyx** consists of 5 sepals, free, usually narrow with valvate aestivation.

The **corolla** consists of 5 petals, usually united, rarely free as in *Luffa* and *Fevillea*. Petals are white or yellow in colour. Aestivation is valvate or imbricate.

The **androecium** consists of 5 stamens that are alternate with petals. There is great complexity in arrangement and union of anthers among the family members. All five stamens are free in *Fevillea*; 2+2+1 in *Thladiantha*; synandria in *Citrullus*; filaments free and anthers are united in *Sicydium*. Anthers are monothecous, sinuous and extrorse. The anther lobe is straight in *Fevillea*; curved in most of the members; completely united in *Cyclanthera*. In *Sechium* one stamen bear monothecous and others are dithecous anthers. Staminodes are seen in pistillate flowers.

The **gynoecium** consists of 3 carpels (rarely 4-5 as in *Cucurbita pepo*), syncarpous. Ovary is inferior, style terminal, usually bear 3 commissural stigmas. Ovary is unilocular with numerous anatropous ovules on parietal placenta. *Sechium* has single pendulous ovule. Usually a pistillode is seen in male flowers.

Pollination is entamophilous. Nectar is secreted by the epigynous disc.

The **fruit** is a pepo. In *Momordica*, fruit breaks to liberate the seeds. Fruits are detached from the stalk at the base in *Ecballium*. The fruit rolls back irregularly in *Cyclanthera*.

The **seed** is dicotyledonous and non-endospermic.

Key Characters of the Family

1. Tendrillar climbers
2. Palmately reticulate veined leaves.
3. Unisexual actinomorphic, epigynous flowers.
4. Synandrium.

5. Inferior ovary with parietal placentation.
6. Fruit pepo.

Important Plants

Several species of this family are cultivated for their vegetable fruits: *Benincasa hispida* (ash gourd; winter melon), *Citrullus lanatus* var. *lanatus* (water melon), *Coccinia grandis* (little gourd), *Cucumis melo* (musk melon), *Cucurbita maxima* (pumpkin), *Cucurbita pepo* (squash), *Lagenaria vulgaris* (bottle gourd), *Luffa acutangula* (ribbed sponge gourd), *Luffa tuberosa, Momordica charantia* (bitter gourd), *Sechium edule, Trichosanthes lobata* (snake gourd). Some common wild plants are medicinal such as *Citrullus colocynthis, Diplocyclos palmatus, Kedrostis rostrata* and *Trichosanthes cucumerina.*

Systematics and Phylogeny

The family Cucurbitaceae exhibits controversial affinities among the flowering plant families and many taxonomists placed it in various positions because of their highly advanced vegetative and floral characters. Bentham and Hooker placed it under order Passiflorales, but their floral character and anatomical characters are closely allied to Campanulaceae. Rendle belevied it to belong the order Peponiferae while Takhtajan (1969) treated under Cucurbitales in between Passiflorales and Begoniales. Cronquist (1981) placed it in the order Violales along with 23 more families whereas Thorne (1983) erected suborder Cucurbitineae of order Violales along with 16 other families. Jeffrey (1990) divided the family into two subfamilies as Zanonioideae and Cucurbitoideae with 8 tribes and 19 subtribes. But recent study on genetics and molecular approaches have not supported the Jeffrey's view. In APG IV system, Cucurbitaceae was placed under clade Rosids, order Cucurbitales with 8 families - Apodanthaceae, Anisophyllacae, Corynocarpaceae, Coroariaceae, Cucurbitaceae, Tetramelaceae, Datiscaceae and Begoniaceae.

Apiaceae (Umbelliferae)

Distribution: This family comprises of about 275 genera and 3,000 species which are distributed mostly in the temperate region of the northern hemisphere. Over 50 genera and 200 species were reported in India. Some of the larger genera are *Eryngium* (230 spp.), *Pimpinella* (150 spp.), *Bupleurum* (150 spp.), *Ferula* (130 spp.), *Peucedanum* (120 spp.) and *Hydrocotyle* (100 spp.). The family is very popular for Hemlock, a poisonous wood that was given to the death of a great Greek philosopher Socretes.

Habit: They are mostly herbaceous annuals. Some are biennials perennating by underground tubers or rhizomes. A few are shrubs, runners and climbing habits.

Vegetative Characters

The **root system** is of the normal tap root type. Sometimes roots stored food material and become tubers as in *Daucus carota.*

The **leaves** are usually compound, pinnately or palmately, dissected many times; rarely simple (as in *Pimpinella mucronata*), alternate (opposite in *Apiastrum*), exstipulate. The leaf base is broad and sheathing. Venation is reticulate but parallel in *Aciphylla*. Foliage is usually aromatic. The chief anatomical characters are schizogenous cavities with ethereal oil glands or resin, in all organs.

Floral Characters

The **inflorescence** is terminal, often supra-axillary, simple or branched umbels, sometime umbels in cymose manner. The main inflorescence is surrounded by caducous or deciduous, simple or compound bracts that form an involucre. The ultimate inflorescence is surrounded by bracts called involucel. The inflorescence is much condensed to form a head in *Eryngium*.

The **flowers** are small, bisexual or unisexual, polygamous or monoecious as in *Echiophora* or dioecious as in *Arctopus*. They are actinomorphic, but in some cases zygomorphic, epigynous, bracteate.

The **calyx** consists of 5 free sepals, aestivation valvate or imbricate, much reduced to teeth-like or obsolete in *Apium*.

The **corolla** consists of 5 petals, free, regular, usually incurved, sometimes outer floral petals exagerated. Aestivation is imbricate or valvate.

The **androecium** consists of 5 stamens that alternate with petals, incurved in bud. Anthers are dithecous, introrse, dehiscing by longitudinal slits.

The **gynoecium** is bicarpellary, syncarpous (posterior carpel aborted in *Lagoecia*), epigynous and ovary bearing free styles on top of the well developed epigynous-disc called stylopodium. The ovary is bilocular with one anatropous unitegmic, pendant ovule on axile placenta in each chamber.

Pollination is through the agency of insects.

The **fruit** is schizocarp (cremocarp), splitting into two, one-seeded mericarp. Both mericarps are held together at the top by a thin, bifid or undivided stalk called carpophore. The mericarps may be flattened laterally or dorsally.

The **seed** is dicotyledonous and endospermic. Only one cotyledon is present in *Conopodium*.

Key Characters of the Family

1. Herbaceous habit.
2. Aromatic leaves with sheathing bases and dissected lamina.
3. Umbel inflorescence.
4. Epigynous flowers.
5. Bilocular ovary with pendant solitary ovule in each chamber.
6. Schizocarpic fruit.

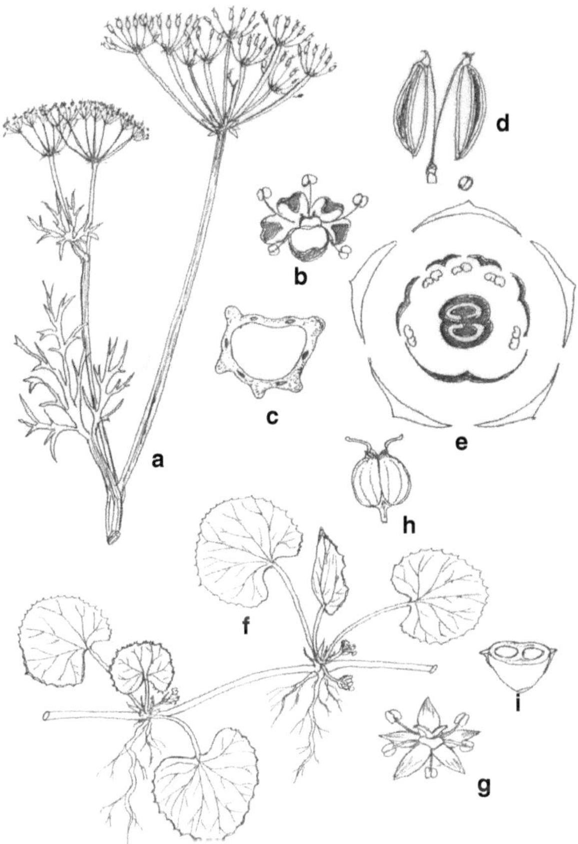

Fig. 15.11: *Foeniculum vulgare* Mill. a. Flowering twig; b. Flower- entire; c. Ovary c.s.; d. Fruit; e. Floral diagram,; f. *Centella asiatica* – plant; g. flower- entire; h. Fruit; i. Ovary c.s.

Important Plants

Several members of the family are useful plants for culinary purpose because of the presence of essential oils and resin. Fruits of *Coriandrum sativum* (coriander), *Carum copticum* (caraway), *Foeniculum vulgare* (fennel), *Cuminum cyminum* (cumin) are used as flavouring materials. The leaves of *Coriandrum sativum*, *Apium graveolens* (celery), *Pimpinella* and *Angelica* are also used for same purposes. Poisonous alkaloids are obtained from *Cicuta virosa* (cowbane), *Aethusa cynapium* (fools parsley) and *Conium maculatum* (hemlock). A gum-resin obtained from the roots of *Ferula asafoetida* (asafoetida). The root of *Daucus carota* (carrot) is of vegetable value. *Centella asiatica* is a common medicinal plant.

Systematics and Phylogeny

Bentham and Hooker placed the family under order Umbellales. Many characters are closely allied to the family Araliaceae, hence Hutchinson believed that Araliaceae is primitive than Apaiceae (Umbelliferae) and placed under Araliales. The floral charcters like reduction of calyx, bicarpellary syncarpous ovary with 2 separate styles shows close affinity with Asteraceae (Compositae). Hutchinson also opined that Umbelliferae is derived from Saxifragales. However, Cronquist opined that it originated from Sapindales. Engler divided the family into 3 subfamilies as Hydrocotyloideae, Saniculoideae and Apioideae with 12 tribes. The Apiaceae is solidly placed within the Apiales order in the APG III classification system. It is closely related to Araliaceae and the boundaries between these families remain unresolved. Traditionally these two families have been delimited largely based on fruit morphology, and the results have not been congruent with the more recent molecular phylogenetic analyses. The subfamilial and tribal classification for the family is currently in a state of flux, with many of the groups being found to be grossly paraphyletic or polyphyletic. In APG IV system the family is placed under the last order Apiales in the clade Asterids along with seven families - Pennantiaceae, Torricelliaceae, Griseliniaceae, Pittosporaceae, Araliaceae, Myodocarpaceae and Apiaceae.

Rubiaceae (Coffee Family)

Distribution: This family is represented by about 500 genera and 6,500 species which are distributed mostly in tropical countries, extending into the temperate regions. A few are found in sub-arctic zone. About 75 genera and 300 species reported in India. Some of the larger genera of the family are *Psychotria* (700 spp.), *Galium* (400 spp.), *Ixora* (400 spp.), *Gardenia* (250 spp.), *Mussaenda* (200 spp.), *Asperula* (200 spp.), *Cephealis* (180 spp.), *Rondeletia* (120 spp.), *Coprosma* (90 spp.), *Guettardia* (80 spp.) and *Rubia* (60 spp.).

Habit: They are usually trees (*Cinchona officinalis*), shrubs, lianes or stragglers and rarely herbs. *Hydnophytum* and *Myrmecodia* are epiphytes. In *Myrmecodia*, the hypocotyl forms a tuberous structure and with numerous chambers. These chambers formed the abode of ants. Similar myrmecophilous species are found in *Nauclea*, *Duroia* and *Cuviera*. The bacterial leaf nodules usually found in most of the species of *Psychotria*. *Uncaria* is a hooked climber.

Vegetative Characters

The **root system** is of the normal tap root type.

The **leaves** are simple, opposite, sometime whorled, anisophyllous as in *Randia* and *Gardenia*, stipulate. The stipules are commonly interpetiolar but in *Psychotria* and *Hamiltonia* stipules are intrapetiolar. In *Spermacoce* they form a tubular structure around the axis. They may be glandular in *Pentas* and much dissected in *Oldenlandia* and *Hedyotis*. The stipules are foliaceous in the tribe *Stellateae*. Consequently each node has four or six leaves with whorled arrangement, because of stipular leaves as in *Galium*. The venation is pinnately reticulate. In *Uncaria* and *Plectronia*, hooks are produced from the leaf axil.

Floral Characters

The **inflorescence** is terminal or axillary, usually cymose type. Dichasial cymes are common but polychasial and monochasial cymes are also seen. The floral axis is condensed in globose head in *Mitragyna* and *Ochrenauclea*. The flowers are united in pairs of their ovaries in *Mitchella* and *Pomax*. Similar union is seen in many flowers of an inflorescence in *Morinda*. In *Vaillantia* and *Sherardia*, involucral bracts surround the flowers.

The **flowers** are bisexual, unisexual in *Coprosma*, dioecious in *Brysophyllum tetrandrum*, polygamo-dioecious in *Tricalysia*, tetramerous or pentamerous, epigynous, actinomorphic, rarely zygomorphic in *Henriquezia*.

The **calyx** consists of 4 or 5 sepals, free, valvate or imbricate aestivation. The sepals are usually small, sometimes reduced to a ring as in *Morinda* or even suppressed as in *Rubia* and *Galium*. In *Mussaenda*, one sepal becomes enlarged, coloured and petaloid.

The **corolla** consists of 4 or 5 petals, united to form a tubular, salver or funnel-shaped structure. It is bilabiate in *Henriquezia*. Aestivation is valvate or twisted.

The **androecium** consists of as many stamens as there are petals, epipetalous, alternate with petals. Anthers are dithecous, introrse, dehiscing by longitudinal slits.

The **gynoecium** is bicarpellary (pentacarpellary in *Hamelia*), syncarpous, epigynous with one to many ovules on axile placentation (parietal placentation in *Gardenia*). Ovary is inferior, half-inferior in *Henriquezia* and superior in *Pagamea*. Style is terminal, simple or forked.

Pollination is usually entamophilous. Heterostyly is seen in *Oldenlandia*, *Morinda*, *Hedyotis* and *Psychotria*.

The **fruit** is a berry (*Coffea*) or a capsule (*Cinchona*). In some species it is a drupe with two pyrenes. The fruit is schizocarp in *Galium* and multiple one in *Morinda*.

The **seed** is dicotyledonous and endospermic except in *Guettarda*. The seeds are winged in *Cinchona* and *Henriquezia*. The hooked fruits of *Galium* are dispersed by animals.

Key Characters of the Family

1. Interpetiolar stipules.
2. Regular epigynous, tetra or pentamerous flowers.
3. Axile placentation.
4. Endospermic seed.

Important Plants

The most important economic plant from the family is coffee, which is obtained from *Coffea arabica*, *C. liberica* and *C. robusta*. Antimalarial drug quinine is obtained from the bark of *Cinchona officinalis* and *C. succirubra*.

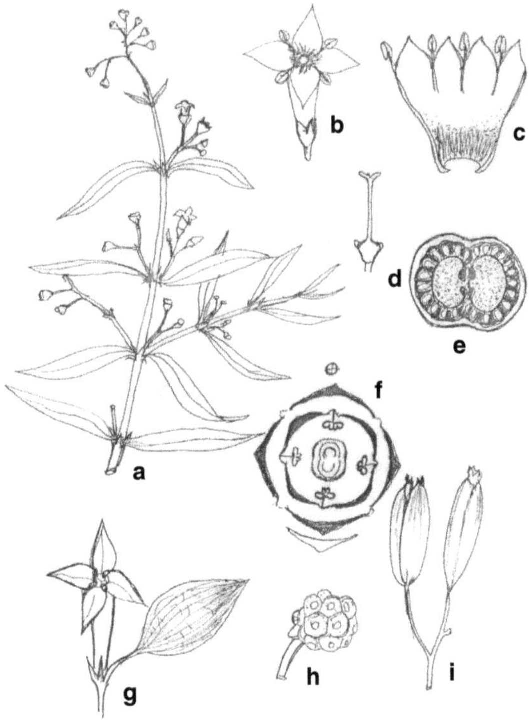

Fig. 15.12: *Oldenlandia umbellata* L. a. Twig; b. Flower-entire; c. Petals with stamens; d. Gynoecium; e. Ovary c.s.; f. Floral diagram; g. Flower of *Mussaenda*; h. Fruit of *Morinda*; i. Fruit of *Cinchona*.

Lancewood is taken from *Calycophyllum candidissimum*. Some species are yielding dycs namely *Rubia cordifolia* (red-dye), *Oldenlandia umbellata* (dull pink) and *Asperula* (pink). Several species of *Mussaenda, Ixora, Pentas, Rondeletia* and *Hamelia* are cultivated in the gardens for their beautiful flowers. *Morinda citrifolia* fruits yields adaptogenic medicine.

Systematics and Phylogeny

Bentham and Hooker placed the family Rubiaceae under the order Rubiales whereas Takhtajan (1969) and Thorne placed it in the order Gentianales. Affinities of Rubiaceae are since now controversial but it is phytochemically related with Loganiaceae. Morphologically it is closer to Caprifoliaceae but only stipular character differs. The family also shows resemblance with Apiaceae and Cornaceae. Several other accepted families are since long included in Rubiaceae, such as Aparinaceae, Asperulaceae, Catesbaeaceae, Cephalanthaceae, Cinchonaceae, Coffeaceae, Coutariaceae, Galiaceae, Gardeniaceae, Guettardaceae, Hameliaceae, Hedyotidaceae, Houstoniaceae, Hydrophylacaceae, Lippayaceae, Lygodisodeaceae,

Naucleaceae, Nonateliaceae, Operculariaceae, Pagamaeaceae, Psychotriaceae, Randiaceae, Sabiceaceae and Spermacoceaceae.

The traditional classification system of Rubiaceae distinguished only two subfamilies as Cinchonoideae, characterized by more than one ovule in each locule, and Coffeoideae, having one ovule in each locule. This view was much criticized because of the distant position of two obviously related tribes, viz. Gardenieae with many ovules in Cinchonoideae and Ixoreae with one ovule in Coffeoideae, and because in species of *Tarenna* the number of ovules varies from one to several in each locule. The introduction of molecular phylogenetics in Rubiaceae research has corroborated or rejected several of the conclusions made in the pre-molecular era. But in APG III classification, there is support for the subfamilies Cinchonoideae, Ixoroideae, and Rubioideae, although differently circumscribed, and Antirheoideae is shown to be polyphyletic. Currently, in most molecular research concerning the Rubiaceae family, the classification is retained three subfamilies (Cinchonoideae, Ixoroideae, and Rubioideae). Molecular studies also have substantial impact on tribal delimitations and taxonomic changes are still being made. Most recent studies have demonstrated the phylogenetic placement of Rubiaceae within the order Gentianales and the monophyly of the family are confirmed. The relationships of the three subfamilies of Rubiaceae together with the tribes Coptosapelteae and Luculieae are recently added in APG IV system.

Asteraceae (Compositae) (Sunflower Family)

Distribution: This is the largest and highly evolved family of the Dicotyledones consisting of about 1,900 genera and 32,900 species, which constitute about 10% of the known flowering plants. They are distributed worldwide and are able to thrive in all places. About 140 genera and 800 species are reported in India. Some of the largely represented genera of the family are *Senecio* (2000 spp.), *Eupatorium* (1200 spp.), *Vernonia* (1000 spp.), *Hieraceum* (1000 spp.), *Centaurea* (600 spp.), *Aster* (500 spp.), *Helichrysum* (500 spp.), *Artemisia* (400 spp.), *Baccharis* (400 spp.), *Mikania* (200 spp.), *Bidens* (230 spp.), *Crepis* (200 spp.), *Inula* (200 spp.), *Achillea* (200 spp.), *Gnaphalium* (200 spp.), *Chrysanthemum* (200 spp.), *Anthemis* (200 spp.), *Erigeron* (200 spp.), *Cirsium* (150 spp.), *Ligularia* (150 spp.), *Helianthus* (110 spp.) and *Hypochoeris* (100 spp.).

Habit: In general, members of this family are herbaceous annuals. Some are shrubs and a few are trees (*Vernonia monosis*). Twiners like *Cissampelopsis* and *Mikania* are not uncommon. Various habits are exhibited by the different species in a single genus *Senecio* (*S. calcadense* – a climber; *S. hastata* – a shrub; *S. articulata* – a succulent; *S. aquatica* – a marshy herb). *Kleinia* and *Notonia* showed remarkable xerophytic adaptations. Schizogenous cavities containing oils or some excretory products are found in various parts of the plant. In the tribe Cichoriae, laticiferous vessels are present. Various types of hairy or glandular outgrowths are produced on the surface.

Vegetative Characters

The **root system** is of normal tap root type. Roots are tuberous in *Dahlia*.

Stem tubers are produced in *Helianthus tuberosus* and *Crepis bulbosa*. The stem is leaf like in *Baccharis*. Radical buds of *Taraxacum* help in vegetative propagation. The stolons of *Achillea* and suckers of *Chrysanthemum* are helpful in a similar way.

The **leaves** are radical or cauline. They are opposite or alternate, rarely whorled (*Zinnia verticillata*), succulent in *Notonia*, scaly in *Haplophyllum*, exstipulate, venation pinnately reticulate. In *Corymbium*, leaves are parallel-veined.

Floral Characters

The **inflorescence** is characteristically a head. Numerous sessile flowers are produced in a centripetal manner on a much condensed axis that is flat (in *Chrysanthemum*), concave (in *Epaltes*), convex (in *Tridax*) or conical (in *Spilanthes*). The flowers are surrounded by a few or numerous bracts that are in one or several whorls. Each flower is in the axil of a scaly bract or palea. Rarely, involucre is absent in which case bracts of the individual flowers or foliage leaves offer protection. The whole inflorescence looks like a single flower. It may be solitary or several inflorescences may be arranged in a racemose or cymose manner or in panicles as in *Solidago*. The head consists of single flower in *Echinops* and a few flowers in *Sphaeranthes*.

The **flowers** of the head may be uniform in structure and function (homogamous) or vary in structure and function (heterogamous). Homogamous head consists of regular flowers as in *Notonia* and ligulate flowers in *Aster*. In the heterogamous head, the flowers of the margin (ray floret) are zygomorphic and usually pistillate, sometime neutral (*Centaurea*), while the central flowers (disc floret) are actinomorphic and usually bisexual. In advanced members, the head bears flowers of only one sex leading to monoecious condition in *Xanthium* and dioecious in *Baccharis*. The flowers are bisexual, unisexual or polygamous. They are epigynous, actinomorphic or zygomorphic, sessile, bracteate (bracts are usually chaffy), achlamydeous in *Xanthium*.

The **calyx** is represented by a rudimentary structure or may be fully developed in the fruit forming numerous hairs called pappus. Pappus hairs are arranged in a ring on the top of the ovary. It is represented in five scaly structure in *Gaillardia*, four in *Bidens* and two or three in *Helianthus*. In *Cichorium*, the calyx formed a ring and completely absent in *Xanthium* and *Sphaeranthus*.

The **corolla** consists of five petals, united, tubular in disc florets and usually ligulate in ray florets. The corolla may be ligulate in *Leontodon* or bilabiate in *Calendula*. Aestivation is valvate. Corolla is absent in the female flowers of *Xanthium*.

The **androecium** consists of 5 stamens that are epipetalous and are alternate with petals. The stamens are syngenesious (anthers are united and filaments are free). In *Xanthium*, the stamens are united only by the filaments (monadelphous). Anthers are dithecous, introrse, dehiscing by longitudinal slits. The connective of the anther is prolonged beyond the anther lobe. Filaments are sensitive in *Centaurea*.

The **gynoecium** is bicarpellary, syncarpous, unilocular, solitary ovule on basal placentation. The ovary is inferior with terminal style and forked stigma.

Pollination is usually entamophilous. Self pollination is observed in *Senecio vulgaris*. Wind pollination occurs in *Artemisia* and *Xanthium*.

The **fruit** is a cypsela which is carried away by wind, as the pappus makes easy to fly. In *Zinnia*, the fruit wall becomes flattened into wings. The corolla covers the fruits and forms a spongy float in fruit of *Diotis*. The calyx is modified into hooked bristles in *Bidens*, spines in *Volutarella* and hooks in *Xanthium*. Fruits are sticky in *Siegesbeckia*.

The **seed** is dicotyledonous and non-endospermic.

Key Characters of the Family

1. Annual herbs.
2. Head inflorescence.
3. Epigynous flowers.
4. Syngenesious anthers.
5. Unilocular ovary with solitary ovule on basal placentation.
6. Cypsela fruit.

Important Plants

This family is commercially important. Oil is extracted from the seeds of *Helianthus annuus*, *Carthamus tinctorius*, *Media sativa* and *Guizotia abyssinica*. Colouring materials are obtained from *Carthamus tinctorius* and *Serrataxia tinctoria*. Leaves of *Artemisia absinthium*, *Cnicus benedictus*, *Eclipta alba* and heads of *Artemisia maritima* are medicinal. Root powder of *Cichorium intybus* is the chicory, an adulterant of coffee. Insecticides are prepared from the heads of *Centrantherum anthelminticum*, *Chrysanthemum coccineum* and *Pyrethrum cinerariifolium*. Tuberous roots of *Cynara scolymus*, *Dahlia tuberosa*, *Helianthus tuberosus* and *Tragopogon porrifolius* are edible. Several species are horticultural value such as *Cosmos*, *Zinnia*, *Helianthus*, *Dahlia*, *Tagetes*, *Rudbeckia*, *Ageratum*, *Vernonia*, etc.

Systematics and Phylogeny

Bentham and Hooker placed the family Asteraceae under order Astreales near to the Rubiales. It is presumed that Asteraceae (Compositae) originated from Rubiaceae. In possessing pentamerous flowers, syngenesious anthers and inferior ovary, the family is closely related to Campanulaceae and Goodeniaceae. In possessing bicarpellary inferior ovary, also comes near to Rubiaceae and Stylidiaceae. In the presence of dense inflorescence with inferior ovary is also closely affine to Dipsacaceae and Valerianaceae.

Traditionally the family Compositae is divided into 13 tribes as Heliantheae, Astereae, Anthemideae, Arctotideae, Inuleae, Senecioneae, Calenduleae, Eupatorieae, Vernonieae, Cynareae, Mutisieae, Liabeae and Lactuceae. The modern approach of the Asteraceae has recognized two subfamilies: Asteroideae (or Tubuliflorae) and Cichorioideac (or Liguliflorae).

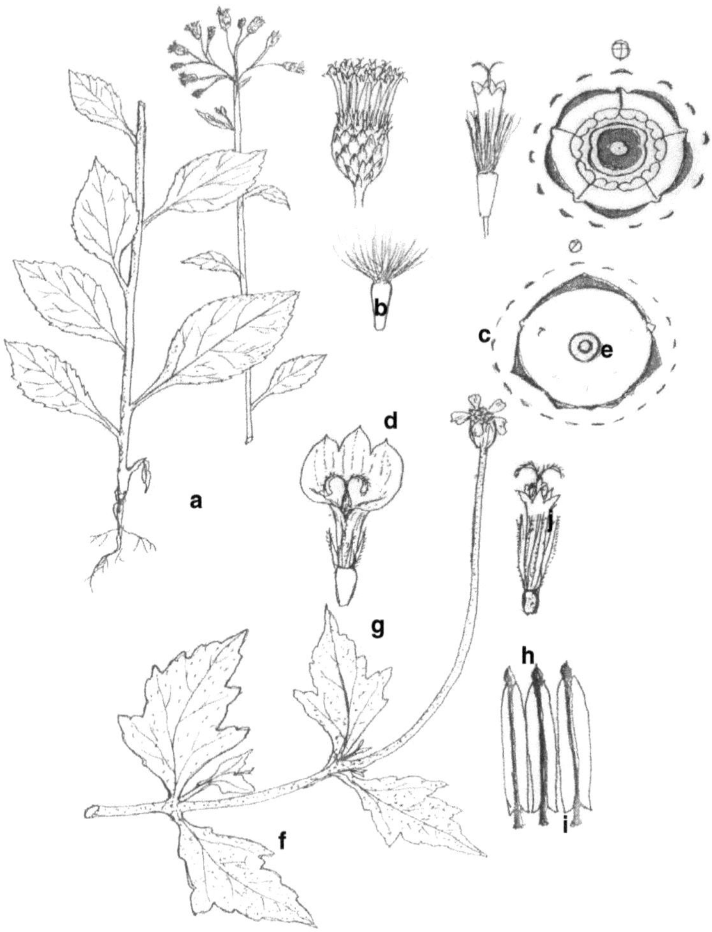

Fig. 15.13: ***Vernonia cinerea*** **(L.) Lees. a. Plant; b. Inflorescence; c. Flower-entire; d. Fruit; e. Floral diagram; f. Twig of** ***Tridax procumbens*****; g. Ray floret; h. Disc floret; i. Stamens; j. Floral diagram for ray floret**

The latter has been shown to be extensively paraphyletic, and has now been divided into 13 subfamilies, but the former still stands. The subfamilies of Asteraceae are Barnadesioideae, Famatinanthoideae, Mutisioideae, Stifftioideae, Wunderlichioideae, Gochnatioideae, Hacastocleiodeae, Carduoideae, Pertyoideae, Gymnarrhenoideae, Cichorioideae, Corymbioideae and Asteroideae (APG IV).

Sapotaceae (Sapota family)

Distribution: This family is comprised of about 65 genera and 800 species which are usually tropical in distribution. About 12 genera and 50 species have been reported in India.

Habit: The members are usually trees and shrubs. Generally they are evergreen in nature with abundant milky latex in the pith and cortex.

Vegetative Characters

Root system is of the normal type and buttress roots are found in some evergreen members.

The **leaves** are simple, alternate, rarely opposite, exstipulate and rarely stipulate (*Madhuca*). They are usually leathery, often with a covering of silky hairs. Venation is pinnately reticulate.

Floral Characters

The **inflorescence** arises sometimes on old stem (cauliflory), flowers solitary (*Chrysophyllum*) or in cymose clusters.

The **flowers** are bisexual, actinomorphic, hypogynous, polycyclic, 2-6-merous, bracteate, usually bracteolate.

The **calyx** consists of one whorl of 5 sepals, free or basally connate. When sepals are in two whorls, each whorl consists of 2, 3 (*Palaquium*, *Mimusops* and *Achras*) or 4 (*Madhuca*) in number. Aestivation is imbricate or valvate.

The **corolla** consists of as many as petals as there are sepals, but they are arranged in one whorl. The petals are united to form a short tube at base. In some cases, the lateral or dorsal appendages are produced, which give the false appearance of corolla lobes as in *Mimusops*. Aestivation is imbricate.

The **androecium** consists of stamens in two whorls. In *Madhuca*, the stamens are in 2 or 3 whorls and all are fertile. The stamens of the outer whorl are usually modified by petaloid staminodes but they are completely suppressed in *Chyrophyllum*. The stamens of inner whorl are fertile, arranged opposite to the petals, epipetalous. The filaments are short and anthers are dithecous, extrorse, dehiscing by longitudinal slits.

The **gynoecium** consists of as many carpels as there are stamens in a whorl, syncarpous with same number of chambers as many as carpels, placentation axile. The ovary is superior, with a terminal style and capitate stigma.

Pollination is by the agency of insects.

The **fruit** is berry with a number of seeds embedded in the fleshy pulp, in which stone cells are common.

The **seeds** are dicotyledonous and albuminous. The reserved food is oily in nature. Sometimes tannins are present in the testa. The seed has a broad hilum and shiny seed coat with stony in nature.

Key Characters of the Family

1. Tree habit.
2. Presence of milky latex.
3. Leathery simple leaves.

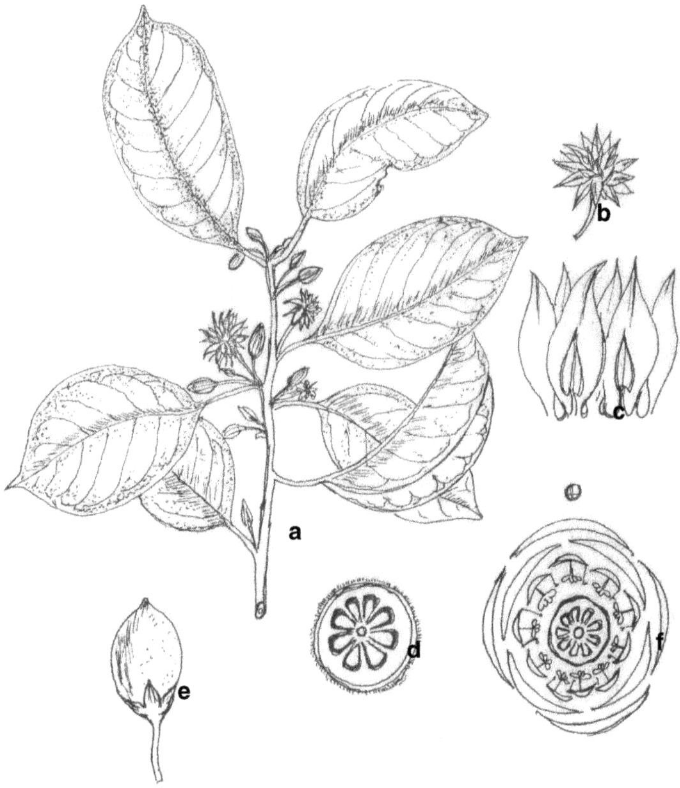

Fig. 15.14: *Mimusops elengi* L. a. Flowering twig; b. Flower-entire; c. Petals with stamen; d. Ovary c.s.; e. Fruit; f. Floral diagram

4. Polycyclic hypogynous flowers.
5. Petaloid appendages of the corolla.
6. Extrorse fertile stamens.
7. Many chambered ovary with solitary ovule.
8. Fruit berry.

Important Plants

The family members have much economic importance. Chewing gum is prepared from the latex of *Achras zapota, Mimusops elengi, Palaqium oblongifolium, P. gutta* and species of *Payena. Achras zapota, Chrysophyllum mamosum, Lucuma nervosa, L. salicifolia, L. obovata* are cultivated for their fleshy edible fruits. Butter-like substance is obtained from *Butyrospermum parkii* which is used in soap industry. Flowers of *Madhuca latifolia* are fermented for the distillation of alcohol. Species of *Sideroxylum* and *Bumelia* are good timber yielding trees.

Systematics and Phylogeny

Bentham and Hooker placed the Sapotaceae under order Ebenales. The family is usually divided into two tribes namely Palaquieae (petal lobes without appendages) and Mimusopseae (petal lobes with appendages). This family is closely related to Ebenaceae by most of the characters. Hallier believed that Sapotaceae have monotypic order Sapotales and Ebenaceae placed under separate order Santalales. Most of the earlier phylogenetists believed that both Sapotales and Santalales originated from common ancestral stock. In recent APG IV system, Sapotaceae is placed under order Ericales with 21 other families, some of the closely related families are Polemoniaceae, Ebenaceae, Primulaceae, Theaceae, Symplocaceae. Tribal classification is not well resolved.

Apocynaceae (Dogbane Family) (Includes Asclepiadaceae)

Distribution: This family consists of 620 genera and 5,100 species which are almost cosmopolitan in distribution. However, they are abundant in tropics. About 80 genera and 350 species have been reported in India. Some of the larger genera are *Hoya* (200 spp.), *Ceropegia* (180 spp.), *Asclepias* (150 spp.), *Oxypetalum* (150 spp.), *Matelea* (130 spp.), *Rauvolfia* (100 spp.), *Tabernaemontana* (100 spp.), *Parsonia* (100 spp.), *Aspidosperma* (100 spp.), *Dischidia* (80 spp.) and *Stapelia* (75 spp.).

Habit and **Habitat**: They are mostly shrubs and herbs, rarely trees. Several members are woody climbers (*Anodendron* and *Hoya*). *Adenium* shows strong xerophytic characters. *Caralluma*, *Brachystelma* and *Ceropegia* showed cactoid habit. *Dischidia* is an epiphytic climber and *Sarcolobus* is a mangrove plant.

Vegetative Characters

The **root system** is of normal tap root type. Some *Ceropegia* and *Brachystelma* species have root tubers. In some *Ceropegia* species roots are associated with mycorrhiza.

Shoot system shows dichasial branching in *Carissa*. Shoots are succulent and photosynthetic in *Caralluma*, *Ceropegia juncea* and *Sarcostemma*. Usually in Asclepiadoideae, stems are weak and twining in nature. Plants have abundant milky latex.

The **leaves** are simple, opposite (rarely alternate in *Cerbera* and *Cascabela*), decussate, sometimes whorled, exstipulate. Leaves are anisophyllous in *Rauvolfia tetraphylla*. Venation is pinnately reticulate. Leaves are reduced in *Caralluma*, *Ceropegia juncea* and *Sarcostemma*. In *Dischidia*, the leaf is transformed into a pitcher.

Floral Characters

The **inflorescence** is axillary or terminal, usually of the cymose type, ultimately tending to be monochasial. In *Caralluma* and *Brachystelma*, inflorescences are terminal and bearing receme type. In *Boucerosia*, flowers are in umbel.

The **flowers** are bisexual, actinomorphic, hypogynous, bracteate, pentamerous with reduction in the inner most whorl.

The **calyx** consists of 5 sepals, free with valvate or imbricate aestivation.

The **corolla** consists of 5 petals, united. The shape of the corolla is variable, rotate, infundibuliform, campanulate or tubular. Aestivation is usually twisted. Coronary outgrowths are often present on the corolla tube, or its mouth.

The **androecium** consists of 5 stamens that are alternating with petals, epipetalous in Apocyneae. Filaments are short, anthers dithecous, introrse, linear, oblong, dehiscing by longitudinal slits. Anthers are adhered to the stigma by connectives. Some members produced feathery out growths in tip of the anthers. Glandular out growths are seen in *Vallaris*. Pollen grains are in tetrads in advanced members like *Condylocarpus*. In Cynanchoideae, the pollen grains of each anther lobe are agglutinated to form a pollinium. From each anther two pollinia are produced. Normally one pollinium from each one of adjacent stamen, is attached to a horny conical body called corpusculum that is situated at the corner of the stigma by means of short or long elastic thread called retinaculum. Such thread like bodies are known as translator and attached to the corner of pentagonal stigma.

The **gynoecium** is bicarpellary, subapocarpous, usually free in the lower part. Stigma is single, dumbbell-shaped or pentagonal with sticky region along the sides. Ovules are usually numerous on ventral placentas of individual carpels.

Pollination is usually entamophilous. The insects are attracted by the size, colour and glandular bodies of the flowers. In *Ceropegia* a special pollination mechanism is adapted by its tubular corolla.

The **fruit** is a pair of follicles or drupes with pyrenes, fibrous drupe in *Cerbera*.

The **seed** is dicotyledonous and endospermic with silky comose.

Key Characters of the Family

1. Woody lianes and cactoid habits.
2. Opposite simple leaves.
3. Latex in the plant body.
4. Regular flower with coronary structures.
5. Sagittate anthers and pollinarium.
6. Subapocarpous pistil.
7. Follicle and drupaceous fruits with silky comose seeds.

Important Plants

Alstonia scholaris and *A. macrophylla* are common avenue trees. Many of them are used in medicinal purposes, these include *Catharanthus roseus, Decalepis hamiltonii, Gymnema sylvestre, Holarrhena pubescens, Hemidesmus indicus, Leptadenia reticulata, Rauvolfia serpentina* and *Wrightia tinctoria*. Many plants like *Allamanda grandiflora, Asclepias curassavica, Cryptostegia grandiflora, Tabernaemontana divaricata* and *Nerium odorum* are cultivated for their ornamental flowers in gardens. The cactioid forms of *Caralluma, Ceropegia, Brachystelma, Cynanchum* and *Hoya* are special ornamental values.

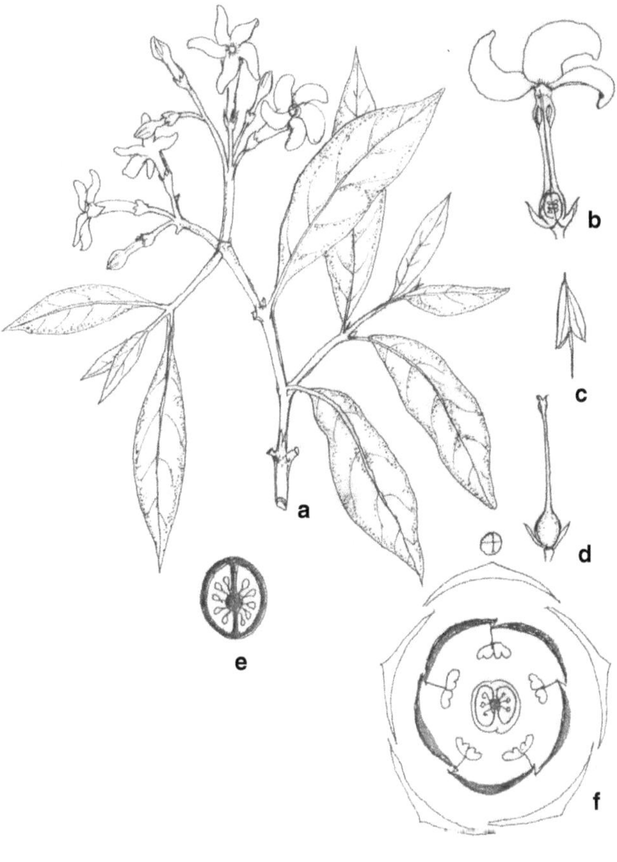

Fig. 15.15: *Tabernaemontana divaricata* (L.) R.Br. a. Plant twig; b. Flower l.s.; c. Stamen; d. Gynoecium; e. Ovary c.s.; f. Floral diagram

Systematics and Phylogeny

Bentham and hooker treated Apocynaceae and Asclepiadacae as separate families under the order Gentianales based on the pollen characters. Many other taxonomists also divided the familyAsclepiadaceae into two sub families, *viz*, Periplocoideae (pollen grains in tetrads, filaments free) and Cynancoideae (pollen grains united in pollinia, filaments united in a tube). In APG IV classification both Asclepiadaceae and Apocynaceae are considered a single family and merged them into a single family Apocynaceae. As of 2012, the family Apocynaceae is divided into five subfamilies, *viz*, Apocynoideae, Asclepiadoideae, Periplocoideae, Rauvolfioideae, and Secamonoideae. The former family Asclepiadaceae is included in Apocynaceae according to the Angiosperm Phylogeny Group IV (APG IV) based largely on molecular taxonomy.

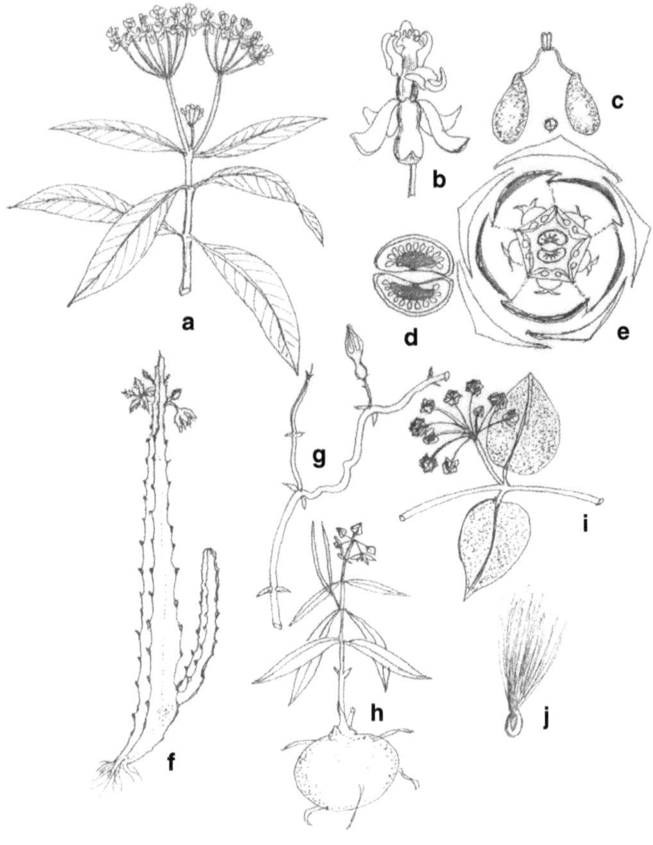

Fig. 15.16: *Asclepias curassavica* L. a. Twig; b. Flower-entire; c. Pollinarium; d. Ovary c.s.; e. Floral diagram; f. *Caralluma adscendens*- habit; g. *Ceropegia juncea* – habit; h. *Brachystelma glabrum*- habit; i. *Hoya wightii*- twig; j. Comose seed.

An updated classification, including 366 genera, 25 tribes and 49 subtribes, was published in 2014. In APG IV system, Apocynaceae is placed under order Gentianales with four other families - Rubiaceae, Gentianaceae, Loganiaceae and Gelsemiaceae.

Convolvulaceae (Glory Family)

Distribution: The family comprises about 50 genera and 1,200 species. The chief centres of distribution are tropical regions of America and Asia, extending the north and south temperate regions. About 20 genera and 150 species have been reported in India. Some larger genera of the family are *Ipomoea* (500 spp.), *Convolvulus* (250 spp.), *Cuscuta* (170 spp.), *Jacquemontia* (120 spp.), *Evolvulus* (100 spp.) and *Calystegia* (25 spp.).

Habit and **Habitat**: They are mostly twining annuals and perennial herbs, shrubs and rarely arborescent form as in *Humbertia*. They are growing in various localities, saline soils (*Cressa cretica*), sand-dunes (*Ipomoea pes-caprae*), floating in water (*I. hastata*), xerophytic (*Humbertia*) and parasitic (*Cuscuta*).

Vegetative Characters

The **root system** is of the normal tap root type. Adventitious roots are formed in *Ipomoea batatas* and they become tuberous.

The **shoot system** is usually slender, twining, herbaceous or woody. Rhizome is produced in *I. scammonia*. Stem is chloroplyllous and filifom in *Cuscuta*. The plant body is usually covered by hairs. In *Cressa critica*, the glandular hairs secrete hygroscopic salt.

The **leaves** are simple or variously lobed, alternate and exstipulate. Venation is pinnately or palmately reticulate. Leaves are absent in *Cuscuta*.

Floral Characters

The **inflorescence** is usually axillary cymose type, dichasial cyme tending to monochasial condition. Sometimes flowers are solitary. In *Porana* panicles are produced.

The **flowers** are bisexual (unisexual in *Hildenbrandtia*), actinomorphic, hypogynous and pentamerous. Bracts and bracteoles are well developed. In *Calystegia*, the bracteoles are large and coloured. Bracteoles are membranous and floating during the fruit formation and it helps to dispersal in *Neuropeltis*.

The **calyx** consists of 5 sepals, free with quincuncial aestivation. The outer sepals are usually bigger.

The **corolla** consists of 5 petals, united, folded marking with twisted aestivation. Scaly appendages are found on the petals of *Cuscuta*.

The **androecium** consists of 5 stamens, epipetalous, alternating with petals, unequal. Anthers are dithecous, introrse, dehiscing by longitudinal slits. Pollen are spinous in *Ipomoea*.

The **gynoecium** is bicarpellary, syncarpous, bilocular but false septum is formed resulting in tetralocular condition, placentation is axile. Ovary is superior with a terminal style. Two distinct styles arise from the base of the ovary in *Dichondra*. Stigma is capitate. In *Erycibe*, there are five carpels and style is absent.

Pollination is through the agency of insects. The flowers are attracted by size, colour and nectar produced by hypogynous disc. Cleistogamous flowers are found in *Dichondra repens*.

The **fruit** is usually 4-6-valved capsule.

The **seed** is dicotyledonous, endospermic with folded cotyledons.

Key Characters of the Family

1. Twining herbaceous habit.
2. Milky latex in the plant body.
3. Actinomorphic flowers with plaited corolla.

4. Axile placentation.
5. Ovules with ventral raphe.

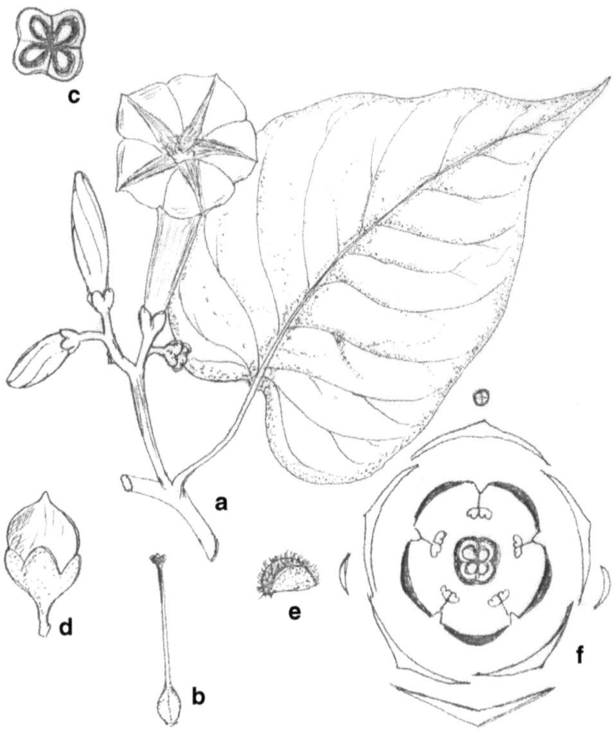

Fig. 15.17: *Ipomoea carnea* Jacq. a. Flowering twig; b. Gynoecium; c. Ovary c.s.; d. Fruit; e. Seed; f. Floral diagram.

Important Plants

The root tubers of *Ipomoea batatas* and root-stock of *Calystegia sapium* are cooked and eaten. Several species are of medicinal value: *Evolvulus alsinoides, Ipomoea turpethum*, etc. Many of them are cultivated in gardens for their ornamental such as *Ipomoea pomacea, Argyreia, Quamoclit, Jacquemontia, Porana paniculata*, etc.

Systematics and Phylogeny

Classically the family Convolvulaceae is divided into 2 subfamilies, *viz.* Convolvuloideae (autophytes, intrastaminal scales absent) and Cuscutoideae (parasites, intrastaminal scales present). Bentham and Hooker classified the family into five tribes i.e. Convolvuleae, Dichondreae, Nolaneae, Cresseae and Cuscuteae. Several other taxonomists placed *Cuscuta* under an unigeneric family Cuscutaceae, and some others also placed *Dichondra* into an independant family Dichondraceae.

Convolvulaceae is usually closer to Solanaceae, Polemoniaceae, Boraginaceae and some extent to Hydrophyllaceae and Nolanaceae. Takhtajan (1969) placed it under order Polemoniales while Cronquist (1981) and Thorne (1983) treated it under Solanales. According to the study of Austin (1997), the family Convolvulaceae can be classified into tribes Ericybeae, Cresseae, Convolvuleae, Merremieae, Ipomoeae, Argyreieae, Poraneae, Dichondreae and Cuscuteae. APG IV system treated the Convolvulaceae under order Solanales in the clade Asterids with four other families, *viz*. Solanaceae, Montiniaceae, Sphenoclaceae and Hydroleaceae.

Solanaceae (Nightshade Family)

Distribution: This family consists of 98 genera and 2,700 species, distributed mainly in the tropical countries. About 40% of the genera are native to American continent. About 15 genera and over 100 species have been reported in India. Some larger genera reported world over are *Solanum* (1700 spp.), *Cestrum* (150 spp.), *Physalis* (100 spp.), *Lycium* (100 spp.), *Nicotiana* (70 spp.), *Capsicum* (60 spp.) and *Petunia* (50 spp.).

Habit: The members are mostly shrubs or herbs, rarely lianes. *Solanum seaforthianum* is a petiole climber.

Vegetative Characters

Root system is of normal type. Root tubers are found in some *Solanum* sp.

Stem is herbaceous or woody, branching usually the cymose type. In *Atropa* and *Scopolia*, the branching is scorpioid. Stem tubers are found in *Solanum tuberosum*.

The **leaves** are simple, often variously lobed, alternate, but opposite in flowering stem, exstipulate. Venation is reticulate. In several members, there is an adnation of the leaves with the axis of the axillary shoots and so the flowers appear to be extra-axillary in position. The leaves are alternate in the vegetative parts and opposite in the flowering parts. In *Datura*, the terminal bud is transformed into a flower, and the axillary buds of the two leaves immediately below it, develop into shoots in dichasial manner.

Floral Characters

The **inflorescence** is usually terminal, sometimes axillary or extra-axillary. It is usually of the cymose type, tending to monochasial condition in the ultimate branches. In *Nicotiana*, it is a panicle.

The **flowers** are bisexual, hypogynous, zygomorphic, rarely actinomorphic, pentamerous with reduction in innermost whorls. In *Nicotiana*, they are pentamerous throughout.

The **calyx** consists of five sepals, united. Aestivation is valvate or imbricate. The calyx is persistent and often overgrowths in the fruits (accrescent) as in *Physalis*, *Withania* and *Nicandra*.

The **corolla** consists of five petals, united, sometimes plicate. Aestivation is valvate or convolute. The petals are uniform or irregular. In *Schizanthus*, the corolla is bilabiate.

The **androecium** consists of 5 stamens that are epipetalous and alternate with petals. They are uniform in size or one may be smaller. In *Hyoscyamus*, the 5th stamen is sterile. Out of the 4 stamens produced, only 2 are fertile in *Schizanthus*. The anthers are dithecous, but monothecous in *Browallia*, longitudinally slits, but porous in *Solanum*.

The **gynoecium** is bicarpellary (pentacarpellary in *Nicotiana*), syncarpous, axile placentation. The ovary is superior with terminal style that ends in capitate stigma. Bicarpellary ovary due to false septum at maturity becomes tetralocular as in *Datura*. The ovary is unilocular with single ovule in *Henoonia*. The arrangement of carpels in this family is peculiar. The posterior carpel is tilted towards the right and the anterior one is tilted to the other end and so the septum occupied on oblique position, whereas in normal position it occupies on horizontal position. So it is often described as an obliquely zygomorphic flower.

Pollination is entamophilous. The flowers are attractive and coloured, in some flowers strongly scented.

The **fruit** is a capsule or berry. It is transversely dehiscent capsule in *Hyoscyamus*. In *Brunfelsia*, the fruit is a drupe like.

The **seed** is dicotyledonous and flattened, endospermic with a curved embryo, but straight embryo in *Nicotiana*.

Key Characters of the Family

1. Adnation of the leaves to the axis.
2. Bilocular ovary with swollen placenta.
3. Oblique septum.
4. Curved embryo.

Important Plants

Several members are economically valuable, stem tubers used as vegetable in *Solanum tuberosum* (tubers), fruit vegetable from *S. melongena* (brinjal), *S. torvum*, *Lycopersicon esculentum* (tomato). Spicy fruits of *Capsicum frutescens* (Chilly) are used as condiment. Many species are yielding medicinal alkaloids such as stramonine from *Datura stramonium*, nicotine from *Nicotiana tabaccum*, belladonna from *Atropa belladonna*, withaferine from *Withania somnifera* and hyoscyamine from *Hyoscyamus niger*. Many members are cultivated in garden ornamental as *Browallia americana*, *Cestrum nocturnum*, *C. diurnum*, *Petunia hybrida*, *Solanum jasminioides*, *Streptosolen jamesonii*, etc.

Systematics and Phylogeny

The family Solanaceae was placed variously under order Polemoniales (Bentham and Hooker), Solanales (Hutchinson, Cronquist, Thorne) and Scrophulariales (Takhtajan). Bentham and Hooker classified this family into 5 tribes, *viz*. Solaneae, Atropeae, Hyoscyaneae, Cestrineae and Salpiglossideae while Wettstein has classified the family differently in 5 tribes as Nicandreae, Solaneae, Datureae, Cestreae and Salpiglossideae.

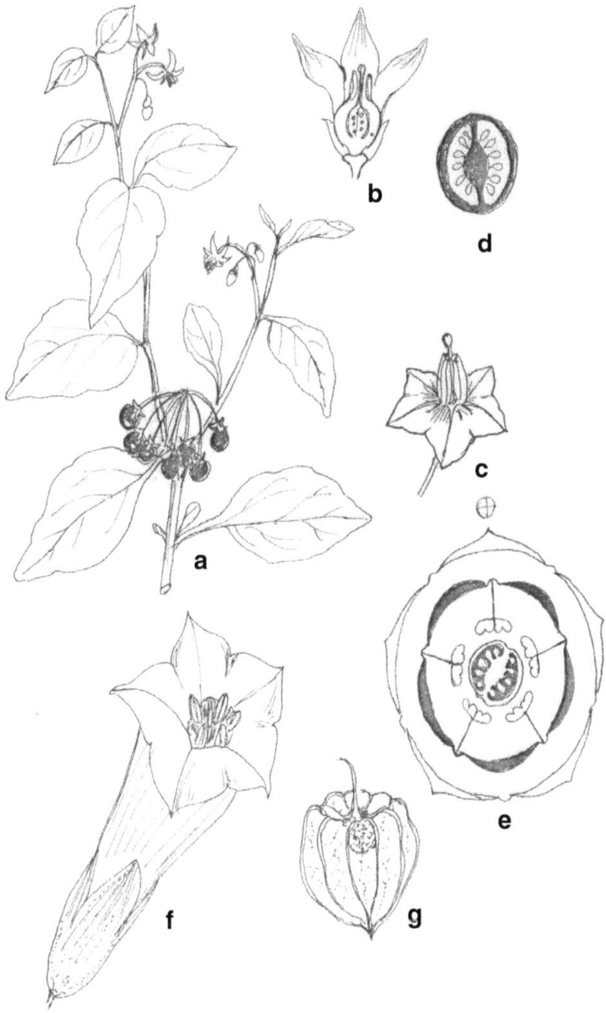

Fig. 15.18: *Solanum nigrum* L. a. Twig; b. Flower-l.s.; c. Flower-entire; d. Ovary c.s.; e. Floral diagram; f. Flower of *Datura*; g. Fruit of *Physalis*.

Taxonomists believed that Solanaceae is allied to Convolvulaceae in possessing pentamerous flowers, persistent calyx and bicarpellary ovary. However, obliquely placed ovary differs with Convolvulaceae. Solanaceae is closely related to Scrophulariaceae, Boraginaceae and Polemoniaceae. Hutchinson erected the tribe Salpiglossideae to the rank of indepentant family Salpiglossidaceae. In APG IV system the order Solanales is included under clade Asterids and placed four other families, *viz*. Convolvulaceae, Montiniaceae, Sphenocleaceae and Hydroleaceae.

Bignoniaceae (Bignonia Family)

Distribution: The family is represented by about 120 genera and 650 species which are distributed mainly in tropical and subtropical, and few are temperate region. About 15 genera and 40 species are found in India. Some of the larger genera are *Tabebuia* (100 spp.), *Arrabidaea* (70 spp.), *Jacarandra* (50 spp.), *Anemopaegma* (30 spp.) and *Tecoma* (18 spp.).

Habit: Mostly woody lianas and tendrillar climbers, trees or shrubs, rarely herbs. Some are climbers with aerial roots (*Campsis radicans*), others twiners (*Pandorea, Tecomaria*).

Vegetative Characters

The **root system** is of normal type. Aerial roots are found in *Campsis radicans*.

The **stem** is woody, branched, anatomically wedges of phloem developed between secondary xylem.

The **leaves** are opposite, pinnately compound, exstipulate, terminal leaflets sometimes modified into tendril; tendrils may be simple or branched, hooked or provided with some adhesive discs, rarely simple leaves in *Diplanthera*.

Floral Characters

The **inflorescence** is usually cymose (dichasial cyme) with bracts and bracteoles.

The **flowers** are bracteate, bracteolate, complete, bisexual, zygomorphic, hypogynous, often showy and campanulate as in *Spathodia*.

The **calyx** consists of 5 sepals, united, 5-toothed, often campanulate and valvate aestivation.

The **corolla** consists of 5 petals, campanulate or funnel-shaped, lobes bilipped with upper lip 2 petals and lower lip 3 petals, imbricate in aestivation.

Androecium is typically of 4 stamens, epipetalous, didynamous, alternate with petals, odd stamen posterior, rarely two stamens in *Catalpa* and 5 in *Oroxylum*. Anthers dithecous, lobes one above the other, introrse, anther dehiscence by longitudinal slits.

Gynoecium is bicarpellary, syncarpous, superior, bilocular, numerous anatropous ovules in each locule, on axile placentation. Unilocular ovary with parietal placentation is found in *Kigelia*. Style is simple, terminal; stigma bilipped and hypogynous nectar glands are found on disc.

Pollination is usually entomophilous.

Fruit is usually capsule, sometime fleshy and indehiscent as in *Kigelia*.

Seeds are dicotyledonous, non-endospermous with straight embryo. Seeds often winged as in *Spathodea*.

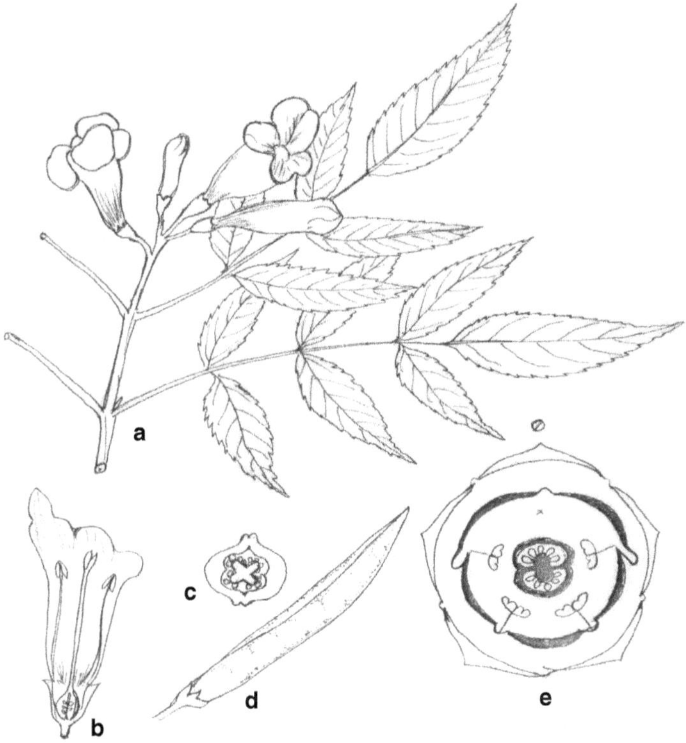

Fig. 15.19: *Tecoma stans* (L.) Juss. a. Tiwg; b. Flower-l.s.; c. Ovary c.s.; d. Fruit-entire; e. Floral diagram.

Key Characters of the Family

1. Woody lianas, trees and shrubs.
2. Leaves opposite and compound.
3. Flowers in cymes and zygomorphic.
4. Fruit capsule.

Important Plants

Some plants yield timber woods such as *Catalpa, Dolichondrone, Haplophragma, Millingtonia, Oroxylum, Radermachera, Spathodea, Stereospermum, Tabebuia* and *Tecoma*. Some are ornamental and avenue trees, *viz. Catalpa bignonioides, Crescentia cujete, Haplophragma adenophyllum, Jacaranda mimosifolia, Kigelia pinnata, Millingtonia hortensis, Spathodea campanulata* and *Tabebuia spectabilis. Bignonia unguis-catii* and *Tecomella undulata* are garden shrubs. *Adenocalymma alliaceum, Pyrostegia venusta* and *Campsis radicans* are ornamental climbers. *Incarvillea* is a herbaceous genus.

Systematics and Phylogeny

The family Bignoniaceae is placed variously under order Personales (Bentham and Hooker), Bignoniales (Hutchinson, Thorne) and Scrophulariales (Takhtajan, Cronquist). Engler and Prantl divided this family into 4 tribes as Bignonieae, Tecomeae, Eccremocarpeae and Crescentieae. This family is closely related with Scrophulariaceae by the presence of didynamous stamens, bicarpellary ovary, zygomorphic flower and capsular fruits. In some other features it is also allied to Acanthaceae and Pedaliaceae. In APG IV system, the family is placed under order Lamiales with 23 other families. The closely related families are Acanthaceae and Lentibulariaceae.

Acanthaceae (Acanthus Family)

Distribution: This is a fairly large family of 246 genera and about 3000 species distributed mostly in the warmer parts of the world. About 75 genera and 350 species have been reported in India. Some larger genera of the family are *Justicia* (300 spp.), *Strobilanthes* (250 spp.), *Barleria* (230 spp.), *Thunbergia* (200 spp.), *Aphelandra* (200 spp.), *Dicliptera* (150 spp.), *Blepharis* (100 spp.), *Crossandra* (60 spp.) and *Acanthus* (50 spp.).

Habit: They are mostly mesophytic perennials, armed or unarmed herbs or shrubs. Some are arborescent forms. Several are woody climbers as *Thunbergia*. Members like *Acanthus ilicifolius* inhabit salt marshes. *Acanthus steraceantha* grows in wet localities. Sessile or stalked cystoliths are characteristically seen in the epidermal cells of the leaves and stems.

Vegetative Characters

The **root system** is of the normal type. In *Ruellia tuberosa*, root tubers are found.

In the **shoot**, nodes are usually swollen and internodes are elongated.

The **leaves** are simple, rarely lobed, opposite, decussate, rarely whorled, anisophyllous in *Strobilanthes anisophyllus*, exstipulate. Sometime leaves are very much reduced and modified into spines as in *Barleria* sp. Venation is reticulate.

Floral Characters

The **inflorescence** is terminal or axillary. Commonly the flowers are on dichasial cymes that are monochasial in the ultimate branches. Spikes and racemes are also common. Rarely flowers are solitary and sometimes crowded on the axis.

The **flowers** are bisexual, hypogynous, zygomorphic, bracteate, bracteolate, pentamerous but reduced in the inner whorls. Bracteoles are well developed and brightly cloured in *Thunbergia*.

The **calyx** consists of 4 or 5 sepals which are usually irregular in size. They are free and aestivation is valvate or imbricate. The calyx is reduced into a ring-like structure in *Thunbergia*.

The **corolla** consists of 5 petals, gamopetalous, usually personate or bilabiate. When it is two-lipped, the upper lip consists of 2 and lower of 3 petals. The upper lip is absent in *Acanthus* and *Blepharis*. Aestivation is imbricate or twisted.

The **androecium** consists of 4 stamens, the posterior stamen being suppressed or represented by a staminode (in *Barleria cristata*). Sometimes the anterior two stamens only are fertile. They are epipetalous, didynamous. Anthers are usually dithecous, basifixed and introrse, sometimes spurred and bearded. The anther cells are unequal in some members.

The **gynoecium** is bicarpellary, syncarpous. Ovary is usually elongated, cylindrical, superior, with a terminal style that bears forked stigma. The stigmatic arms are usually unequal in size. Ovary is bilocular (often the posterior one is smaller) with axile placenta. The base of the ovule from the funicle formed hook or papilla-like structure. At maturity it is helpful for seed dispersal mechanism as jaculator or retinaculum in the fruit.

Pollination is entomophilous. The sensitive stigmas of *Strobilanthes*, when touched, move downwards and become pressed against lower lip. Cleistogamy is also recorded in some members.

The **fruit** is a few- or many-seeded loculicidal capsule, rarely a 2-seeded drupe. In capsules seeds are dispersed by jaculator or retinaculum mechanism.

Seeds are dicotyledonous and spherical or lens-shaped, non-endospermic (endospermic in *Nelsonioideae*). Sometimes seeds are with scaly or unicellular hairy out-growths which become mucilaginous on wetting.

Key Characters of the Family

1. Swollen nodes.
2. Presence of cystoliths on epidermal cells.
3. Well developed bracts and bracteoles.
4. Zygomorphic flowers.
5. Didynamous stamens.
6. Bilocular ovary with serially arranged ovules.
7. Loculicidal capsular fruit.
8. Jaculator or retinaculum mechanism of seed dispersal.

Important Plants

Andrographis paniculata, Justicia adhatoda, Rhinacanthus nasutus and *Hygrophila* are medicinally important plants. Several of the members are garden favourites: species of *Aphelandra, Barleria, Crossandra, Jacobina, Justicia, Eranthemum, Ruellia, Thunbergia, Meyenia*, etc.

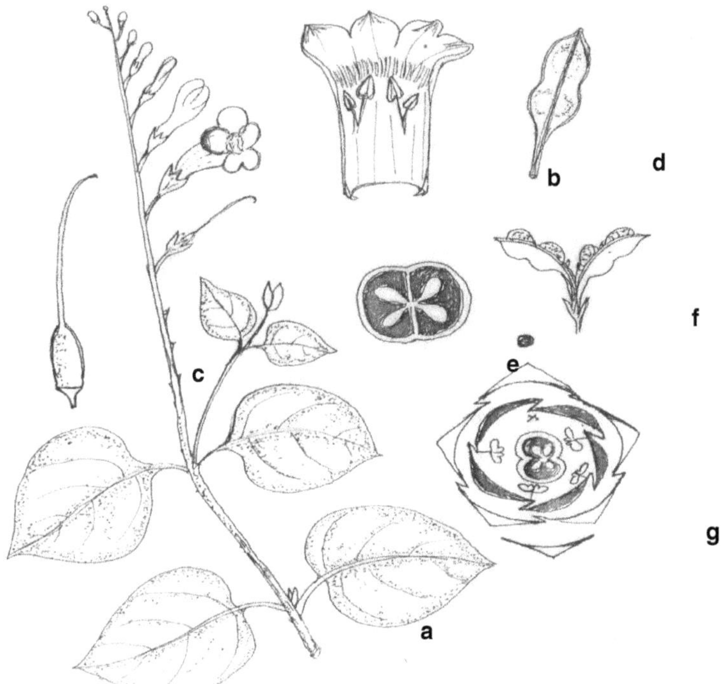

Fig. 15.20: *Asystasia gangetica* (L.) T.And. a. Plant twig; b. Petals with stamens; c. Gynoecium; d. Fruit-entire; e. Ovary c.s.; f. Fruit- split open; g. Floral diagram.

Systematics and Phylogeny

The family Acanthaceae is placed variously under order Personales (Bentham and Hooker, Hutchinson), Tubiflorae (Engler and Prantl) and Scrophulariales (Takhtajan, Cronquist). The family is classically divided into four subfamilies, viz. Nelsonioideae, Mendoncioideae, Thunbergioideae and Acanthoideae. Some members of Acanthaceae are closely related to Scrophulariaceae and others are closer to Bignoniaceae. Taxonomists believed that the family Acanthaceae is derived from Scrophulariaceae. In APG IV system, the family is placed under order Lamiales with 23 other families. The families Pedaliaceae and Bignoniaceae are closely related to the Acanthaceae.

Lamiaceae (Labiatae) (Mint Family)

Distribution: This family consists of 200 genera and 3,600 species distributed in warm and temperate regions. About 65 genera and 400 species have been reported in India. Some of the largely represented genera are *Salvia* (700 spp.), *Hyptis* (400 spp.), *Thymus* (400 spp.), *Teucrium* (300 spp.), *Stachys* (300 spp.), *Scutellaria* (200 spp.), *Coleus* (200 spp.), *Ocimum* (150 spp.), *Nepeta* (150 spp.), *Lavendula* (30 spp.) and *Mentha* (25 spp.).

Habit: They are mostly annual or perennial herbs, or rarely shrubs. Species of *Hyptis* and *Leucosceptrum* are exceptionally arborescent. Some species of *Scutellaria* are lianes.

Vegetative Characters

The **root system** is of normal type, some species of *Coleus* have root tubers.

The **stem** is usually four-sided in younger parts and is covered by hairs or sometimes glands that produce aromatic oils.

The **leaves** are simple, rarely lobed, opposite, decussate, sometimes whorled, usually exstipulate and commonly aromatic.

Floral characters

The **inflorescence** are usually terminal, sometimes axillary, dichasial cymes leads to monochasial in the ultimate branches and form a close cluster, around the stem, giving false appearance of whorls called verticillaster. Sometimes, it is a thyrsus as in species of *Ocimum* or diffuse panicle. In *Scutellaria*, flowers are solitary.

The **flowers** are bisexual, hypogynous, zygomorphic, bracteate and bracteolate. Bracts are highly differentiated in *Monarda*.

The **calyx** consists of 5 sepals, united to form a tube. They are irregular or with variable number of ribs. Aestivation is valvate or imbricate. The calyx is persistent often accrescent, rarely fleshy as in *Hoslundia*. Calyx is two-lipped in *Salvia*.

The **corolla** consists of 5 petals, united, irregular and bilabiate. The upper lip is usually two and the lower three petals. In *Ocimum*, the lower lip consists of only one petal. All the five petals are in anterior position in *Teucrium*. Aestivation is imbricate.

The **androecium** consists of 4 stamens, epipetalous and didynamous. Generally posterior stamen is completely suppressed by a staminode. In *Salvia*, only two stamens are fertile. The stamens are monadelphous in *Coleus*. Anthers are dithecous, rarely monothecous as in *Geniosporum*, introrse, dehiscing longitudinal slits.

The **gynoecium** is bicarpellary, syncarpous, tetralocular due to false septum and axile placentation. The ovary is superior, with a gynobasic style that ends in forked stigma. Terminal style present in *Ajuga* and *Teucrium*. A hypogynous nectariferous gland is produced.

Pollination is entomophilous, Protandry is common adaptation, Cleistogamous flowers occur in the species of *Salvia*, *Ajuga* and *Lamium amplexicaule*.

The **fruit** is a schizocarp breaking into 4 nutlets. It is a drupaceous in Prasioideae. Geocarpy is seen in *Lycopus virginicus*. Fruits are dispersed by wind as the calyx forms a chaffy membranous float. Sometimes hairs, bristles or hooks produced on surface of the persistent calyx helps in dispersal through the animals.

The **seed** is dicotyledonous, non-endospermic, rarely endospermic as in Prostantheroideae.

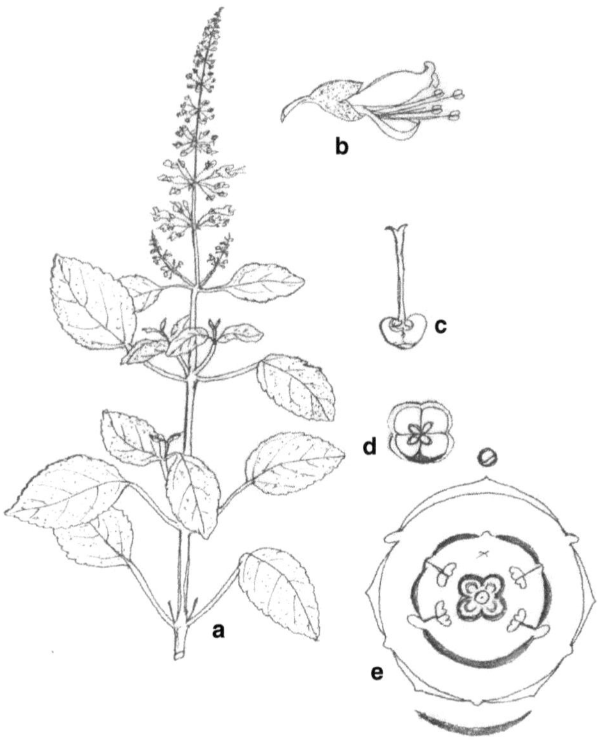

Fig. 15.21: ***Ocimum tenuiflorum*** **L. a. Plant twig; b. Flower-entire; c. Gynoecium; d. Ovary c.s.; e. Floral diagram**

Key Characters of the Family

1. Herbaceous habit.
2. Quadrangular stem.
3. Aromatic foliage.
4. Thyrsus and verticillaster inflorescence.
5. Irregular calyx.
6. Bilabiate corolla.
7. Didynamous and epipetalous stamens.
8. Tetralocular ovary with gynobasic style.
9. Forked stigma.
10. Schizocarpic nutlets.

Important Plants

Many of the members are aromatic and medicinal, menthol is obtained from *Mentha arvensis*, *M. viridis* (spearmint) and *M. crispus* (curleymint). *Origanum vulgare* and *Lavendula officinalis* (lavender) are used in perfumery industries. *Coleus aromaticus* is cultivated for its aromatic oil. *Ocimum tenuiflorum*, *Leucas aspera*, *Anisomeles malabarica* are common medicinal plants. *Coleus blumei*, *C. thyrsoideus*, *Lavendula bipinnata*, *Rosmarinus officinalis*, *Salvia officinalis*, *S. leucantha*, *S. splendens*, *Thymus vulgaris* are common garden ornamentals.

Systematics and Phylogeny

Lamiaceae is placed under order Lamiales in majority of the classification systems. Engler and Prantl divided the family into 8 subfamilies, *viz.* Ajugoideae, Prostantheroideae, Prasioideae, Scutellarioideae, Lavenduloideae, Lamioideae, Ocimoideae and Catopherioideae. Many taxonomists believed the family is closely related to Verbenaceae and Boraginaceae. Some features are closely allied to Scrophulariaceae also. Hutchinson suggested that Lamiaceae (Labiatae) is the most highly evolved family among Dicotyledons. APG III systems in 2004 Lamiaceae were divided into seven subfamilies with ten genera not placed in any of the subfamilies. The unplaced genera are: *Tectona*, *Callicarpa*, *Hymenopyramis*, *Petraeovitex*, *Peronema*, *Garrettia*, *Cymaria*, *Acrymia*, *Holocheila* and *Ombrocharis*. The subfamilies are Symphorematoideae, Viticoideae, Ajugoideae, Prostantheroideae, Nepetoideae, Scutellarioideae, and Lamioideae. The subfamily Viticoideae is probably not monophyletic. Prostantheroideae and Nepetoideae are divided into tribes. *Callicarpa* and *Tectona* which are unplaced in APGIII classification are now placed in Lamiaceae in APG IV Classification. Phylogenitically the family Lamiaceae is rooted from Verbenaceae.

Amaranthaceae (*Amaranthus* or Pigweed Family)

Distribution: The family is of about 165 genera and 2,040 species distributed in tropical world. About 18 genera and 50 species have been reported in India. Some of the larger genera in the family are *Alternanthera* (200 spp.), *Gomphrena* (100 spp.), *Iresine* (80 spp.), *Amaranthus* (60 spp.) and *Celosia* (60 spp.).

Habit: They are mostly annual or perennial herbs or shrubs. Many of them are mesophytes. A few species of *Aerva* are xerophytes, species of *Alternanthera* inhabit in damp localities. *Deeringia amaranthoides* is a climbing shrub.

Vegetative Characters

The **root system** is of normal type.

The **shoot system** is usually herbaceous, branching monopodial or sympodial (*Alternanthera*).

The **leaves** are simple, alternate in *Aerva*, opposite in *Alternanthera* and *Achyranthes*, exstipulate and sometimes variegated.

Floral Characters

The **inflorescence** is terminal or axillary. It is spike in *Achyranthes*, fasciculated in species of *Celosia*, globose head in *Gomphrena*. In *Pupalia*, the lateral flowers are sterile, whereas central flowers alone are fertile.

The **flowers** are bisexual, rarely unisexual in *Amaranthus*, sometimes polygamous, actinomorphic, hypogynous, monochlamydous, bracteate and bracteolate. Bracts and bracteoles sometimes coloured.

The **calyx** consists of 5 sepals, free and persistent. The sepals are membranous and petaloid. Aestivation is quincuncial. Usually **corolla** is absent.

The **androecium** consists of 5 stamens that are opposite to the sepals. Usually they are united by filaments to form a tube around the gynoecium. In *Achyranthes*, alternating with the fertile stamens, are fringed scaly structures which are taken to represent stamens in second whorl. The anthers are dithecous, introrse, dehiscing by longitudinal slits.

The **gynoecium** is 2- or 3-carpellary, syncarpous, unilocular with basal placentation. Ovary is superior with terminal style and 2- or 3-lobed stigma. In *Celosia*, the ovules are many in ovary chamber.

Pollination is through the agency of insects. Bracts and bracteoles are coloured in some genera.

The **fruit** is usually an achene or rarely nut. It is a cicumscissile capsule in *Celosia*.

The **seed** is curved, docotyledonous and endospermic. Usually seeds are with hard seed coat.

Key Characters of the Family

1. Herbaceous habit.
2. Monochlamydous flowers.
3. Monadelphous stamens.
4. Unilocular ovary.
5. Basal placentation.

Important Plants

Most of the members are used as vegetable greens and commonly called amaranths; *Amarathus viridis*, *A. caudatus*, *A. spinosus*, *Alternanthera sessils*, *A. tenella*, *Celosia argentia*, *Allmania nodiflora*, *Digera muricata*, etc. Many of them are cultivated for their ornamental value; *Gomphrena globosa* (bachelor's button), *Celosia cristata* (cock's comb), *Irisine herbestii*, *Telanthera ficoidea*, etc. *Aerva lanata* and *Achyranthes aspera* are used as medicinal. Seeds of *Amaranthus caudatus* are edible.

Systematics and Phylogeny

Many taxonomists placed the family Amaranthaceae in different positions, under series Curvembryae (Bentham and Hooker), order Centrospermae (Engler and Prantl), Chenopodiales (Hutchinson, 1959; Thorne, 1983) and Caryophyllales (Takhtajan, 1969; Cronquist, 1981).

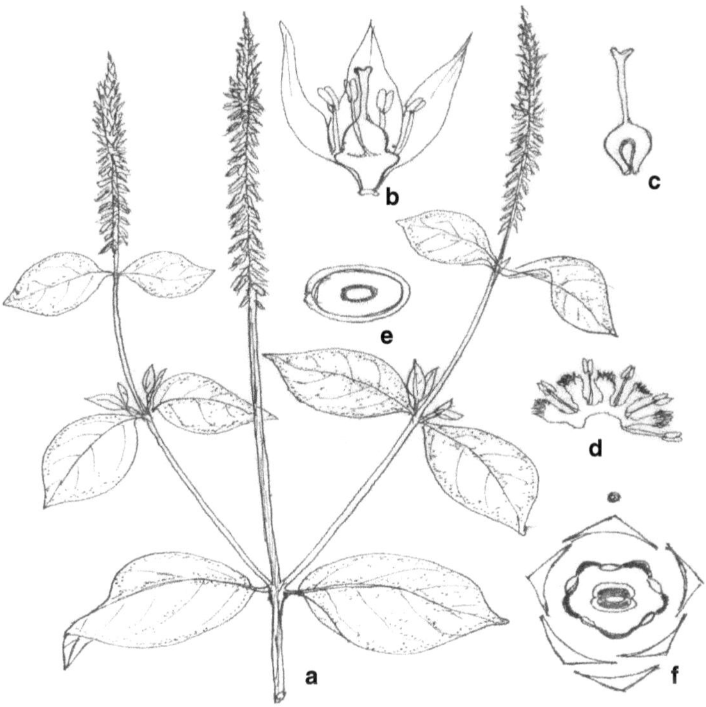

Fig. 15.22: *Achyranthes aspera* L. a. Plant twig; b. Flower l.s.; c. Gynoecium; d. Staminal column; e. Ovary c.s.; f. Floral diagram.

Bentham and Hooker classified the family into three tribes, *viz*. Celosieae, Amarantheae and Gomphreneae. This family is closely related to Chenopodiaceae by possessing most of the characters. Hutchinson believed that this family has evolved from Caryophyllaceae.

In the APG II system of 2003, the family was placed in the order Caryophyllales. It includes the plants formerly treated as the family Chenopodiaceae. Phylogenetic research revealed the important impact of the subfamily Polycnemoideae on the classification, if Polycnemoideae are considered being part of Chenopodiaceae, then Amaranthaceae *sensu stricto* have to be included, too, and the name of the extended family is Amaranthaceae. If Polycnemoideae would be separated as an independent family, Chenopodiaceae and Amaranthaceae *s.str.* would form two distinct monophyletic groups and could be treated as two separate families. Amaranthaceae Juss. (*s.l.*) include the former families Achyranthaceae, Atriplicaceae, Betaceae, Blitaceae, Celosiaceae, Chenopodiaceae, Corispermaceae, Deeringiaceae,

Dysphaniaceae, Gomphrenaceae, Polycnemaceae, Salicorniaceae, Salsolaceae and Spinaciaceae. The systematics of Amaranthaceae are the subject of intensive recent research, which also revealed the traditional classification, based on morphological and anatomical characters, often did not reflect the phylogenetical relationships. Current studies classified the species of former Chenopodiaceae to eight distinct subfamilies, *viz.* Polycnemoideae, which are regarded as a basal lineage, Betoideae, Camphorosmoideae, Chenopodioideae, Corispermoideae, Salicornioideae, Salsoloideae and Suaedoideae (APG IV).

Aristolochiaceae (Birthwort Family)

Distribution: The family is represented by 10 genera and 600 species distributed in tropical and temperate America, Asia, Europe and Africa. Some larger genera are *Aristolochia* (500 spp.) and *Asarum* (70 spp.).

Habit: Herbs or shrubs and usually twinig lianas.

Vegetative Characters

The **root system** is of normal tap roots.

The **stem** is woody, branched with broad medullary rays.

The **leaves** are simple, alternate, petiolate, palminerved and exstipulate.

Floral Characters

The **inflorescence** is axillary clusters (*Aristolochia*), solitary or arranged in racemes or cymes.

The **flower** is bisexual, actinomorphic (*Asarum*) or Zygomorphic (*Aristolochia*), epigynous.

Perianth is usually a petaloid gamosepalous calyx, variously 3-lobed teeth, often differently coloured; corolla usually absent but sometimes inner whorl of 3 minute teeth present. There are three united perianth lobes formed a tube and limb in *Aristolochia*.

The **androecium** consists of 6-36 stamens, free or united with style to form staminal column (gynostegium). Filament is short and thick. Anthers are dithecous, free or adnate to the style, dehisce longitudinally.

The **gynoecium** consists of 4-6 united carples, mostly 4-6 locules with numerous anatropous ovules, horizontal or pendulous in each locule on axile placentation. Ovary is inferior or half-inferior, style single with 4-6 stigmas.

Pollination is through the agency of insects.

Fruit is many-seeded or septicidal capsule, often dehiscing basally as in *Aristolochia*.

Seeds are dicotyledonous, endospermic and with small embryo.

Important Plants

Asarum and *Aristolochia* are important genera and their species are of ornamental and medicinal value. *Aristolochia clematis, A. gigas, A. grandiflora, A. ornithocephala, A. macrophylla* and *Asarum europaecum* are gardern ornamental plants. *Aristolochia serpentina* (Virginia snakeroot) and *A. indica* are medicinally important usually for antidote purposes.

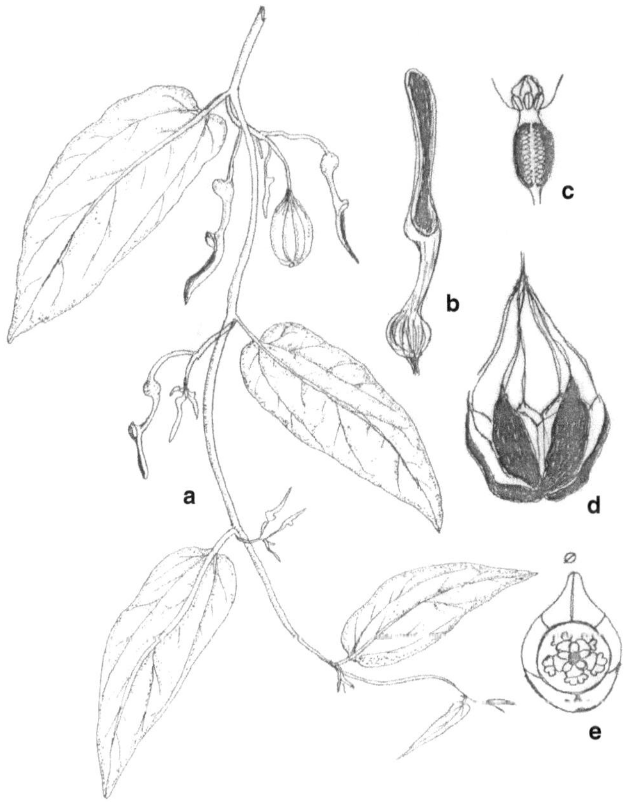

Fig. 15.23: *Aristolochia indica* L. a. Plant twig; b. Flower-entire; c. Ovary with androecium; d. Fruit; e. Floral diagram.

Key Characters of the Family

1. Terrestrical herbs, shrubs and lianas.
2. Simple alternate leaves.
3. Gamosepalous calyx often trumpet-shaped or bell-like perianth tube and lobes.

4. Stamens with 6-36 in number and adnate to style.
5. Septicidal capsule.

Systematics and Phylogeny

The family is placed variously under series Multiovulatae Terrestris of Monochlamydeae by Bentham and Hooker, uner order Aristolochiales of subclass Archichlamydeae by Engler and Prantl, under the same order in the class Magnoliopsida by Takhtajan and Cronquist. Hickey and King (1988) suggested that the family is very difficult to place, however it is related to Annonaceae of Magnoliales. Engler divided the family into three tribes, *viz*. Sarumeae, Bragantieae and Aristolochieae. Hutchinson (1948) treated the family Aristolochiaceae as the descendants of Ranales via Berberidaceae. Recent classification based on molecular phylogeny, Aristolochiaceae are magnoliids, a basal group of angiosperms which are not part of the large categories of monocots or eudicots. As of APG IV (2016), the former families Hydnoraceae and Lactoridaceae are included, because exclusion would make Aristolochiaceae in the traditional sense paraphyletic. Some newer classification schemes, such as the update of the Angiosperm Phylogeny Group, place the family Aristolochiaceae in the order Piperales, but it is still quite common, though superseded, for the Aristolochiaceae to be assigned, sometimes with some other families, their own order Aristolochiales.

Loranthaceae (Mistletoe Family)

Distribution: This is a tropical family of about 75 genera and 1,100 species. The important genus *Loranthus* is represented by about 500 species. About 10 genera and 60 species have been reported in India. Some other larger genera are *Phoradendron* (190 spp.), *Struthanthus* (75 spp.), *Viscum* (70 spp.), *Dendrophthoe* (55 spp.) and *Psittacanthus* (50 spp.).

Habit: This is interesting family of semiparasitic shrubs or herbs. *Nuytsia* and *Guiadendron* are autotrophic trees found in Australia.

Vegetative Characters

The **root system** is absent and their nourishment from the host with help of specially modified roots called haustoria.

The **shoots** are normal or greenish, nodes swollen, with dichasial branching, in some cases flattened and photosynthetic as in *Viscum articulatum*.

The **leaves** are simple, opposite, sometimes alternate, rarely whorled, entire, leathery and exstipulate. In some species of *Viscum*, leaves are scaly and stem becomes cladode.

Floral Characters

The **inflorescence** is usually terminal, sometimes axillary, simple or dichasial cyme, rarely raceme. Flowers are solitary or fascicle in axillary and sometimes solitary flowers sunk into the axis in *Phoradendron*.

The **flowers** are bisexual, unisexual in *Viscum*, regular with tendency to zygomorphy, epigynous, bracteate and bracteolate.

Perianth is in 2-merous or trimerous (*Viscum*) whorls. It consists of 5 lobes in *Loranthus*. Perianth lobes are free in *Loranthus* and united in *Viscum*. Perianth is sepaloid in *Viscum* and petaloid in *Loranthus*. In Loranthoideae, there is a toothed or irregular rim called calyculus below the perianth.

The **androecium** consists of stamens equal in number to the perianth lobes and opposite to them. Sometimes they are attached or completely fused with perianth lobes (*Viscum*). Anthers are dithecous, rarely monothecous, introrse or sometimes with numerous chambers, dehisce by longitudinal slits or transverse or pores. Pistillode is commonly seen in male flowers.

The **gynoecium** is usually 2-3-carpellary, syncarpous with inferior ovary, terminal, short or long style and capitate stigma. Ovary is unilocular with basal placentation. In *Viscum* the locule is indistinct.

Fig. 15.24: *Dendrophthoe falcata* (L.f.) Ettingsh. a. Flowering twig; b. Flower-entire; c. Stamens with floral tube; d. Gynoecium; e. Fruits; f. Floral diagram.

Pollination is anemophilous or entomophilous as in *Loranthus*. Flowers of *Viscum album* are not attractive, but flies and bees visited for the nectar.

The **fruit** is a drupe, or a pseudocarp (pome). The fruits are much liked by birds that are fond of the fleshy delicious outer part of the fruit. Mesocarp layer is with viscid substances which is called bird-lime. It helps to disperse the seeds through the birds.

The **seeds** are albuminous, but exalbuminous in *Psittacanthus* with 2-3 embryos in certain species.

Key Characters of the Family

1. Semiparasitic herbs or shrubs.
2. Epigynous flowers.
3. Unilocular ovary with ill-defined basal ovule.

Important Plants

The members are commonly called mistletoe, not much economic importance. Common genera are *Loranthus, Viscum, Korthalsella, Scurrla*, etc. are wild growing on many host trees.

Systematics and Phylogeny

Many taxonomists treated the family Loranthaceae under order Santalales (Lawrence, 1951; Takhtajan, 1969; Cronquist, 1981; Thorne, 1983). But the family was placed under series Achlamydosporeae of Monochlamydeae by Bentham and Hooker. Engler divided the family into 2 subfamlies, *viz*. Loranthoideae (calyculus present) and Viscoideae (calyculus absent). The family shows close affinity towards Santalaceae. The APG IV system (2016) assigns the family to the order Santalales in the clade Core Eudicots.

Euphorbiaceae (Spurge Family)

Distribution: It is a quite large family of about 300 genera and above 7,500 species with cosmopolitan distribution. Several species of them are distributed in tropical forests and some are characteristic of salt marshes. The chief genus *Euphorbia* with about 20% of the species of the family is cosmopolitan in distribution. About 60 genera and 350 species have been reported in India. Some of the larger genera are *Euphorbia* (2,000 spp.), *Croton* (750 spp.), *Acalypha* (450 spp.), *Jatropha* (175 spp.), *Manihot* (170 spp.), *Sapium* (120 spp.), *Tragia* (100 spp.), *Clutia* (70 spp.) and *Manadenium* (50 spp.).

Habit and **Habitat**: They are mostly shrubs and trees and a few are herbs. Species with various habits can be seen in the same genus. *Tragia* is a climber. Most of the species of *Euphorbia* are xerophytic. *Excoecaria* and *Coperonia* inhabit salt-marshes.

Vegetative Characters

The **root system** is of the normal type. Tuberous roots are found in *Manihot utilissima* and *Euphorbia fusiformis*.

The **shoot system** is mostly adapted to xerophytic condition. Stem is flattened or water reservoir and photosynthetic organ when leaves are scaly or poorly developed. Stem becomes succulent and ribbed or angular in *Euphorbia antiquorum, E. tortilis* and *E. vajraveluei*. In *Xylophylla* stem is flattened green leaf-like cladode.

The **leaves** are simple or rarely compound, alternate, often opposite, rarely whorled and variously palmately lobed (in *Ricinus*). The leaves are stipulate, stipules may be hairy, glandular or spinous. In *Phyllanthus*, leaves on dwarf branches look-like a compound leaf. Some xerophytes leaves are reduced into scales, spines and bristles. Sometimes rudimentary leaves formed and become caducous as in *Euphorbia tirucalli*. Plant body usually have milky or watery sap.

Floral Characters

The **inflorescence** is terminal or axillary, often racemose and subsequently branching cymose-dichasial. The inflorescence in *Euphorbia* looks like a bisexual flower and is highly specialised called cyathium. There is a terminal naked female flower which is surrounded by many male flowers in scorpoioid manner with basal involucral bracts. The bracts sometimes are petaloid and they bear glandular bodies on the outer surface. Male flower is represented by a single stamen. It is taken as a reduced flower because of the presence of a joint between the stamen and lower pedicel. Usually inflorescence showed regular fashion but in *Pedilanthus*, it is bilateral symmetry. Cyathium inflorescence may be produced in racemose or cymose manner.

The **flowers** are unisexual, monoeciuos or dioeccius (*Excoecaria*), hypogynous (perigynous in *Bridelia*), usually monochlamydeous, rarely dichlamydeous, pentamerous (throughout in *Wielandia*), sometimes reduction in number of carpels.

The **perianth** is in one or two whorls, sometimes distinguished into calyx and corolla. The lobes are sepaloid or petaloid, free or united with valvate or imbricate aestivation. Perianth is absent in *Euphorbia*.

The **androecium** consists of single (in *Euphorbia*) to many (in *Croton*) stamens, usually free. They are branched in *Ricinus*, united in *Euphorbia reticulata*. Synandrium is found in *Phyllanthus cyclanthera*. Anthers are dithecous, introrse or extrorse, dehiscing by longitudinal or transverse slits or apical pores.

The **gynoecium** is 3-carpellary, rarely bicarpellary in *Mercurialis*, pentacarpellary in *Weilandia* and multicarpellary in *Gleidion*, syncarpous, axile placentation. Ovule is single or two in a chamber with distinct obturator.

Pollination is through the agency of insects. Bracts are brightly coloured in *Euphorbia pulcherrima*. Extrafloral nectaries usually present.

The **fruit** is usually schizocarpic regma. It is a drupe in *Hippomane* and a berry in *Bischofia*.

The **seed** is endospermic, endosperm is usually oily. Caruncle is usually characteristically present.

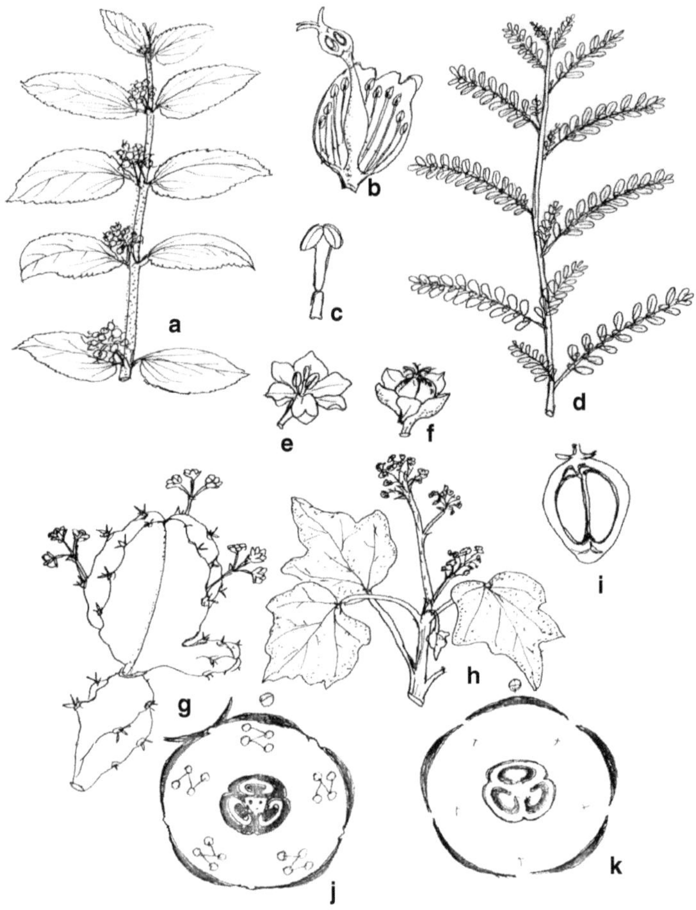

Fig. 15.25: *Euphorbia hirta* L. a. Twig; b. Cyathium l.s.; c. Staminal flower; d. *Phyllanthus amarus* – Twig; e. Male flower-entire; f. Female flower-entire; g. *Euphorbia antiquorum* – Habit; h. *Jatropha curcas* – Twig; i. Fruit l.s.; j. Floral diagram for *Euphorbia* sp.; k. Floral diagram – *Phyllanthus* pistillate flower.

Key Characters of the Family

1. Xerophytic habit.
2. Presence of latex.
3. Dwarf branches.
4. Reduced unisexual flowers.
5. Tricarpellary, tri-locular, superior ovary with axile placenta.
6. Schizocarpic fruit.
7. Caruncled seeds.

Important Plants

This is an important family for extraction of valuable economic products. Rubber is extracted from the latex of *Hevea brasiliensis*, *Manihot glaziowii* and *Siphonia elastica*. Oils are obtained from the seeds of *Aleurites fordii* (tung oil), *A. moluccana* (candlenut oil), *A. montana* (chinawood oil) and *Ricinus communis* (castor oil). A medicinal oil is obtained from *Croton tiglium* (croton oil). Biodiesel is obtained from the seed oil of *Jatropha curcas*. Tapioca starch is prepared from the root tubers of *Manihot utilissima*. Raw materials for soap industry are obtained from *Stillinga sebifera*. *Phyllanthus amarus* is a good remedy for jaundice. Edible fruits are obtained from *Emblica officinalis* and *Phyllanthus acidus*. Some are garden ornamentals such as *Acalypha wilkesiana*, *Euphorbia pulcherrima*, *Jatropha podagrica* and *Codiaeum variegatum*. *Euphorbia grantii*, *E. tirucalli* and *Pedilanthus tithymaloides* are used for live fence.

Systematics and Phylogeny

Euphorbiaceae is placed under Unisexuales of Monochlamydeae by Bentham and Hooker, order Geraniales by Engler and Prantl, under Euphorbiales by Hutchinson (1959), Takhtajan (1969), Cronquist (1981) and Thorne (1983). Hickey and King (1988) divided the family into four subfamilies, *viz.* Phyllanthoideae, Euphorbioideae, Porantheroideae and Ricinocarpoideae. Euphorbiaceae resembles Geraniaceae with flower characters and it is close to Sterculiaceae by their fruit and seed characters. Cronquist (1981) believed that this family is closely related to Sapindales, Geraniales, Linales and Polygalales and all these plant groups may have derived from common root stock. In APG II (2006) the family Phyllanthaceae is separated and it is placed somewhat distant from the core Euphorbiaceae. In APG IV (2016), the family Euphorbiaceae is placed under common uncertain clade of order Malpighiales with 35 other families including Phyllanthaceae.

Orchidaceae (Orchid Family)

Distribution: This is one among the largest flowering plant families in the world with about 1,000 genera and 20,000 species which are cosmopolitan in distribution. In the tropics, majority of the orchids are epiphytes and in the subtropical region they may be terrestrial. It is a second largest family in India represented by about 130 genera and 900 species distributed mainly in Eastern Himalayas, Western Ghats and Khasi hills. Some of the larger genera are *Dendrobium* (1,500 spp.), *Habenaria* (600 spp.), *Epidendrum* (400 spp.), *Oncidium* (350 spp.), *Odontoglossum* (200 spp.), *Cattelya* (60 spp.), *Vanda* (60 spp.) and *Cymbidium* (50 spp.).

Habit and **Habitat**: They are usually perennial or annual herbaceous forms that show great variety of design and mode of life owing to diversity condition in which they grow. Terrestrial foms inhabit humus, peaty or rarely boggy soils. Some are non-green bearing scale leaves and are saprophytic (*Epipogium*). Several terrestrial forms showing the roots associates with mycorrhiza in various degress either ectotrophic or endotrophic. Some *Bulbophyllum* and *Vanilla* are climbers.

Vegetative Characters

The **roots** are usually adventitious type. Several terrestrial forms showed mycorrhizal roots (*Aenhenrya, Anoectochilus, Zeuxine*). Some genera have root tubers (as in *Habenaria* and *Brachycorythis*). Tendrillar roots in *Vanilla* help in climbing. In *Taeniophyllum*, roots are green and flattened photosynthetic organ. Similar roots are found in some epiphytic orchids. In some epiphytic orchids, roots have different function. Usually climbing roots attach to the substratum and clinging roots become specialized to absorb moisture from the air because of the presence of the velamen tissue.

The **stem** is aerial or underground. The aerial stem is slender and grows slowly in a sympodial or monopodial manner. Underground stem may be rhizome, corm, tuber or bulb. The rhizome may be slender and branched forming coral-like body in *Corallorhiza*. Sometimes stem forms pseudobulbs below the leaves which is used to hybernate the dry season. Pseudobulb is in different sizes, shapes and arrangement that may be small pin head size in *Bulbophyllum minutissimum*, whereas, in *Peristeria elata* it is of a child's head size. In some *Dendrobium*, it is elongated and cane-like structure with several internodes.

The **leaves** are small and scaly in saprophytes. They are radical or cauline, usually simple, distichously or pentastichously arranged. The leaf base is sheathing often covering the stem and venation is parallel. The leaves are thick, succulent with sunken stomata, well-developed water storage tissue and thick cuticle over the body.

Floral Characters

The **inflorescence** is terminal or axillary. Scape is produced in terrestrial forms. Commonly it is racemose types. In *Renanthera lawii*, the inflorescence is several feet long. Sometime spike like racemose because the slender ovary of the flower is mistaken for its pedicel. Solitary flowers are sometimes produced as in *Cypripedium*.

The **flower** is epigynous, rarely unisexual due to abortion of one sex (monoecious or dioecious), zygomorphic. They are constructed on the trimerous plan with the odd sepal placed in the anterior position; but due to twisting of the floral axis on its axis through 180^0 (resupination), the posterior labellum is brought to the anterior position such that it forms a convenient landing place for their pollinators. The flowers are dichlamydous, usually monochlamydous, bracteate and bracteolate.

The **perianth** is in trimerous with two whorls, all the members may be uniform or not. When there is a distinct calyx, its members are smaller than petals, but in *Masdevallia*, the sepals are much larger than petals. All the petals are not always uniform. The posterior petal is much different from the other two and forms the labellum or lip. It is variously cut, highly coloured and may be bear warts or out-growths. It is slipper-shaped in *Cypripedium* and trumpet-shaped in *Cattleya*. In *Bulbophyllum tremulum*, cloured bristles are bearing on the labellum. Labellum forms sometimes long spur with nectar for attracting pollinators. The lateral petals are absent in *Corysanthes* but in *Epicranthes* they are filiform and are ribbon like in *Cyperipedium*. Beyond the perianth in the flower, the central elongated body is called column. It bears at the tip anther lobes and at a lower level the stigma spots. This is interrupted as the axis elongated to raise the position of the stamens.

The **androecium** consists of 1-2 stamens and shows a great complication due to the adnation of the stamen with the parts of the gynoecium to form gynostegium. In *Diandrae*, two fertile stamens of the lateral position in the inner whorl are fertile, while odd stamen in outer whorl is sterile. But in *Nieuwiedia* the odd stamen of the outer whorl is also fertile. In *Monandrae* group, odd stamen of the outer whorl alone fertile and others are sterile while others may produce in the form of appendages and glands. Anthers are usually dithecous and pollens are granular. The pollen cannot be properly called pollen grains, but sometimes represents tetrads. Pollen grains are united to form a sac like structure called pollinia. The adjacent pollinia of two anthers are united by a translator and joint through caudicle. The whole structure is called pollinarium. In *Vanda,* the two pollen masses called stipes and pedicel which bear sticky glands at base. The pollinarium may be attached in the base of rostellum (basitonous) or by the apical part (acrotonous). The number of pollinia in the anther varies in different genera. The basic number is two or four or eight as in *Spathoglottis*.

The **gynoecium** is tricarpellary, syncarpous. Ovary is inferior and unilocular with 3 or bifid parietal placentation. In *Apostasia*, column is usually stout, the stigma or stigmatic lobes are originally three and they are confluent from one in *Cypripedium*. The stigmas are smooth, viscid or flat cushion like structures produced or at the tip of the column. In monandrous forms only two of them are fertile and receptive, while third one is sterile and forms a proliferated structure called rostellum. In diandrous forms all three stigmas are receptive and fertile.

Pollination is usually entamophilous. Majority of the orchids are usually big and highly coloured for attracting various types of pollinators. Each and every orchid flower mimic different kinds of their pollinators such as insects, bees, wasps, birds and other agencies.

The **fruit** is a dry dehiscent capsule breaking by 6 slits.

The **seed** is minute, powdery, monocotyledonous and nonendospermic. Usually the seeds contain undifferentiated embryo and hygroscopic elaters. The seed germination takes place only when associated with specific mycorrhiza either from parental or from neighbourhood.

Key Characters of the Family

1. Epiphytic herbaceous nature.
2. Raceme like inflorescence or scape.
3. Zygomorphic flowers with labellum and often with spur.
4. Reduced number of stamens (usually pollinium).
5. Gynostegium (column).
6. Inferior ovary.

7. Parietal placentation.
8. Minute, powdery, undifferentiated, non-endospermic seeds.

Fig. 15.26: *Vanda tessellata* (Roxb.) Hook.f. a. Plant with inflorescence; b. Column with pollinium; c. Column-entire; d. fruit; e. Floral diagram; f. *Habenaria rariflora* – Habit; g. *Vanilla planifolia*- Habit.

Important Plants

Plants of this family are called jewels of plant kingdom and flowers are highly showy with great ornamental values and much contributed to horticulture and floriculture industries. Ornamental genera are *Anoectochilus, Bulbophyllum, Cattleya, Cypripedium, Dendrobium, Eulophia, Habenaria, Ophrys, Orchis* and *Spathoglottis*. From the capsules of *Vanilla* an essence is obtained and it can be used in flavouring material in food items.

Systematics and Phylogeny

Bentham and Hooker placed the family under series Microspermae while other taxonomists treated them in separate order Orchidales by Hutchinsom (1959), Takhtajan (1969) and Cronquist (1981). However, Thorne (1983) ranked it under the suborder Orchidineae of the Liliales. Willis (1973) divided the Orchidaceae into 3 subfamilies as Apostasioideae, Cypripedioideae and Orchidoideae. These subfamilies are also further divided into 6 tribes, *viz*. Apostasieae, Cyperipedieae, Orchideae, Neottieae, Epidendreae and Vandeae. Most of the taxonomists consider this family as most advanced and highly evolved family among monocotyledons due to the presence of herbaceous habit with several epiphytes, reduced number of stamens, resupinate epigynous ovary, presence of rostellum and non-endospermic seeds. Many viewed that the family originated from Liliaceous stock (Hutchinson, 1959). The Orchidaceae is currently placed in the order Asparagales with 13 other families in APG IV system of 2016. It was supported by morphological studies, but never received strong support in molecular phylogenetic studies. The family is divided into 5 tribes, *viz*. Apostasioideae, Vanilloideae, Cyperipedioideae, Epidendroideae and Orchidoideae. Vanilloideae is considered as the primitive tribe of the family.

Liliaceae (Lily Family)

Distribution: The family comprises of 705 genera and 3,700 species distributed throughout the world. They are abundant in warmer temperate and tropical regions. About 35 genera and 200 species have been reported in India. Some larger genera in the family are *Allium* (450 spp.), *Smilax* (350 spp.), *Aloe* (300 spp.), *Lilium* (80 spp.), *Colchicum* (65 spp.) and *Asphodelus* (15 spp.).

Habit: The members are mostly herbaceous or shrubby forms, perennating by underground stems. Several of them show straggling habit and some are arborescent forms as in *Dracaena* and *Aloe*. Many of them are strong xerophytic nature. *Petrosaevia* is a partial root parasite.

Vegetative Characters

The **root system** is of adventitious type. Contractile roots are produced commonly when the stems are underground. Fasciculate roots are found in *Asparagus*.

The **stem** is underground or aerial. The rhizome of *Polygonatum* grows in sympodial manner and *Gloriosa* has plough like structure. The underground stem is corm in *Colchicum*. Several of them produce bulbs as in *Allium*. In arborescent forms the stem shows anomalous secondary thickening due to formation of extra fascicular cambium. The stem may be cladode in *Asparagus* and leaf-like flattened in *Ruscus*. Bulbils are produced in the axils of leaves in *Allium* and *Chlorophytum*.

The **leaves** are cauline or radical as in several herbaceous forms. They are alternate, closely arranged, opposite in *Scolyopus* and whorled in *Lilium canadense*. The leaves are scaly in *Asparagus* and succulent in *Aloe*. Venation is parallel but reticulate in *Smilax*. The leaf tip function as a tendril in *Gloriosa*. Epiphyllous buds are produced in *Drymia* and the leaf tip when touch the soil.

Floral Characters

The **inflorescence** is terminal or axillary, usually scape. Sometimes flowers are protected by large spathaceous bracts. Flowers are solitary in *Tulipa*. The arrangement of the flowers may be in umbel, panicles or monochasial cyme. The flowers are born along the margin of the flattened phylloclade in *Semele androgyna*.

The **flowers** are usually actinomorphic, bisexual, hypogynous, typically trimerous (tetramerous in *Paris*), bracteate and bracteolate. In *Asparagus* by the reduction of one sex the flowers become unisexual. They are dioecious in species of *Smilax*. The flowers are slightly zygomorphic in *Aloe* and *Haworthia*.

The **perianth** is in trimerous whorls. The lobes are uniform in size and structure. They are free or united as in *Aloe*. Perianth is two-lipped in *Haworthia*. Aestivation is valvate or imbricate.

The **androecium** consists of 6 stamens in two whorls. In *Smilax*, the members of one whorl is absent. In *Gillesia*, the stamen in the posterior part is sterile. The anthers are dithecous, dehiscing by pores or longitudinal slits.

The **gynoecium** is tricarpellary, syncarpous with terminal style and capitate stigma. Ovary is superior, ovules two in each chamber on axile placentation. Rarely ovary is unilocular with parietal placenta.

Pollination is entamophilous. The flowers of *Allium* are protandrous and *Smilax* is protogynous. Flowers of *Yucca* are specially adapted for insect pollination.

The **fruit** is loculicidal capsule, rarely septicidal. In *Dracaena*, the fruit is berry.

The **seed** is monocotyledonous and endospermic. Polyembroyony is seen in some members.

Key Characters of the Family

1. Underground stem.
2. Typical trimerous, homochlamydous flowers.
3. Axile placentation.

Important Plants

Several of the members are economically valuable plants such as *Allium cepa* (onion), *Allium sativum* (garlic) and *Asparagus* are vegetable value. Drug yielding plants are *Aloe barbadense*, *Veratrum*, *Colchicum autumnale*, *Smilax glabra*, *S. officinalis*, *S. rotundifolia*, *Convallaria majalis*, etc. Resins are obtained from *Dracaena* and *Xanthorhoea*. Some are garden ornamentals like *Hemerocallis*, *Chlorophytum*, *Dracaena*, *Yucca*, *Ornithocalum*, *Sansevieria*, *Corydaline*, *Gloriosa*, *Lilium*, etc.

Systematics and Phylogeny

Liliaceae is placed under the series Coronariae by Bentham and Hooker, under Liliflorae by Engler and Prantl (1931) and Liliales by Hutchinson (1959), Takhtajan (1969), Cronquist (1981) and Thorne (1983). Cronquist merged Amaryllidaceae under Liliaceae and some other taxonomists included certain genera of Liliaceae under separate families, viz. *Aloe* under Aloeaceae and *Smilax* under Smilacaceae.

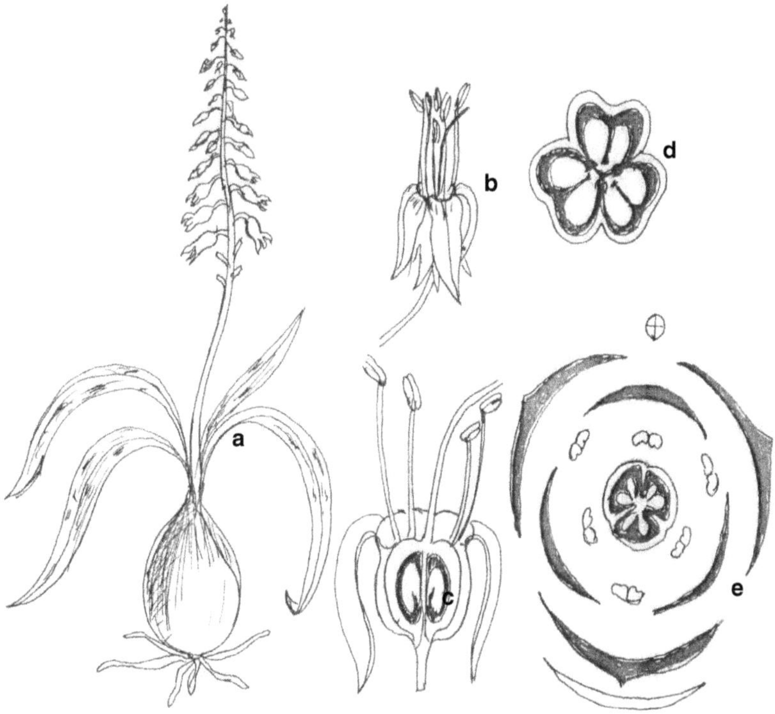

Fig. 15.27: *Drimia indica* (Roxb.) Jessop. a. Habit; b. Flower-entire; c. Flower l.s.; d. Ovary c.s.; e. Floral diagram

However, Bentham and Hooker divided the family Liliaceae into 20 tribes whereas Engler and Prantl treated only into 12 tribes. Family Liliaceae represents most primitive monocotyledons stock.

Modern taxonomic systems, such as the APG which is based on phylogenetic principles using molecular biology, have redistributed many of these genera resulting in the relatively small family that is currently recognised. Despite establishing this relative degree of monophyly (genetic homogeneity) for the Liliaceae family, their morphology remains diverse and there exists within the Liliaceae clade with a number of subclades (subgroups). But until now some enigmatic genere were *Clintonia, Medeola, Scoliopus* and *Tricyrtis.* They were not properly placed in the system. Of the fifteen genera within the Liliaceae, the ten genera of the Lilioideae subfamily form one morphological group that is characterised by contractile bulbs and roots, and a Fritillaria-type embryo-sac. Within the Lilioideae, *Clintonia* and the closely related *Medeola* form a subclade, and are now considered a separate tribe Medeoleae. The other major grouping consists of the five genera constituting the Streptopoideae (including *Scoliopus)* and Calochortoideae (including *Tricyrtis)* subfamilies characterised by creeping rhizomes, styles which are divided at their apices, and by megagametophyte development of the Polygonum-type of embryo sac.

Arecaceae (Palm Family)

Distribution: The family comprises of about 180 genera and 2,600 species which are exclusively tropical in distribution. About 30 genera and 100 species have been reported in India. Some larger genera of the family are *Calamus* (300 spp.), *Chamedorea* (130 spp.), *Areca* (55 spp.), *Coccothrinax* (50 spp.), *Copernicia* (30 spp.) and *Raphia* (30 spp.).

Habit: They are mostly attaining size of trees and form characteristic vegetation. *Cocos* and many members are found along the coastal region. *Nipa* is an esturial element particularly in Sundarbans of our country.

Vegetative Characters

Root system is of adventitious type and to maintain in position the tall erect trunk in some forms buttress roots are produced. The leaves are usually coriaceous and show strong xerophytic characters. The venation is parallel. In the bud condition the leaves are characteristically folded and when expand is torn along the lines of folding and hence it forms a compound structure.

Floral Characters

The **inflorescence** is usually an axillary, but in monocarpic forms like *Corypha umbraculifera* and *Metroxylon rumphii* it is terminal. It is usually covered by one or several broad, leathery bracts that are called spathes. They enclose and protect the simple or much branched inflorescence axis in the young condition. The sessile flowers are sometimes embedded in the axis.

The **flowers** are usually sessile, unisexual, monoecious (*Cocos*) or dioecious (*Borassus* and *Phoenix*). They are bisexual in *Pritchardia* and polygamous in *Chamerops*. The flowers are usually small, trimerous, hypogynous and actinomorphic. Sterile structures are seen in the flowers of the opposite sex.

The **perianth** is in 2-merous whorls with distinct lobes. It may be petals and sepals by their size alone. They are usually leathery but are scaly in *Phytelephas*. The perianth lobes are united or free, valvate or imbricate and persistent in female flowers.

The **androecium** consists of six stamens. They are in 2-merous whorl, but in *Areca* in one whorl. They are free with dithecous, introrse, versatile anthers that dehisce longitudinally. In the female flowers of the *Cocos* and *Borassus*, the staminodes are represented as a ring with six teeth.

The **gynoecium** is tricarpellary (bicarpellary in *Caryota*), syncarpous (apocarpous in *Phoenix*). The ovary is superior, terminal style with capitate stigma.

There is one ovule in each chamber and usually one carpel is functional. The placentation is axile or basal.

Pollination is anemiphilous but in *Sabal* it is entomophilous.

The **fruit** is a berry in *Phoenix* and a fibrous drupe in *Cocos*.

The **seed** is monocotyledonous and endospermic. Endosperm is oily in *Cocos* and of cellulose as in *Phytelephas*. Ruminate endosperm is seen in *Areca*.

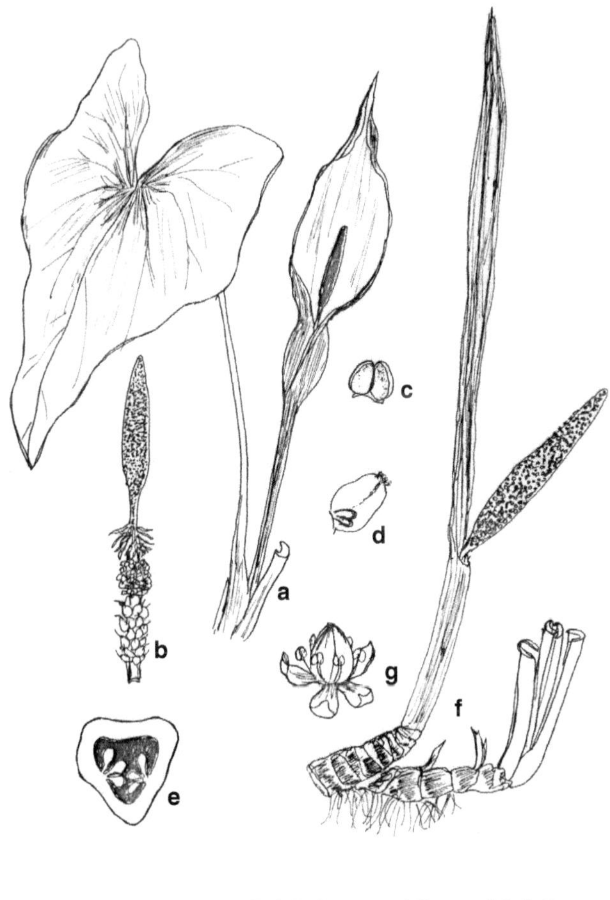

Fig. 15.28: *Colocasia esculenta* (L.) Schott. a. Plant with inflorescence; b. Inflorescence stype; c. Stamen; d. Ovary l.s.; e. Ovary c.s.; f. *Acorus calamus* – Habit; g. Flower-entire.

Key Characters of the Family

1. Arborescent unbranched habit.
2. Fan-like or feather-like leaves.
3. Unisexual flowers produced on spadix inflorescence.

Important Plants

The wood of *Borassus* and *Cocos* is used for poles and rafters. The reedy stem of *Calamus rotang* (cane) is used in the making of baskets, caskets, cradles, chairs, etc. Sugar containing sap (toddy) is obtained from the inflorescence of *Cocos nucifera*, *Borassus flabellifer*, *Mauritia vinifera*, *M. flexuosa*, *Arenga wightii*, *Nypa fruticans* and *Phoenix sylvestris*. Fibres are obtained from the leafbases and mesocarp of the fruits of *Borassus flabellifer*, *Leopoldenia piassaba*, *Cocos nucifera*, *Attalea* and *Trachycarpus*. Wax is obtained from the leaves of *Copornicia cerifera* and the stem of *Ceroxylon andicolum*. Edible oil is extracted from the fruits of *Elaeis guineensis* and *Cocos nucifera*. The endosperm of *Phytelephas* is vegetable ivory. Young buds of *Euterpe*, *Cocos* and *Attalea* are used as cabbage.

Systematics and Phylogeny

Arecaceae (Plamae) is variously placed under series Calycinae (Bentham and Hooker), order Principes (Engler and Prantl), order Palmales (Hutchinson) and order Arecales (Takhtajan, Cronquist and Thorne). The family is closely related to Araceae, and it was treated under Spadiciflorae (Rendle) along with Araceae and Lemnaceae. Hutchinson believed that this family originated from Agavaceae whereas Takhtajan and Cronquist opined that Arecaceae and Araceae are closely related in possessing spadix with unisexual flowers.

Palms are a monophyletic group of plants, meaning the group consists of a common ancestor and all its descendants. Extensive taxonomic research on palms began with botanist H.E. Moore, who organized palms into 15 major groups based mostly on general morphological features. N.W. Uhl and J. Dransfield (1987) have revised Moore's classification that organizes Arecaceae into six subfamilies, *viz.* Coryphoideae, Calamoideae, Nypoideae, Ceroxyloideae, Arecoideae and Phytelephantoideae. Currently, few extensive phylogenetic studies of the Arecaceae exist. Baker *et al.* (1997) explored subfamily and tribe relationships using chloroplast DNA from 60 genera from all subfamilies and tribes. The results strongly showed the Calamoideae are monophyletic, and Ceroxyloideae and Coryphoideae are paraphyletic. The relationships of Arecoideae are uncertain, but they are possibly related to the Ceroxyloideae and Phytelephantoideae. In APG IV system, the family Arecaceae is placed under order Arecales which contains only two families Arecaceae and Dasypogonaceae.

Cyperaceae (Sedge Family)

Distribution: This family consists of 55,500 species classified under 90 genera. They are cosmopolitan in distribution. *Carex* and *Scirpus* are abundantly found in cold temperate climates and are frequently towards equator. *Cyperus* is the genus mostly distributed in tropics. About 25 genera and 400 species have been reported in India. Some of the larger genera are *Carex* (2,000 spp.), *Cyperus* (700 spp.), *Fimbristylis* (300 spp.), *Scirpus* (250 spp.), *Rhynchospora* (230 spp.), *Eleocharis* (200 spp.), *Scleria* (200 spp.), *Schoenus* (100 spp.) and *Uncinia* (40 spp.).

Habit and **Habitat**: These are grass-like or rush-like perennial herbs and are generally confined to sour, swampy localities. Some are characteristically growing in sand dunes (*Carex arvensis*). *Schoenodendron* is an arborescent form.

Vegetative Characters

Root system is of the adventitious type, generally fibrous. Root tubers are found in some species of *Cyperus*.

Stem is usually perennial modified into rhizome or tuberous. Growth is sympodial, new shoots of each year is adnate to the parent shoot for one internode, branching seems to be extra-axillary. The stem is slender or trigonal, solid with elongate rhizome. Internodes are usually long.

The **leaves** are simple, radical or cauline; three-ranked. Leaf base is sheathing and closed the stem.

Floral Characters

The **inflorescence** is spikelet, usually terminal, sometimes axillary. They are arranged in simple spikes, racemes, panicles or umbels with commonly subtended to foliaceous bracts. The spikelet is the taxonomic character in this group.

The **flowers** are much reduced in size and structure and are enclosed by imbricate bracts. They are bisexual, sometimes unisexual (monoecious or dioecious), hypogynous, actinomorphic, dichlamydeous or achlamydeous.

The **perianth** is in 2 whorls of 3 each as in *Scirpus sylvaticus*, free. Perianth is absent in *Carex*. Perianth lobes are scaly or subpetaloid but commonly represented by hairs or completely suppressed.

The **androecium** consists of six stamens in 2 trimerous whorls. Sometimes only outer whorl is present. Stamen is single in *Hemicarpha*. The filament is long with dithecous, introrse basifixed anther that dehisce by longitudinal slits.

The **gynoecium** is tricarpellary, sometimes bicarpellary, syncarpous. Ovary is superior with terminal style that bears 3 or 2 stigmatic branches. Ovary is unilocular bearing a single ovule on basal placenta.

Pollination is through the agency of wind.

The **fruit** is an achene or nut.

The **seed** is monocotyledonous and endospermic. Embryo is situated in the basal part of the endosperm. Plumule is often covered by a sheath.

Key Characters of the Family

1. Grass-like stem
2. Three-sided solid stem.
3. Leaves with closed sheathing base.
4. Inflorescence is spikelet.
5. Scaly perianth.

6. Unilocular ovary with basal placenta.
7. Fruit – achene or nut.

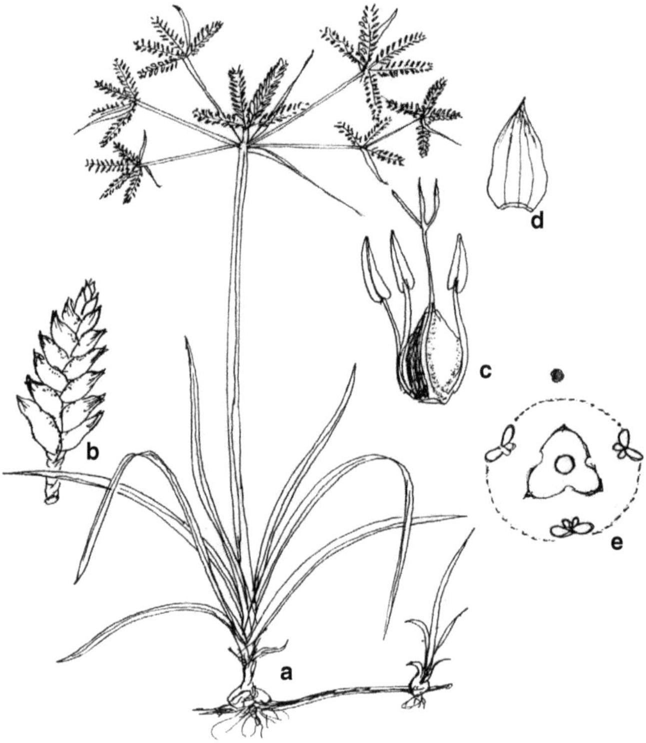

Fig. 15.29: *Cyperus rotundus* L. a. Habit; b. Spike; c. Flower-entire; d. Glume; e. Floral diagram

Important Plants

The members are common occurrence with less economic importance. Starch is stored in rhizomes of *Cyperus esculentus, Eleocharis tuberosus* and certain species of *Scirpus*. *Isolepis* and *Cyperus alternicifolius* are horticultural importance. *Scirpus lacustris* and *Lepironia mucronata* are used in weaving mats. The stems of *Cyperus papyrus* were used by the ancient Egyptians for making papers.

Systematics and Phylogeny

Cyperaceae is closely related to Poaceae and Bentham and Hooker placed both the families under the same series Glumaceae while Engler and Prantl placed them under order Glumiflorae. Cronquist (1981) also treated both the families under Cyperales. Thorne (1983) included these two families under Commelinales. However, Hutchinson (1959) erected two separate orders for these families, *viz.* Cyperales and Graminales respectively. Takhtajan (1969) believed that Cyperaceae belongs to independent order Cyperales and it is evolved from Juncaceous stock. Cyperaceae is divided into 3 subfamilies, *viz.* Cyperoideae, Rhynchosporioideae and Caricoideae by Hickey and King (1988). Radford (1986) has classified the family into 8 tribes, *viz.* Cypereae, Dulichieae, Rhynchosporeae, Scirpeae, Hypolytreae, Sclerieae, Cryptangieae and Cariceae. In APG IV system, the family is placed under order Poales with 12 other families including Poaceae and Juncaceae. This system is also evidenced that the family is rooted with Juncaceae.

Poaceae (Grass Family)

Distribution: This is one of the largest flowering plant family comprising about 780 genera and 12,000 species that are distributed in all parts of the world. They formed important constituent of vegetations of prairies, steppes and savannas. About 240 genera and 1,200 species have been reported in India. Some of the larger genera in the family are *Panicum* (500 spp.), *Digitaria* (400 spp.), *Aristida* (380 spp.), *Poa* (300 spp.), *Eragrostis* (300 spp.), *Stipa* (300 spp.), *Paspalum* (250 spp.), *Agrostis* (200 spp.), *Arundinaria* (150 spp.), *Pennisetum* (150 spp.), *Setaria* (140 spp.), *Agropyron* (150 spp.), *Avena* (70 spp.) and *Sorghum* (60 spp.).

Habit: They are usually herbaceous annuals or perennials and they perennate by means of underground rhizomes. A few are growing more than 100 meter tall as in Bamboos. Some species are characteristically found on sand dunes as sand binder (*Spinifex arvensis* and *Psamma*).

Vegetative Characters

The **root system** is of adventitious type and usually fibrous. In *Maize* and *Sorghum* stilt-roots are formed.

The **shoot system** is sub-terranean or aerial. Commonly it is a rhizome. Aerial stem is erect or trailing culm. Sometimes they are runners. The stem is usually cylindrical, hollow with long internodes. It is woody in bamboos. The nodes are swollen at the base of the internode and intercalary meristem is situated.

The **leaves** are radical or cauline. Usually in the sub-terranean stem leaves are scaly. In the aerial stems, leaves are simple, distichous, leaf-base sheathing and covered the internodes. On the upper part of the sheath is produced ligule. The liguels may be scaly, membranous or hairy. Sometimes it is represented by a ring. Ligule is absent in *Echinochloa*. Auriculate structure is produced in *Oryza*. The lamina is usually linear or lanceolate with parallel venation. In *Bambusa*, petiole is cylindric.

Floral Characters

The **inflorescence** is terminal or axillary. In grass flowers the essential organs are reduced and the flowers are produced in racemose manner on a condensed axis to form spikelets. The spikelets are of taxonomic importance. They may be sessile or stalked. They are usually produced on single or many-fold branched inflorescence axis in a racemose or rarely cymose manner.

The **spikelet** bears a few to several flowers in an acropetal order in the axils of closely fitting, imbricate, distichously arranged bracts on the slender, condensed, often jointed axils called rachilla. The first two bracts are empty and hence called sterile glumes. In the axil of each bract excepting top most one there is a flower. Such a bract is called the flowering glume or the lemma. On the axis of the flower is produced a thin structure called palea. It is binerved and often envelope the flower and is pressed between the flower and rachilla. The flower bears two membranous lodicules, 3 or 6 stamens and the gynoecium.

The **flowers** are bisexual or unisexual, sometimes polygamous. The unisexual flower may be produced on the same inflorescence or on different inflorescences on the same plant (monoecious) as in maize or on different plants (dioecious) as in *Spinifix*. They are polygamous, zygomorphic and trimerous. The sterile glumes of the flowers are unequal in size and sometimes they are reduced in different degrees in many genera. It may bear an awn on the back (as in *Avena*) or at the tip (as in *Chloris*). The palea is a thin and binerved structure that commonly covers the flower. The lodicules are usually two (three in *Bambusa*), arranged laterally. Lodicules are numerous in *Ochlandra*.

The **androecium** consists of usually 3 stamens, rarely six in *Bambusa* and *Oryza* in trimerous whorls. The numerous stamens of *Ochalandra* are united. Single stamen is found in *Uniola*. The stamen has a slender filament and a dithecous, introrse anther that is apparently versatile and dehisce through longitudinal slits.

The **gynoecium** is fundamentally tricarpellary, syncarpous with unilocular ovary in which single anatropous ovule is there on basal placentation. The ovary is superior with terminal style and plumose stigma. In *Bambusa*, three styles are present whereas in *Nardus* there is single style.

Pollination is facilitated by wind.

The **fruit** is a caryopsis, a dry indehiscent one. Seeds usually are monocotyledonous and endospermic.

Key Characters of the Family

1. Cylindrical hollow stem with distichously arranged leaves.
2. Open leaf-sheath and ligule.
3. Spikelet inflorescence.
4. Lodicules and glumaceous floral parts.
5. Versatile anthers.
6. Feathery stigma.
7. Caryopsis fruit.

Important Plants

The family members yield food grains for the whole human population of the world. The common members are wheat (*Triticum aestivum, T. vulgare, T. turgidum,*

T. amylum), rye (*Secale cereale*), barley (*Hordeum vulgare*), maize (*Zea mays*), oat (*Avena sativa*), ragi (*Eleusine coracana*), jower (*Sorghum vulgare*) and rice (*Oryza sativa*).

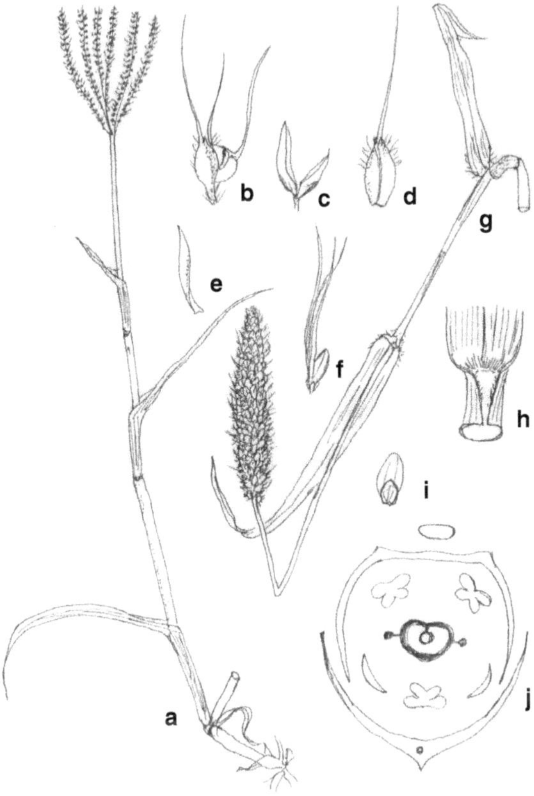

Fig. 15.30: *Chloris barbata* Sw. a. Plant; b. Flower-entire; c. Glumes; d. Glume with awn; e. Bract; f. Lodicules; g. *Setaria pumila* – Habit; h. Auricle; i. *Setaria* seed; j. Floral diagram

The leaves of *Stipa tenacissima*, *Psamma arenaria*, *Elymus arenarius* and *Zea mays* are used for making paper. Essential oil is obtained from the leaves of *Cymbopogon citratus* and roots of *Vetiveria zizanioides*. *Saccharum officinarum* is a source of sugar. *Distichlis maritima* and *Spinifix arvensis* are sand binders.

Systematics and Phylogeny

Poaceae (Gramineae) is variously placed under series Glumaceae (Bentham and Hooker), order Glumiflorae (Engler and Prantl, 1931), order Graminales (Hutchinson, 1959), order Poales (Takhtajan, 1969), order Cyperales (Cronquist, 1981) and order Commelinales (Thorne, 1983). Traditionally the family is divided into 6 subfamilies, *viz*. Festucoideae, Panicoideae, Egragrostoideae, Bambusoideae, Oryzoideae and Arundinoideae with 25 tribes. Poaceae resembles Cyperaceae by their small inconspicuous flowers with subteneded bracts, presence of spikelet inflorescence. However, Poaceae differs with Cyperaceae by terete, hollow, jointed stems, distichous leaves and structure of spikelets. APG IV system treated this family as highly evolved among monocotyledons and advanced family also. Poaceae is placed in last position in monocotyledons under the order Poales and furher it is divided into 12 subfamilies, *viz.* Anomochlooideae, Pharoideae, Puelioideae, Pooideae, Bambusoideae, Ehrhartoideae, Aristidoideae, Arundinoideae, Chloridoideae, Panicoideae, Micrairoideae and Danthonioideae.

16

Economic Botany

Sl. No.	English Name	Botanical Name	Family	Plant Part Used
I.	**Cereals**			
1.	Macaroni wheat	*Triticum durum* Desf.	Poaceae	Fruit
2.	Bread wheat	*Triticum* aestivum L.	Poaceae	Fruit
3.	Emmer wheat	*Triticum dicoccum* (Schank) Schubles	Poaceae	Fruit
4.	Rice, Paddy	*Oryza sativa* L.	Poaceae	Fruit
5.	Barley	*Hordeum vulgare* L.	Poaceae	Fruit
6.	Oats	*Avena sativa* L.	Poaceae	Fruit
7.	Rye	*Secale cereale* L.	Poaceae	Fruit
II.	**Millets**			
8.	Pearl millet	*Pennisetum typhoides* Stapf & Hubb.	Poaceae	Fruit
	Bulrush millet	(=*P. typhoideum* L.,		
	Spiked millet	*Pennisetum glaucum*)		
9.	Finger millet	*Eleusine coracana* Gaertn.	Poaceae	Fruit
10.	Great millet	*Sorghum bicolor* (L.) Moench	Poaceae	Fruit
11.	Italian millet	*Setaria italica* Beauv.	Poaceae	Fruit
12.	Barnyard millet	*Echinochloa colonum* (L.) Link	Poaceae	Fruit

Sl. No.	English Name	Botanical Name	Family	Plant Part Used
13.	Common millet Proso millet Hog millet	*Panicum miliaceum* L.	Poaceae	Fruit
14.	Kodo millet	*Paspalum scrobiculatum* L.	Poaceae	Fruit
15.	Little millet	*Panicum miliare* L.	Poaceae	Fruit
16.	Maize	*Zea mays* L.	Poaceae	Fruit
III.	**Pulses**			
17.	Black gram	*Vigna mungo* (L.) Hepper	Fabaceae	Seed
18.	Green gram	*Vigna radiata* (L.) Wilczek	Fabaceae	Seed
19.	Horse gram	*Vigna unguiculata* (L.) Walp. ssp. *unguiculata* (=*Dolichos biflorus* L.)	Fabaceae	Seed
20.	Cow pea	*Vigna unguiculata* (L.) Walp. ssp. *cylindrica* (L.) Walp. (=*V. catjang* Wall.)	Fabaceae	Seed
21.	Chick pea Bengal gram	*Cicer arietinum* L.	Fabaceae	Seed
22.	Pea	*Pisum sativum* L.	Fabaceae	Seed
23.	Pigeon pea Red gram	*Cajanus cajan* Millsp.	Fabaceae	Seed
24.	Chickling vetch	*Lathyrus sativus* L.	Fabaceae	Seed
25.	Lentil	*Lens culinaris* Medik.	Fabaceae	Seed
IV.	**Oil Seeds**			
26.	Ground nut Pea nut	*Arachis hypogaea* L.	Fabaceae	Seed
27.	Coconut	*Cocos nucifera* L.	Fabaceae	Endosperm
28.	Palm oil	*Elacis guinensis* Jacq.	Arecaceae	Fruit
29.	Black mustard	*Brassica nigra* Koch	Brassicaceae	Seed
30.	Brown sarson	Brassica campestris L. var. *dichotoma*	Brassicaceae	Seed
31.	Rape seed	*Brassica napus* L.	Brassicaceae	Seed
32.	Gingelly oil Sesame	*Sesamum orientale* L. (=*S. indicum* L.)	Pedaliaceae	Seed
33.	Sunflower	*Helianthus annuus* L.	Asteraceae	Fruit
34.	Safflower	*Carthamus tinctorius* L.	Asteraceae	Fruit
35.	Niger seed	*Guizotia abyssinica* Cass.	Asteraceae	Fruit
36.	Soy bean oil	*Glycine max* Merr.	Fabaceae	Seed
37.	Linseed	*Linum usitatissimum* L.	Linaceae	Seed
38.	Castor	*Ricinus communis* L.	Euphorbiaceae	Seed

Sl. No.	English Name	Botanical Name	Family	Plant Part Used
V.	**Vegetables**			
39.	White gourd	*Benincasa hispida* Cogn.	Cucurbitaceae	Fruit
40.	Bitter gourd	*Momordica charantia* L.	Cucurbitaceae	Fruit
41.	Spine gourd	*Momordica dioica* Roxb. ex Willd.	Cucurbitaceae	Fruit
42.	Bottle gourd	*Lageneria siceraria* (Molina) Standl. (=*L. vulgaris Ser.*)	Cucurbitaceae	Fruit
43.	Little gourd	*Coccinia grandis* Voigt (=*C. indica* Wt. & Arn.)	Cucurbitaceae	Fruit
44.	Ridge gourd	*Luffa acutangula* (L.) Roxb.	Cucurbitaceae	Fruit
45.	Smooth gourd	*Luffa cylindrica* (L.) Roem.	Cucurbitaceae	Fruit
46.	Pumpkin	*Cucurbita pepo* L.	Cucurbitaceae	Fruit
47.	Red pumpkin	*Cucurbita maxima* Duch.	Cucurbitaceae	Fruit
48.	Snake gourd	*Trichosanthes anguina* L.	Cucurbitaceae	Fruit
49.	Cucumber	*Cucumis sativus* L.	Cucurbitaceae	Fruit
50.	Brinjal Egg plant	*Solanum melongena* L.	Solanaceae	Fruit
51.	Potato	*Solanum tuberosum* L.	Solanaceae	Stem
52.	Tomato	*Lycopersicon esculentum* Mill.	Solanaceae	Fruit
53.	Broad bean	*Vicia faba* L.	Fabaceae	Fruit
54.	Cluster bean Field vetch	*Cyampopsis tetragonaloba* Taub.	Fabaceae	Fruit
55.	Indian bean	*Lablab purpureus* (L.) Sweet	Fabaceae	Fruit
56.	Sword bean	*Canavalia gladiata* DC. (=*C. ensiformis*)	Fabaceae	Fruit
57.	Velvet bean	*Mucuna cochinchinensis*	Fabaceae	Fruit
58.	Drum stick	*Moringa oleifera* Lam. (=*M. pterygosperma Sajina* & Gaertn.)	Moringaceae	Fruit
59.	Beet Root	*Beta vulgaris* L. var. *esculenta*	Chenopodiaceae	Stem-root
60.	Carrot	*Daucus carota* L.	Apiaceae	Root tuber
61.	Knol-khol	*Brassica oleracea* L. var. *caulocarpa* L.	Brassicaceae	Stem tuber
62.	Cabbage	*Brassica oleracea* L. var. *capitata* L.	Brassicaceae	Stem tuber
63.	Cauliflower	*Brassica oleracea* L. var. *botrytis* L.	Brassicaceae	Inflore-escence
64.	Turnip	*Brassica campestris* L. var. *rapa* L.	Brassicaceae	Root tuber
65.	Lady's finger	*Abelmoschus esculentus* (L.) Moench (=*Hibiscus esculentus* L.)	Malvaceae	Fruit
66.	Onion	*Allium cepa* L.	Liliaceae	Bulb

Sl. No.	English Name	Botanical Name	Family	Plant Part Used
67.	Radish	*Raphanus sativus* L.	Brassicaceae	Root
68.	Elephant ear	*Colocosia antiquorum* Schott.	Araceae	Corm
69.	Elephant's foot Blume	*Amorphophalus campanulatus*	Araceae	Corm
70.	Sweet potato	*Ipomoea batatas* L.	Convolvulaceae	Root
71.	Bladder dock	*Rumex vesicarius* L.	Polygonaceae	Leaf
72.	Amaranth	*Amaranthus tricolor* L.	Amaranthaceae	Leaf
73.	Amaranth	*Amaranthus oleracea*	Amaranthaceae	Leaf
74.	Amaranth	*Amaranthus paniculatus* L.	Amaranthaceae	Leaf
75.	Red vine spinach	*Basella rubra* L.	Basellaceae	Leaf
76.	Gongura	*Hibiscus cannabinus* L.	Malvaceae	Leaf
77.	Spinach	*Spinacea oleracea*	Chenopodiaceae	Leaf
VI.	**Fruit Crops**			
78.	Apple	*Malus pumila* Miller (=*Pyrus malus* L.)	Rosaceae	
79.	Apricot	*Prunus armeniaca* L.	Rosaceae	
80.	Pear	*Pyrus communis* L.	Rosaceae	
81.	Peach	*Prunus persica* (L.) Batsch	Rosaceae	
82.	Plum	*Prunus domestica* L.	Rosaceae	
83.	Almond	*Prunus dulcis* (Miller) D.Web.	Rosaceae	
84.	Cherry	*Prunus avium* (L.)	Rosaceae	
85.	Strawberry	*Fragaria vesca* L.	Rosaceae	
86.	Raspberry	*Rubus idaeus* L.	Rosaceae	
87.	Litchi	*Litchi sinensis*	Sapindaceae	
88.	Fig (Anjur)	*Ficus carica* L.	Moraceae	
89.	Avocado	*Persea americana* Mill.	Lauraceae	
90.	Walnut	*Juglans regia* L.	Juglandaceae	
91.	Pistachio	*Pistacia vera* L.	Anacardiaceae	
92.	Banana	*Musa paradisiaca* L.	Musaceae	
93.	Mango	*Mangifera indica* L.	Anacardiaceae	
94.	Papaya	*Carica papaya* L.	Caricaceae	
95.	Jack fruit	*Artocarpus integrifolia* L.	Moraceae	
96.	Grape	*Vitis vinifera* L.	Vitaceae	
97.	Guava	*Psidium guajava* L.	Myrtaceae	
98.	Cashew	*Anacardium occidentale* L.	Anacardiaceae	
99.	Custard apple	*Annona squamosa* L.	Annonaceae	
100.	Orange	*Citrus reticulata* Blanco	Rutaceae	

Sl. No.	English Name	Botanical Name	Family	Plant Part Used
100.	Sweet orange	*Citrus sinensis* Osbeck	Rutaceae	
101.	Citron	*Citrus medica* L.	Rutaceae	
102.	Lemon	*Citrus limonia* Osbeck	Rutaceae	
103.	Acid lime	*Citrus aurantifolia* Swingle	Rutaceae	
104.	Rough lemon	*Citrus limon* Burm.f.	Rutaceae	
105.	Shaddock	*Citrus grandis* (L.) Osbeck	Rutaceae	
106.	Jujube	*Ziziphus jujuba L.*	Rhamnaceae	
107.	Pineapple	*Ananas comosus* Merr. (= *A. sativa* Schutt.)	Bromeliaceae	
108.	Pomegranate	*Punica granatum* L.	Punicaceae	
109.	Watermelon	*Citrullus lanatus* (Thunb.) Matsumura & Nakai	Cucurbitaceae	
110.	Muskmelon	*Cucumis melo* L.	Cucurbitaceae	
111.	Indian cherry	*Syzygium cumini* (L.) Skeels	Myrtaceae	
112.	Sapota	*Manilkhara zapota* (L.) P. Royen (=*Achras sapota* L.)	Sapotaceae	
113.	Dogwood apple	*Limonia acidissima* L. (= *Feronia elephantum* Correea)	Rutaceae	
114.	Indian almond	*Terminalia catappa* L.	Combretaceae	
115.	Date palm	*Phoenix dactylifera* L.	Arecaceae	
VII.	**Spices and Condiments**			
116.	Indian mustard	*Brassica juncea* (L.) Czern & Coss	Brassicaceae	Seeds
117.	Poppy seeds	*Papaver somniferum* L.	Papaveraceae	Seeds
118.	Black pepper	*Piper nigrum* L.	Piperaceae	Fruit
119.	Cardamon	*Elettaria cardamomum* Maton	Zingiberaceae	Fruit
120.	Chilli, Red pepper	*Capsicum frutescens* L.	Solanaceae	Fruit
121.	Coriander	*Coriandrum sativum* L.	Apiaceae	Fruits & leaves
122.	Cumin	*Cuminum cyminum* L.	Apiaceae	Fruit
123.	Fenugreek	*Trigonella foenum-graecum* L.	Fabaceae	Seed
124.	Garlic	*Allium sativum* L.	Liliaceae	Bulb
125.	Ginger	*Zingiber officinale* Roscoe	Zingiberaceae	Rhizome
126.	Nutmeg	*Myristica fragrans* Houtt.	Myristichaceae	Aril
127.	Turmeric	*Curcuma longa* L.	Zingiberaceae	Rhizome
128.	Mint	*Mentha piperata* L.	Lamiaceae	Leaf
129.	Long pepper	*Piper longum* L.	Piperaceae	Fruit
130.	Cloves	*Syzygium aromaticum* (L.) Merr. & Perry	Myrtaceae	Flower

Sl. No.	English Name	Botanical Name	Family	Plant Part Used
131.	Bishop's weed	*Trachyspermum ammi* Sprague (=*Carum copticum*)	Apiaceae	Fruit
132.	Anise	*Pimpinella anisum* L.	Apiaceae	Fruit
133.	Cinnamon Blume	*Cinnamomum zeylanicum*	Lauraceae	Stem bark
134.	Asafoetida	*Ferula asafoetida* L.	Apiaceae	Gum resin
135.	Curry leaf	*Murraya koenigii* L.	Rutaceae	Leaf
136.	Vanilla	*Vanilla planifrons* Ander.	Orchidaceae	Leaf
137.	Tamarind	*Tamarindus indica* L.	Fabaceae	Mesocarp
138.	Fennel	*Foeniculum vulgare* Mill.	Apiaceae	Fruit
139.	Saffron	*Crocus sativus* L.	Iridaceae	Stigma & Style
VIII.	**Rubber and Latex**			
140.	Para Rubber	*Hevea brasiliensis* Muell.-Arg.	Euphorbiaceae	Latex
141.	Guyayule rubber	*Parthenium argentatum*	Asteraceae	Root
142.	Guttapercha	*Palaquium gutta* (Hook.)	Sapotaceae	Latex
143.	Chewing gum	*Manilkhara zapota* (L.) P. Royen (=*Achras sapota* L.)	Sapotaceae	Latex
IX.	**Bevarages**			
144.	Coffee	*Coffea arabica* L.	Rubiaceae	Fruit
145.	Tea	*Thea sinensis* O. Kuntze (=*Camellia sinensis*)	Theaceae	Leaf
146.	Cacao	*Theobroma cacao* L.	Sterculiaceae	Fruit
X.	**Fumicatories and Masticatories**			
147.	Areca nut	*Areca catechu* L.	Arecaceae	Endosperm
148.	Tobacco	*Nicotiana tabacum* L.	Solanaceae	Leaf
149.	Beedi Leaf	*Diospyros melanoxylon* Roxb.	Ebenacae	Leaf
150.	Betel leaf	*Piper betle* L.	Piperaceae	Leaf
XI.	**Drugs and Narcotics**			
151.	Indian Hemp	*Cannabis sativa* L.	Cannabinaceae	Leaf
152.	Poppy	*Papaver somniferum*	Papaveraceae	Latex
XI.	**Dyes**			
153.	Indigo	*Indigofera tinctoria* L.	Fabaceae	Leaf
154.	Indian privet	*Lawsonia inermis* L.	Lythraceae	Leaf
155.	Madder	*Rubia tinctoria* L.	Rubiaceae	Root
XI.	**Fibre crops**			
156.	Cotton	*Gossypium herbaceum* L. *Gossypium barbadense* L. *Gossypium hirsutum* L.	Malvaceae	Epidermal

Sl. No.	English Name	Botanical Name	Family	Plant Part Used
157.	Jute	*Corchorus capsularis* L. *Corchorus olitorius* L.	Tiliaceae	Phloem fibre
158.	Kenaf	*Hibiscus cannabinus* L.	Malvaceae	Phloem fibre
159.	Rozelle	*Hibiscus sabdariffa* L.	Malvaceae	Phloem fibre
160.	Hemp	*Cannabis sativa* L.	Cannabinaceae	Phloem fibre
161.	Sun hemp	*Crotalaria juncea* L.	Fabaceae	Phloem fibre
162.	Sisal hemp	*Agave sisalina* Pers.	Agavaceae	Leaf xylem
163.	Flax	*Linum usitatissimum* L.	Linaceae	Phloem fibre
164.	Ramie	*Boehmeria nivea* Goud.	Musaceae	Leaf fibre
165.	Manila hemp	*Musa textilis* Nees	Musaceae	Leaf fibre
166.	Coir	*Cocos nucifera* L.	Arecaceae	Mesocarp of fruit
XII.	**Sugars and starches**			
167.	Sugar cane	*Saccharum officinarum* L.	Poaceae	Stem
168.	Sugar beat	*Beta vulgaris* L. var. *saccharifera*	Chenopodiaceae	Stem-Root tuber
169.	Tapioca	*Manihot utilissima* Pohl	Euphor biaceae	
170.	Sago	*Metroxylon sago* Rottl.	Arecaceae	Stem exudate
171.	Arrow root	*Maranta arundinacea* L.	Marantaceae	Rhizome
XIII.	**Woods**			
172.	Teak	*Tectona grandis* L.f.	Verbenaceae	
173.	Sandalwood	*Santalum album* L.	Santalaceae	
174.	Red sanders	*Pterocarpus santalinus* L.f.	Fabaceae	
175.	Sal	*Shorea robusta* Gaertn.f.	Dipterocar paceae	
176.	Deodar	*Cedrus deodora* (Roxb. ex D. Don) G.Don	Coniferae	
XIV.	**Miscellaneous**			
177.	Camphor	*Cinnamomum camphora* (L.) Nees & Eberm. Quercus suber	Lauraceae	Young shoots
178.	Bottle cork	*Quercus suber*	Fagaceae	Bark
179.	Olive oil	*Olea europaea*	Oleaceae	Leaves
180.	Haematoxylin	*Haematoxylon compechianum*	Caesalpini aceae	Dye from wood
181.	Soapnut tree	*Sapindus emarginatus* Vahl.	Sapindaceae	Fruit
182.	Shika kai	*Acacia concinna* (Willd.) DC.	Mimosaceae	Fruit
183.	Leek	*Allium ampeloprasum* L.	Liliaceae	Bulbs
184.	All spice	*Pimenta dioica* (L.) Merr.	Myrtaceae	Fruit

17

Botanical Explorations in India and Taxonomists of India

The science of Botany in India had its beginning in the remote past. Vrikshayurveda, the Science of medicinal plants was written by Parasara probably before the beginnings of the Christian era. Extensive systematic work based on plant exploration must have preceded the writing of this work.

The Portuguese were the first to carry on botanical explorations in India on modern lines.

Garcia da Orta (1490-1570)

Garcia da Orta was a Portuguese Pharmacologist, Colonial administrator, explorer and botanist. He held the positions of Physician for Viceroy of Goa for 30 years. His most significant work is "Colloquios dos simples a drogashe consas medicinais da India Combostosalle Doutor Garcia da Orta 1563 (1565)" (This work is famous in translated forms : Aromatum et Simplicium Historia).

He was the first European to describe the famous double coconut fruit and referred it as 'Cocoeus de Maldiva' and gave information on medicines used in India that were new to west. In his book he described Bel fruit (*Aegle marmelos*), nirgundi (*Vitex negundo*), nim (*Melia azadirachta*), harshingher (*Nyctanthes arbortristis*), Conessi bark tree (*Holarrhena antidysenterica*). His work 'Colloquios' is a landmark in the history of Civilization being not only the 3rd European book to be printed in the East Indies in 1563. He provided information on Eastern Drugs sent by Arabs to West, local fruits, borax, the betel quid, narcotics from hemp and Datura.

C. Acosta

A Portuguese who published the next book on Indian Plants in 1578, entitled *"Tractado de las Drogas"*.

Hendrik Van Rheede (1660-1699)

He was the Governor of Malabar, a Dutch by origin and an amateur botanist who made a large collection of Indian plants. Between 1663 and 1669 Van Rheede also travelled over most of the Coastal regions from Goa to Tuticorin and made notes on vegetation. Van Rheede wrote a book *"Hortus Malabaricus"*. The first volume of *"Hortus Malabaricus"* was published in 1678 and it's final, 12th volume in 1693. His book contains the descriptions and uses of over 700 different species of plants growing in the region. All the plants are beautifully illustrated. Names of all plants are written in four scripts, Malayalam, Roman, Arabic and Nagari. The entire text was in Latin, covering 1534 pages. The text as well as the 793 illustrations in double folio size in 12 volumes was printed with copper engraved plates. Van Rheede's descriptions on the richness of the vegetation of the forests of Malabar are unique for his times, for their scientific accuracy of observations.

John Burmann (1707-1779)

He published a book, entitled *"Thesaurus Zeylanicus"* in 1737. This contained an account of the plants of Ceylon and Peninsular India.

Paul Hermann (1646-1695)

He collected a large number of plants from Ceylon and India. He published his *"Paradiscus botanicus"* in 1687. Later many of his collections were dealt with by Linnaeus in 1747 as *Flora Zeylanica* and by Nicholaus Burman in 1768 as *Flora Indica.*

John Gerard Koenig (1728-1785)

A Danish botanist, an earlier pupil of Linnaeus arrived in India in 1768. He organized a society called "The United Brothers" in order to promote the study of Indian plants. He made explorations in the Madras region and sent many specimens of Indian plants to botanists such as A.W. Roth, A.J. Retz and H.A. Schrader in Europe who described them.

Lieutenant Colonel Robert Kyd (1746-1793)

He was the founder (1787) and first Superintendent of Royal (now Indian) Botanic Garden, Howrah. The main objective of the garden was to establish a stock for disseminating such articles as may prove beneficial to the inhabitants.

William Roxburgh (1751-1815)

He succeeded Col. Kyd as the next Superintendent of the Royal Botanic Garden, Sibpur. Roxburgh toured India extensively for plant explorations. He made extensive plant explorations in Nepal and the adjoining areas but returned to England in 1815.

His collections were studied and described by D. Don who in 1825 published his, *"Prodromus Florae Nepalensis"*.

William Roxburgh made a survey of the Coromandel Coast from 1795 to 1820 and published a book *"The plants of the Coast of Coromandel"* in three folio volumes (1795, 1798, 1819). It contained 300 coloured plates of plants with vivid description and names in Latin, Telugu and English languages in 12 volumes which formed the basis for his subsequent work *Flora Indica*. He also wrote a book *'Hortus Benghalensis'* which have names of plants as *'nomen nudum'*.

Roxburgh's *"Flora Indica"* was issued in 1820-1824. It was a classical work in which a systematic account of Indian plants was presented for the first time. 2533 coloured drawings of Indian plants were left by Roxburgh.

Dr. Francis Buchanan (1763-1829)

He succeeded Roxburgh (1814) as the third Superintendent of the Royal Botanic garden, Sibpur, Calcutta.

Nathaniel Wallich (1786-1854)

He succeeded Buchanon as the fourth Superintendent of the Royal Botanical Garden, Calcutta and held this post from 1815 to 1835.

Wallich made extensive plant explorations in Nepal, Western India and Lower Burma. His collections were also sent to Europe and were studied among others by de Candolle, Bentham, Kunth, Lindley and others. Wallich Cat. is a catalogue of 9148 dried specimens published by Wallich from 1828-1849. His major contribution was the three volume flora *"Plantae Asiaticae Rariores"* published in 1830, 1831, 1832.

Robert Wight (1796-1872)

Robert Wight, a medical man, reached India in the year 1819. He was very much attracted by the plants around him and made extensive collections especially in the southern region of Peninsular India. In the year 1826 Wight was appointed as the Madras Government Botanist, which position he held for two years, till the post was abolished. Even after this unexpected incident, Wight continued his interest in this field and made collections with great enthusiasm. During his thirty-five years in India, Wight has described about 38 genera and more than 3000 species of Indian plants. His contributions to Indian Botany include about 28 publications of which *"Illustrations to Indian Botany"* (1831), *Icones Plantarum Indiae Orientalis* (1838-1853), *Specilegium Neilgherrense* (1846-1851), are invaluable for the students of taxonomy of India. In 1834 he published conjointly with G.A. Walker Arnott the *'Prodromus Florae Peninsulae Indiae Orientalis'*. Wight's name will also be immortalized in the history of Indian taxonomy for it is through him the natural system of classification reached India.

Wight in his monumental work *"Icones Plantarum Indiae Orientalis"* (1838- 1853) presented 2101 plates of Illustrations in 6 volumes.

Hugh Francis Clarke Cleghorn (1820-1895)

He became Professor of Botany and Materia Medica at Madras in 1851, Conservator of Forests in 1857 and Inspector General of Forests in 1867. He established a Herbarium of Madras Museum, Madras city around 1853. His herbarium specimens along with those of Dr. Drew and Sir E. Elliot bearing the year 1853 under the Madras Museum Herbarium labels appear to have formed the nucleus of the present Madras Herbarium located at Coimbatore under the Botanical survey of India. Cleghorn's collections enriched the herbaria at Edinburgh and Howrah.

William Griffith (1810-185)

An English botanist who made large collections for thirteen years in Assam valley, Burma, Bhutan, Sikkim, Central India and Malacca, collecting 9000 species and described and published posthumously.

Thomas Thomson (1817-1878)

An English naturalist who made extensive collections in Punjab (1847-1849), Himalayas and Tibet.

Colonel Richard Henry Beddome (1830-1911)

He entered the Forest Service of Madras in 1856. He immortalized his name in the field of Indian Botany particularly in the studies of South Indian Ferns, by the valuable and commendable work he did for about 60 years. He followed the path of Wight in illustrating the plants by the use of lithography. He had an insatiable appetite for the study of Ferns as he had for flowering plants. His prolific writing started in 1858 with the *Flora and Vegetable products of Pulney hills* followed by the *Flora of Nilgiris* (1876, 1880) and the *Flora of Tirunelveli* (1877). In addition to these, his illustrations with descriptions like "*The Flora Sylvatica for Southern India*" (1869-1874) and "*Icones Plantarum Indiae Orientalis*" (1868-1874) are the significant contributions. His other works include the genus *Impatiens* (1858), *Contributions to the botany of Southern India* (1861-62), "*Trees of Madras Presidency*" (1863) etc. He made remarkable contributions to the study of ferns through a series of illustrated publications spread over 60 years. '*The Ferns of Southern India*'(1863-1864)', '*Ferns of British India*' (1865-1870), '*Supplement to the Ferns of Southern India and British India*' (1876), *Handbook to the Ferns of British India, Ceylon and Malay Peninsula* (1883) and its supplement (1892) are still very valuable standard reference for any taxonomic study on Ferns.

Joseph Dalton Hooker (1817-1911)

He was a British botanist, explorer, pioneer plant geographer, ethnobotanist, traveller, surgeon and naturalist. He held the positions Assistant naturalist and Surgeon, HMS Erebus (1839-1843), Assistant Professor of Botany at Edinburgh (1843-1845), Botanist, Geological Survey (1846-1847), Assistant Director (1855- 1865), Director, Royal Botanical Gardens and Museum, Kew (1865-1885).

Hooker started writing *Flora of British India* on the basis of Griffith's and Thomson's collections which were sent to him at Kew. This work appeared in seven volumes from 1872-1897. Hooker himself visited India in 1847-1851 and along with Thomson made extensive explorations in the Himalayas. Here he discovered various species of *Rhododendron* on which he provided a monograph (1849).

Contributions of other Botanists

The *Flora of British India*, by Hooker (1872- 1897) focussed the urgent need for intensive exploration of underexplored and unexplored regions as well as the publication of regional floras. The history of early taxonomic work in any part of the country must mention the establishment of Botanical Survey of India and Indian (Royal) Botanic Garden at Sibpur, Howrah. Robert Kyd started the Howrah Garden. Botanical Survey of India was established in 1890 under the directorship of Sir George King with its headquarters at Calcutta and three regional offices at Madras, Bombay and Saharanpur. During the first three decades of the 20th Century intensive exploration was done all over the country by Botanical Survey of India. This resulted in the publication of a number of provincial floras, such as the '*Flora of the Presidency of Bombay*' (T. Cooke, 1901-1908), *Bengal plants* (D. Prain, 1903), '*Flora of the Upper Gangetic Plain and the adjacent Siwaliks and Sub-Himalayan Tracts*' (J.F. Duthie, 1903-1922), '*Flora of the Presidency of Madras* (J.S. Gamble and C.E.C. Fischer, 1915-1936), *The Botany of Bihar and Orissa* (Haines, 1921-1925), '*The Flora of Indus Delta*' (Blatter et al., 1929) and '*Flora of Assam*' (Kanjilal *et al.*, 1934- 1940).

Burkill (1965) has dealt with the history of Botany in India up to the year 1900 in his book "*Chapters in the History of Botany in India*".

Other Floras in first half of 20th Century include, *Flora of South Indian Hill Stations* (Fyson, 1932), *Handbook of some South Indian Grasses* (Rangachari and Tadulinga Mudaliar, 1921), *A Handbook of some South Indian weeds* (Tadulinga Mudaliar and Venkata Narayana, 1932), *Flora Simlensis* (Collect, 1902) etc.

J.S. Gamble (1847-1925)

James Sykes Gamble name in India is always associated with the publication of '*Manual of Indian Timbers*' (1881) and '*Monograph on the Bambuseae of British India*' (1896) and Cecil Erneste Claude Fischer's (1874-1950) with the Survey of Anaimalai (1922) and Lushai hills (1938). But the prominent role played by both of them in compiling the '*Flora of the Presidency of Madras*' (1915-1936) immortalized their names in the annals of Botany of India. '*The Flora of the Presidency of Madras*' was received by botanical world as one of the best regional floras of India, and the system of descriptive keys which could be referred to as a sort of full description of species was carried out to a great efficiency here. Gamble collected plants intensively throughout the Presidency of Madras between February 1883 and 1889. His collections are housed in K, CAL, MH and DD. Fischer in 1923 gave a descriptive list of the Forest flora of the East Central Districts.

H. Santapau (1903-1970)

Born at La Galera, Tarragonna, Spain on December 5, 1903, Fr. Santapau joined the Society of Jesui Gandia, Valencia at the age of 16. After his schooling in classics and languages, mainly in Spain and Latin, he proceeded to Rome and from the Gregorian University he took his Ph.D. (Philosophy) in 1927. He left for India in 1928 to serve out his regency period and to complete his Theological studies. In 1936 he became a student of Royal College of science and subsequently of the Imperial College of Science and Technology, London. He obtained the B.Sc. Hons. and Ph.D. degrees in Botany of the London University as well as the A.R.C.S (Associated Diploma of the Royal College of Science) and the D.I.C. (Diploma of the Imperial College) in Botany. He spent the following two years at the Herbarium of the Royal Botanic Gardens, Kew, Surrey, England before joining the teaching staff of St. Xavier College, Bombay in 1940.

Fr. Santapau had begun his Botanical explorations as early as in 1926, in the Central Cevennes, France. He continued it in the Eastern Pyrenees (1934), The Italian Alps (1936), Baluchistan (1946), Kathiawar (1946-1954), The Dangs forest (1950-1955). The Western Ghats from Karjat and Khandala down to Goa and the Eastern Ghats from Waltair to Vijayawada, the Pulneys and the Nilgiris came under his exploration. He then turned his attention to the North-East and the northern regions —the Khasia and Jaintia hills in Assam, Darjeeling and Kalimpong in eastern Himalayas, the Dehra Dun and Mussori. He collected over a lakh botanical specimens from these regions. Most of which are kept in Bombay, Representative collections were sent also to Kew (London), The Arnold Arboretum of Harvard University, the Missouri Botanical Garden, and many other National and International Herbaria. The results of these explorations have been published in several books and journals in India and abroad. His practice was to go on Botanical explorations during his vacations and lecture during academic sessions, mainly at St. Xavier College and the University of Bombay. He had delivered lectures at all the important universities in India on invitation

His boundless energy and tireless activity can be gauged from his efforts in the educational field and from his publications. The more important of his publications are: *The Flora of Khandala in the Western Ghats of India* (1953), *Botanical Collectors Manual* (1955), *The Flora of Purandhar* (1958), *The Flora of Saurastra Part I* ((1962), *The Acanthaceae of Bombay* (1952), *The Asclepiadaceae and Periplocaceae of Bombay* (1962), *Publication Dates* (1965), *The Orchids of Bombay* (1966) and *Common Trees* (1966).

He was associated with numerous learned Societies and Scientific bodies. He was a Fellow of the Linnean Society, London (1947) and the Indian Botanical Society (The Society, of which he was the President in 1955, had bestowed on him the Birbal Sahni Medal in 1964).

The Government of India nominated Fr. Santapau as Chief Botanist in 1954 for the revival of the Botanical Survey of India. In 1961 he was appointed as the Director of Botanical Survey of India and continued till June 1968.

In recognition of his services to the country in education and research the Government of India conferred upon him the Padmashri Award in 1967. In the same year Spain, his mother country, also honoured him with the "Order of Alphonsox, the Wise".

Fr. Santapau spent the major portion of his life in India and later became Citizen of India. He died on January 13, 1970.

Dr. E.K. Janki Ammal (1897-1984)

Dr. Edavaleth Kakkat Janaki Ammal was born on November 4, 1897 at Tellicherry in the Kerala State. She took her B.A. (Hons.) Degree in Botany in 1921 and M.A. Degree in 1923 from Madras University. After spending a couple of years in teaching profession at the Women's Christian College, Madras, Janaki Ammal went over to the University of Michigan, USA and took her M.Sc. degree in 1925 and D.Sc. degree in 1931. She joined Sugarcane Breeding Station at Coimbatore as Geneticist in 1934 and continued up to 1939. She joined John Imes Horticultural Institution, London in 1939 as Assistant Cytologist and continued till 1945. She then moved to the Royal Horticultural Society, Wesley, London as Cytologist and was there for five years (1946-51).

Government of India invited her to work as the Officer-on-Special Duty (1952-54) for reorganizing Botanical Survey of India and in 1954 she became the first Director, Central Botanical Laboratory, Allahabad and continued up to 1959. She then moved to the Regional Research Laboratory, Jammu where she worked in various capacities from 1959 to 1969. She then moved down to Madras University where she continued to work till her death on February 7, 1984.

Janaki Ammal worked mainly in Cytogenetics of *Saccharum* and *Zea mays*. She also worked on polyploidy in *Solanum, Cymbopogon, Terminalia, Datura* etc.

The publication "*Chromosome Atlas of Cultivated Plants*" by Darlington and Janaki Ammal is well known to Botanists and Agricultural Scientists all over the world. The compilation speaks of the pains taken by Janaki Ammal. She was responsible for establishment of Division of Ethnobotany at Allhabad (BSI).

Index

A

Acanthaceae (Acanthus Family) 272
Actinocytic 137
Agamodeme 208
Aizoaceae 26, 46, 65, 72, 138, 145, 153, 154, 169, 171, 226
AlgaeBase 197
Algorithmic approaches 195
American code 86
Amino acid sequencing 174, 175
Amplified fragment length polymorphism (AFLP) 137, 184, 185
Amsterdam code 86
Anatomical 2
Androecium 218
Andrographis nallamalayana 91, 94
Andrographis paniculata 273
Angiosperm phylogeny group (APG) 63
Angiosperm phylogeny group classification 66, 67
Anisocytic 136
Annonaceae (Custard apple family) 217
Anomocytic 136
Anther 145
Apertural patterns 155
Apiaceae (Umbelliferae) 249
Apocynaceae (Dogbane family) 261
Arbitrary primed PCR (AP-PCR) 185
Archesporium 145
Arecaceae (Palm family) 294
Aristolochiaceae (Birthwort family) 280
Artabotrys hexapetalus 219
Assemble and align dataset 189
Asteraceae (Compositae) (Sunflower family) 255
Austrobaileya scandens 141
Autodeme 208
Automated bee identification system (ABIS) 198

B

Banding patterns 162
Barcoding 191, 193
Basellaceae 151
Bataceae 71, 145, 223

Bennettitalean ancestry 77
Bentham and Hooker's system 37
Berlin code 87
Bignoniaceae (Bignonia family) 270
Biodiversity informatics 203
Biological diversity 183, 184, 197
Biosystematic categories 206
Biosystematic phase 2
Biosystematics 205
Bootstrap analysis 190
Botanical explorations 311
Botanical gardens 131
Botanical Survey of India 101, 116, 118, 129, 211-216, 315-317
Brachyscome dichromosomatica 159
Brassicaceae (Cruciferae) (Cabbage family) 220
Brassica chinensis 143
Breeding systems 2

C

Callistemon linearis 246
Cambridge code 86
Cananga odorata 219
Capparis cynophallophora 143
CAPS 186
Caryophyllaceae 38, 46, 57, 72, 137, 145, 153, 169, 171, 174, 223, 226, 279
Caryophyllaceae (Pink or carnation family) 223
Cassia auriculata 243
Caytonialean ancestry 79
CBOL (Consortium for the barcode of life) 199
Central botanical laboratory 212, 213, 317
Cephalanthera ensifolia 158
Ceropegia pullaiahii 96
Chemotaxonomy 162, 168, 172
Chromosomal homologies 209
Chromosome behaviour 162
Chromosome number 159
Chromosome size 161
Cleaved amplified polymorphic sequences 186
Clinodeme 208
Cluster analysis 178
Coenospecies 207
Collection methods 118
Colpodium versicolor 159
Comparium 207
Computer applications 195
Computer stored keys 113
Coniferalean ancestry 78
Convolvulaceae (Glory family) 264
Crossing programmes 209
Cryptogamic unit 213
Cucurbitaceae (Gourd family) 247
Cyanogenic compounds 171
Cyclocytic 137
Cyperaceae 147
Cyperaceae (Sedge family) 296
Cyprepedium cordigerum 158
Cytodeme 208
Cytogenetic analysis 209
Cytological 2

D

DAF 185
Darwin's theory 44
Database resources 197
DELTA (Description language for taxonomy) 198
DELTA system 113
Deme rerminology 207
Diacytic 137

Dichrostachys cinerea 241

Dicotyledons 18, 20-29, 31, 37, 43, 63, 78, 81, 135, 137, 140, 141, 146, 158, 176, 277

Digital automated identification system (DAISY) 198

Digital floras 202

Digital taxonomy 197, 201

Dimorphocalyx kurnoolensis 91, 95, 96

Discrete data methods 190

Distance-matrix method 189

DNA amplified fingerprinting 185

DNA barcoding 183, 191, 192, 199

DNA sequences 63, 184, 186, 189

Durian theory 80

E

Economic botany 303

Ecospecies 206

Ecotype 206

Edinburgh code 87

eFloras 202

Electronic resources for taxonomy 196

Electrophoresis 175

Embryo 145

Embryological 2

Embryological data 146, 149

Endodeme 208

Endosperm 145

Environment impact assessment 201

Epidermal cells 139

Ethnobotanical studies 118

ETI bioinformatics 198

Euphorbiaceae (Spurge family) 284

Evolutionary 1

Evolutionary potential 2

Exine sculpturing 156

Experimental taxonomy 208

F

Fertilisation 145

Field and herbarium methods 115, 129

Floral characters 217, 221, 224, 226, 229, 231, 234, 236, 239, 241, 242, 244, 245, 248, 250, 253, 256, 259, 261, 265, 267, 270, 272, 278, 280, 282, 285, 288, 292, 294, 297, 299

Flowering plant index 103

Fouquieriaceae 53, 72, 145

Frankeniaceae 48, 53, 72, 145

Fruit wall 142

Fumigation 128

G

General botanical gardens 132

General taxonomic indexes 103

Gene sequencing 186

Gentianaceae 152

Geographic information system (GIS) 200

Global biodiversity information facility (GBIF) 197, 203

Gnetalean ancestry 78

Gray herbarium card index 103, 104

Gross morphology 2

Gymnosperms 18-20, 23, 24, 28, 43, 78, 81, 141, 146

Gynoecium 218

Gyrostemonaceae 57, 71, 145, 155, 233

H

Herbarium knowledge 2

Herbarium label 126

Hexacytic 137

Hu card index 104

Hutchinson's system 28, 29

I

Index Holmiensis 103

Index Kewensis Plantarum Phanerogamarum 103
Index Londonensis 85, 103
Index Nominum Genericorum 104, 203
Internal structure 139
International plant names index 104, 197
Inter simple sequence repeats (ISSR) 184, 185
Iseilema anthephoroides 160
Isoetalean ancestry 78

J

Jatropha heterophylla 99

L

Lamiaceae (Labiatae) (Mint family) 274
Last common ancestor (LCA) 187
Lectotype 91
Leiden botanic garden 133
Leningrad code 87
Liliaceae (Lily family) 291
Loranthaceae 148
Loranthaceae (Mistletoe family) 282
Lycopersicon esculentum 87, 98, 99, 268, 305

M

MADS-box genes 82
Malvaceae (Mallows family) 226
Mature embryo sac 145
Maximum parismony 190
Megasporogenesis 145
Melaleuca leucodendron 246
Melbourne code 88, 95
Menyanthaceae 152
Mesogenous 137
Mesoperigenous 138
Minute plants 123
Missouri botanical garden 66, 134, 201, 316
Molecular markers 184, 185
Molecular phylogenetic trees 188, 189
Molecular plant taxonomy 183
Molecular sequence-based analysis 199
Molluginaceae 57, 65, 66, 72, 145, 169, 171
Monocotyledons 21-25, 28, 29, 31, 33, 42, 43, 59, 78, 81, 135, 137, 141, 144, 146, 156, 158, 171, 176
Montreal code 86
Multiple arbitrary amplicon profiling (MAAP) 185
Myrtaceae (Myrtle family) 245

N

Nanopore genome sequence 184
National Institute for Biotechnology Informatics (NCBI) 200
New York botanical garden 33, 134, 202
NPC system 155
Numerical taxonomy 17, 176, 180, 181, 199, 200
Nyctaginaceae 41, 46, 53, 72, 145, 169, 171

O

Operational taxonomic units (OTUs) 177
Orchidaceae (Orchid family) 287
Orchidarium 132, 214
Ovule 145

P

Padua botanic garden 133
Paeonia 149
Palermo botanic garden 133
Palynological aspects 2
Paphiopedilium villosum 158
Paracytic 137
Paris code 85, 86
Pentoxtlalean theory 79

Perigenous 137
Phenodeme 208
Phylogenetic 1
Phylogenetic analysis 186, 189
Phylogenetic sequence 44
Phylogenetic systems 7, 22
Phylogenetic trees 187, 189
Phytolaccaceae 19, 41, 46, 57, 72, 145, 153-155, 169, 171, 226
Pisa botanical garden 133
Plant classification 7, 30
Plant molecular taxonomy 191
Plant nomenclature 83, 84
Plant presser 121
Plant taxonomy 3
Plumbaginaceae 49, 58, 72, 145
Poaceae (Grass family) 299
Podostemaceae 146
Pollen association 156
Pollen grain 145
Pollen nuclear number 158
Polygonaceae 20, 41, 46, 57, 72, 145, 146, 306
Polymerase chain reaction 184
Polystachya flavescens 158
Portulacaceae 38, 46, 57, 72, 138, 145, 151, 153, 154, 169, 171
Prosopis juliflora 241
Pteridophytes 23, 78, 93, 103, 141
Pteridospermean ancestry 79

R

Random amplified polymorphic DNA (RAPD) 185
Raw drug repository 203
Restricted fragment length polymorphism (RFLP) 185
Rhabdodendraceae 72, 145
Rhamnaceae (Buckthorn family) 236
Rhynchospora tenuis 159
RNA hybidization 174
Rochester code 85
Root system 217, 226, 229, 231, 233, 236, 239, 240, 245, 248, 259, 267, 294, 297
Royal botanic garden 133, 134, 312
Rubiaceae (Coffee family) 252
Rutaceae (Citrus family) 233

S

Sapotaceae (Sapota family) 258
Scanning electron microscopy (SEM) 143
Seattle code 87
Seed coat 146
Seed coat anatomy 142
Sequence characterized amplified regions (SCAR) 186
Sequence tagged microsatellites (STMS) 185
Sequence tagged sites (STS) 185
Shenzhen code 88
Shoot system 217, 261
Simultaneous character-set identification 113
Single molecule real time (SMRT) sequencing 186
Single primer amplification reactions (SPAR) 185
Slender aquatic plants 124
Spiranthes australis 158
Sterculiaceae 228
St. Louis code 87
Stochastic models of evolution 190
Stockholm code 86
Stomata 136
Succulent plants 123
Sydney code 87
Systematic phase 2

Syzygium cumini 246, 247, 307

T

Taxonomic literature 101

Tetracytic 137

Theligoniaceae 145

Tiliaceae 231

Time of origin 80

Tokyo code 87

Transmission election microscopy (TEM) 143

Trichomes 138

Triticum aestivum 14, 87, 98, 176, 300, 303

U

Unisexual flowers 23, 44, 79, 81, 286, 296

UPGMA method 190

V

Variable number of tandem repeats 185

Variation patterns 2

Vasculum 119

Vegetative characters 217, 220, 226, 229, 233, 239-245, 248, 252, 255, 259, 261, 265, 267, 270, 275, 277, 280, 284, 288, 291, 294, 297, 299

Vienna code 85, 87

Vivianiaceae 53, 145

W

Web tools 198

X

Xylopia aromatica 219

Z

Zingeria biebersteiniana 159

www.ingramcontent.com/pod-product-compliance
Ingram Content Group UK Ltd.
Pitfield, Milton Keynes, MK11 3LW, UK
UKHW020238290926
15470UKWH00002B/603

9 789387 057807